AF520148

MANAGING WATER SCARCITY

Experiences and Prospects

Managing Water Scarcity

Experiences and Prospects

Edited by
A. VAIDYANATHAN
H.M. OUDSHOOM

IDPAD

MANOHAR
2004

First published 2004

© Individual Contributors, 2004

All rights reserved. No part of this publication may be reproduced or transmitted, in any form or by any means, without the prior permission of the editors and the publisher

ISBN 81-7304-553-4

Published by
Ajay Kumar Jain for
Manohar Publishers & Distributors
4753/23 Ansari Road, Daryaganj
New Delhi 110 002

Printed at
Lordson Publishers Pvt Ltd
Delhi 110 007

Contents

Preface

Water is essential for sustaining life. Historically, the availability of fresh water has played a key role in the evolution of agriculture and human settlements. In modern times, a large quantity of water is used for non-agricultural purposes. With an ever-growing population and rise in living standards, the demand for water for irrigation, energy generation, industrial production, and urban consumption has also increased rapidly. The total supply of fresh water is, however, limited by nature. Technological progress has enabled us to overcome the constraints on sourcing water imposed by gravity and terrain and increased our capacity to harness water for human use. However, in many parts of the world, growth in water requirements has outstripped the increase in water supply. At the same time, droughts, depletion of aquifers, deforestation, and pollution have cut into the availability of fresh water of good quality. This has resulted in intensified competition and conflicts over access to the available supplies of water.

These conflicts of interest are manifest in both industrial and non-industrial countries, market and centrally planned economies, as well as in arid and humid climates. The specific problems may vary and change, but everywhere water related issues are becoming increasingly widespread. Issues concerning water resources management have become matters of global concern, leading to several international initiatives to address them.

The United Nations convened a high level World Water Conference in Mar del Plata, Argentina, in March 1977. This conference, the first of its kind ever held to discuss water related issues at a global level, drew attention to the importance of a responsible use of the resource, and the need to view water issues and their remedies, even when they appear specific to particular regions or localities, in a wider perspective. The 1980s were declared as the International Water Supply and Sanitation Decade and the conference adopted a limited Action Plan stressing the importance of ensuring the supply of clean water to every citizen of the world, and the need for concerted efforts by the world community to ensure this.

The ensuing years have witnessed a growing concern with the environmental impact of water resource development and a widening of the scope of the discussion and actions on water related matters. This became evident in the United Nations Conference on Environment and Development held in 1992 in Rio de Janeiro. Fresh water figured prominently in the deliberations. The Rio Declaration on Environment and Development, adopted by the conference, emphasized that water resource development and its use must be informed by a recognition of the integral and interdependent nature of the earth, respect for the interests of all concerned stakeholders, and protecting the integrity of the global environmental and developmental system.

Specifically, in relation to water, measures to ensure adequate supplies of water of good quality for various human needs must also take care to preserve the hydrological, biological, and chemical functions of ecosystems, and adapt human activities within the capacity limits of nature and combating vectors of water related diseases. This implies that the present fragmented approach, largely dominated by hydrologists, engineers, and economists, to the development and management of water resources must be replaced by an integrated, multidisciplinary effort.

After the Rio Conference 'Integrated Planning for a Sustainable Use of Water Resources' has gained wide currency among water professionals. But it still remains largely a slogan. The literature on this subject is still scarce and the publications available so far are often of a very general and academic nature. Examples of concrete applications of 'integrated planning' and 'multidisciplinary approach' in specific projects or systems to specific projects are rare. This is not surprising. Apart from the resistance of disciplines currently dominating water resource planning, getting specialists from several disciplines to work as interactive teams is difficult. The constitution of such teams is still rare. And even when such teams are formed, differences in perspectives, theoretical frameworks, and analytical tools, not to mention availability of relevant data, present formidable hurdles to meaningful dialogue.

There is also a tendency to see 'integrated planning' as a method for finding an all-embracing solution satisfactory to all concerned interests. Of course, this is not true—integrated planning does not create new water—and the problem of resolving competing claims will remain. What it can do, and this is certainly very valuable, is to spell out alternative possibilities and their implications so that politicians and decision makers can make more informed and reasoned choices between different objectives (maximizing overall benefits, equity in distribution, preventing environmental damage) and competing claims of different uses and users' conflicting interests.

Developing satisfactory mechanisms and procedures for integrated water resource planning is obviously a huge and difficult task. Perhaps a practical way to set the process in motion would be select a few specific issues and organize a focused discussion across different disciplines on how to address them. Water scarcity and attendant conflicts, being widespread and often serious, is one such issue around which the possibilities and problems of the multidisciplinary approach could be explored.

It is on this understanding that the Indo-Dutch Programme on Alternatives in Development (IDPAD) took the initiative in organizing the seminar 'Managing Water Scarcity: Experiences and Prospects' held at Amersfoort, Netherlands in October 1997. The intention was to focus on the sources and nature of water scarcity and conflicts in selected projects/systems in India and in Europe and the USA, the manner in which they have been handled, the mechanisms used and their effectiveness. It was expected that a discussion of the diverse experiences in these respects from different disciplinary perspectives would help to better understand the issues involved and hopefully come up with some concrete, practical ideas for action and further research. For this purpose, persons (academics and officials) known to have considerable knowledge and experience in studying or handling the problem were requested to prepare papers on pre-selected projects/themes and subjects well in advance of the seminar. Experts from different backgrounds interested and knowledgeable about the issues involved from a broader, more general, perspective were invited to be keynote speakers and discussants. The success of the seminar in assembling a truly multidisciplinary group is reflected in the composition of the participants who included six engineers, six economists, four social scientists, three environmentalists, two lawyers, and senior administrators, many of whom combined social activism in water issues with their professional work.

The papers having been circulated to participants ahead of the seminar, there was ample time for detailed and focused interaction among the participants. The discussions showed that meaningful trans-disciplinary dialogue is possible when it is centred on a well-defined issue/theme and that it can lead to meaningful and practical conclusions on the ways in which different disciplines can contribute to evolving solutions based on an integrated perspective. Participants were unanimous in recognizing the possibility and utility of such dialogues and the need for furthering them.

This volume includes a selection of papers presented at the seminar and revised subsequently in the light of the discussions at the seminar and editorial comments. Though the publication comes six years after the seminar, we would like to emphasize that the authors have updated information, to the extent possible and that, in any case, events in the intervening period have not made a significant difference to the substantial issues raised or conclusions reached.

We would like to thank the authors of the chapters of this book who managed, in the first instance, very successfully, to produce the initial papers well in advance of the seminar. This gave the participants the opportunity to familiarize themselves with the complex nature of the subject of the seminar and helped to focus the discussion in the seminar. After the seminar they again willingly revised and updated their texts for publication.

Thanks to the keynote speakers and discussants who were very helpful in initiating and illuminating the discussions with their observations and to all the participants who engaged enthusiastically in the discussions and widened the framework of the discussions with inputs from their disciplines. Our main regret is that, for logistical reasons, it was not possible to prepare and publish a detailed resume of their interventions and the rich discussion they stimulated.

We owe a special word of thanks to the Dutch IDPAD Coordinator, Huub Coppens, who was in the first phase the inspired initiator of the seminar, Marc Verhagen who later managed to implement the concept into realization, and Astrid van Esch who put together the individual papers into the final manuscript. Last but not least we would like to express our sincere thanks to Sonia van Bost and Kirsten Oleson who took the full burden of organizing the seminars as well as keeping constantly in touch with the convenors, authors, discussants, and participants.

A. VAIDYANATHAN
H.M. OUDSHOOM

Introduction

In many areas of the world the demand for fresh water for human use is fast approaching the available supply and even outstripping it in some. This situation results from inadequate availability, excessive withdrawals, and mismanagement.

The total amount of rainfall in a given region is determined by nature and is beyond human control. Availability of water can, however, be increased by diverting or storing river flows outside the region and by lifting water from local aquifers. These in turn are dependent upon the geography of the area and technological and economic considerations. Developments in civil engineering, pumping, and agricultural sciences have progressively reduced the physical and economic constraints on the depth from which water can be extracted, the size of storages, and the distances over which it can be transported. The volume of water impounded in reservoirs, as well as the volume extracted from underground, has increased at a rapid rate during the past century.

In relatively dry and densely populated areas irrigation is an important factor in enhancing the productivity of land. It is observed worldwide that under such situations, there is intense competition for access to irrigation water between different parts of a river basin and within the command of particular systems. Population growth, expansion of irrigation for increased agricultural production, urbanization, and industrialization have led to a rapid growth in the overall demand for water. In many parts of the world the harnessed supplies are inadequate to meet the demand, leading to a widespread incidence of water scarcity. The relatively faster growth of non-agricultural demand for water have led to a change in the pattern of demand for water, and given the water scarcity, to increased conflicts between different uses.

Overexploitation and misuse of water resources aggravate the problem. Where groundwater exploitation exceeds the recharge rate in a given aquifer, the quantum of fresh water available from it is bound to decline over time. Excessive diversion of surface flows adversely affects the volume of water available to satisfy the needs for minimum regime flows, navigation, and for the conservation of scenic values and riverine/estuarine ecosystems. Increasingly, misuse (for example, due to over-irrigation and discharge of sewage, sludge, and industrial effluents) is compounding the scarcity of fresh water and, by causing deterioration of water quality, endangering the utility of water for various uses.

These phenomena are, however, area-specific. Comparisons of water availability and its use at the regional level, let alone on a national or global level, are not meaningful. River basin comparisons are more meaningful than inter basin relationships. By analysing variations in the nature and intensity of shortages and conflicts across basins, it is possible to focus on basins where the problem is serious and on solutions appropriate to specific contexts. This is what the present collection of papers seeks to do.

CASE STUDIES

The 16 papers in this volume have been divided into four broad sections. The first group, consisting of seven papers, describes the nature and causes of water scarcity and conflict in selected systems across India and the responses of different stakeholders. Chapters 1, 2, and 5 focus on overexploitation, extensive pollution of water sources, and the resultant conflicts between different segments and between different sectors in three river basins—one in Gujarat and two in Tamil Nadu. These chapters are: Chapter 1, 'Managing Conflict in the Bhavani River Basin' by Paul P. Appasamy; Chapter 2, 'Competing Demands for Water in the Sabarmati Basin: Present and Political Conflicts' by Vishwa Ballabh and Katar Singh; and Chapter 5, 'Conflict Over the Use of Groundwater: Some Evidence from Tamil Nadu' by S. Janakarajan. Chapter 4, 'Water Use in the Punjab Region: Conflicts and Frameworks for Resolution' by Kanchan Chopra and Seema Bathla, deals with the overexploitation of groundwater in Haryana state and explore strategies to contain it. In Chapter 6, 'Managing and Utilizing the Kosi Waters: Experience and Prospects', T. Prasad describes the experience in implementing a strategy to control extensive flooding by the river Kosi during the monsoon and to extend irrigation in the fertile plains of Bihar. His paper highlights the complex geological, technical, and managerial problems of water resource development in an international river basin. Chapter 3, 'Urban Water Supply and Management: Delhi' by Ashis Banerjee and Suresh K. Rohilla, deals with the problems of water shortage, falling groundwater levels, and deteriorating water quality faced by the city of Delhi and outlines a strategy for increased groundwater recharge and conservation. Chapter 7, 'The Legal Framework for Water Management and its Working' by K.V. Raju, provides a critical review, backed by concrete examples, of the lacunae in the legal framework governing water management in India, the nature of rules and regulations governing water allocation in irrigation systems, and weakness of enforcement. All these chapters give some idea of how various concerned interests have (or have not) responded to the emerging problems, how effective they have been, and what could be done in the future to better cope with them.

NON-GOVERNMENTAL INITIATIVES

The second group of papers describes non-governmental, community-based initiatives to overcome water shortages due to droughts and promote equitable sharing of water and public action to prevent and mitigate the effects of pollution. In Chapter 8, 'The Krishna Valley Development: An Approach to Equitable and Efficient Use of Water', K.R. Datye et al. describe the emergence of grass-roots initiatives in a water scarce region of Maharashtra to augment water supplies through local mobilization, developing cost effective technologies for harvesting rainfall, and making more efficient use of water as well as its equitable distribution. Chapter 9, 'The Deepening Divide: Diverse Responses to the Challenge of Groundwater Depletion in Gujarat' by Tushar Shah, describes a widespread and apparently successful community-based movement for rain harvesting and groundwater recharge in Gujarat—a region marked by chronic water scarcity. In Chapter 10, 'Water Pollution Control: Role of Community and Public Action', Himanshu Thakkar provides a critical review of the role of governments, non-governmental organizations (NGOs), the affected communities, and the courts in dealing with industrial pollution related problems in different parts of the country. With numerous concrete examples, he highlights the indifference and ineffectiveness of the government, the difficulties in getting effective judicial redress, and the key role of public action (through mobilization of affected communities, agitations, and public interest litigation) to tackle these problems.

Another example of NGO activity at the international level is provided by Arthur van Norden in Chapter 11, 'The International Water Tribunal', which describes the activities of the International Water Tribunal (IWT). A non-governmental organization, IWT seeks to provide an informal public forum in which specific water-related complaints can be subjected to professional legal scrutiny on the basis of evidence. While it simulates the judicial procedures, its findings have no sanction. Nevertheless it helps to increase public awareness of the issues involved as well as the remedies available and, thereby, generate pressures for a fair resolution in formal forums.

THE INTERNATIONAL EXPERIENCE

The third group of papers deals with the working of laws, regulations, and institutions in managing water and resolving water conflicts in the international context mainly in Europe. In Chapter 13, 'The Rhine Research Project: The Port of Rotterdam Scour Control Programme', Tiedo Vellinga describes the largely informal, but remarkably successful, strategy adopted by the Rotterdam Port Authority to deal with the adverse effects of upstream pollution in the Rhine basin on its operations. His paper describes, in a concise and telling manner, how the Port Authority went about getting a satisfactory settlement without formal involvement of the governments of riparian states or of the judiciary. The Port Authority succeeded, with the support of NGOs concerned with the environment and non-official experts, and through informal negotiations, in persuading the industries to accept independent and objective measurements of their contribution to the pollution load and to agree to a time-bound, targeted programme for bringing down pollution level to acceptable levels. The circumstances and mobilization tactics used to accomplish this—through negotiated contracts rather than adversarial adjudication—is highly instructive.

Complementing this is Chapter 14, 'The Rhine Experience: The Development of Integrated River Basin Management in North-West Europe' by Koos Wieriks, which gives an account of the establishment of a formal organization—based on a formal agreement negotiated by the governments of the riparian states—for integrated planning of water resource exploitation and use in the entire Rhine basin.

In Chapter 12, 'The Increasing Role of Public Policy in Integrated Water Management', Wim van Leussen describes the growing awareness of the necessity for integrated water management and the increasing complexity of public policy-making in relation to water. He reviews recent, and ongoing, efforts in the USA and the Netherlands to develop better methods of analysis and institutional arrangements. They seek to build 'cooperative partnerships' of relevant stakeholders and a process of decision-making in which all important concerned interests participate in determining development priorities and resolving competing claims and conflicting interests through mutual negotiation. His account of the structure and functioning of water resource planning in the Netherlands is supplemented by an extensive discussion of the structure and working of Water Boards in the Netherlands and the French experience of incorporating integrated basin planning into the institutional structure of the government.

THE NETHERLANDS WATER BOARDS

In the Netherlands, the Water Boards are the oldest democratic form of government. Like the provincial and municipal authorities, these Boards are instruments of the government. Originally meant to manage the collaborative task of protection against inundation and sea intrusion, they are now responsible for all the essential aspects of regional water planning and management. Their functions include management of the water related infrastructure in their jurisdiction; providing protection against flooding; managing the amount of water and its allocation to match the demands of various sectors; resolution of conflicts of interest; combating pollution; implementing the Water Pollution Act; maintaining water quality; and improving the quality of surface waters. Organized on the basis of coherent water basins or hydrological units, they operate on the basis of Basic Action Plans covering all sectors, prepared and decided by a process involving all parties concerned.

The Water Boards have a governing body and an executive council headed by a professional chairman and a high level technical and administrative staff. Land and property owners as well as residents are represented on the general council. The former elect their representatives directly. Representatives of resident groups, including household polluters, are chosen indirectly by the municipal executive. Chambers of Commerce select representatives of industries discharging their waste into rivers and water bodies. The Boards are financed entirely from two taxes—namely the general Water Board charge and the pollution levy—levied on those concerned.

RIVER BASIN COMMITTEES OF FRANCE

The French 'River Basin Committees' are composed of 60-110 persons, representing interested parties; the national, regional, and local governments; industrial and agricultural groups, and citizens. The Committees approve the long-term (20 to 25 year) schemes for developing water resources. Every five years they vote on an Action Plan to improve water quality. In addition, they vote annually on two fees to be paid by water users within the basin: one based on the level of water consumed and the other one based on the level of pollution.

The policies set by the Committees in their basins form the basis for the preparation of Action Plans for the implementation of the long-term development plan for the basin's water resources, a five-year programme, and the level of fees and incentives for various users. The responsibility for proposing the Action Plans, and implementing them, vests with the Financial Agencies. For this purpose the Agencies generate and use extensive data on the current and targeted quality and quantity of water and industrial effluents.

During the last three decades these Committees have become the medium and the focal point for negotiations and policy-making at the river basin level. The fact that all 'stakeholders' are involved in the process facilitates coordination among all the parties involved and securing the general consensus necessary for managing of the water resources in accordance with the agreed overall plan.

WOMEN'S PARTICIPATION

The importance of ensuring greater and more effective participation of women in water management has been increasingly emphasized in discussions relating to water. This is particularly relevant in those spheres, notably domestic water use, where women play a central and active role. There is also a view that even in irrigation, women should be given a larger role and more rights over water. These aspects are not covered in any depth by the contributions summarized above though the issue did figure in the discussions on these papers. A fuller, though general discussion of the issue is provided in Chapter 15, 'Some Gender Dimensions in Managing Water Scarcity' by Loes Schenk-Sandbergen, and Chapter 16, 'Gendered Poverty Alleviation Under Water Scarcity' by Barbara van Koppen, which comprise the fourth section of the volume.

SUMMARY OF DISCUSSIONS

Conflicts over water between different river basins, between agricultural and non-agricultural uses, and among agriculturists using water for irrigation are evidently pervasive. The growing menace of water pollution from untreated industrial effluents and urban sewage in many areas has also been highlighted. At the same time it is clear that the nature and severity of conflicts varies considerably from case to case.

It is evident from the Indian case studies that in general the responses on the part of government and water managers to emerging scarcity and conflicts have been inadequate and ineffective. Fragmented planning of water resource development; little concern for efficiency of use, conservation, and other aspects of demand management; overemphasis on augmentation (more storages, conjunctive use, encouragement of groundwater exploitation), and encouragement of excessive and wasteful use by means of subsidized supplies are serious weaknesses in water resource planning.

Projects are planned and designed on incomplete and unreliable data. Efforts to fill this gap are inadequate and weak. The sensitivity of decisions to inadequacies data is rarely examined in a systematic manner. The technical complexity of problems, the difficulty of balancing competing claims, and inadequate access to facts and expert assessment of possibilities and implications limit the generation of options for solutions.

Allocation principles and rules are fuzzy, incomplete, and often internally inconsistent. The affected stakeholders have to rely largely on informal, individual/community efforts to secure favourable action from officials (through bribery) and/or for exploiting advantageous locations, local political power or access to power centres.

Resort to formal procedures such as petitions to government, organized political pressure, and seeking redress through courts does not happen as much as one would expect because the costs of mobilizing and sustaining the necessary effort are very high (because of large numbers, limited resources, and internal conflicts) and the probability of success low. Adjudication (through courts and tribunals) is costly and time consuming, the outcomes uncertain, and enforcement of judgements weak.

Governments have been unwilling or unable to confront these problems. The responses in terms of legislative changes have been tardy and the laws are often defective. Weak enforcement of regulations compounds the problem.

INTEGRATED PLANNING AND MANAGEMENT

The complexity and difficulty of water management came into sharp relief in the course of the discussion. With the scope for supply augmentation being limited in many areas, it has become necessary to find ways to conserve water and use it more efficiently and less wastefully. There is also the intricate task of striking a viable balance between the interests of different claimants and ensuring long-term sustainability of the resource. There are several sources of water and numerous claimants over limited supplies, and they are all interrelated in complex ways. Deciding how available water from a given source(s) is to be allocated between various claimants involves balancing considerations of social costs and benefits and their equitable distribution among stakeholders. Water resource planners have also to accommodate the demand for water for uses other than domestic needs, agriculture and industries (such as recreation, and fisheries), the need to maintain minimum regime flows in rivers, preserve riverine and estuarine eco-systems, and concern for long sustainability in water resource management. Matching multiple sources and uses to meet multiple objectives underscores the importance and the necessity for integrated planning and management.

Clearly, this is not possible if, as is presently the case in India, water resource planning is dominated by civil engineers and hydrologists. Even in their limited domain, civil engineers tend to be far too preoccupied with the design and construction of dams and canals and do not give due importance to the treatment of upper catchments and drainage in the commands essential for effective and sustained use of harnessed water. Irrigation project design does not take ino account the scope for conjunctive use of surface water and groundwater or of large and small storages. Civil engineers and hydrologists are not equipped, even if they are interested and concerned, to address issues relating to the relation between irrigation regime, crop patterns, and productivity, the impact of alternative crop patterns on overall productivity and its distribution across regions and groups, socio-economic implications of alternative source-use combinations, and issues relating to environmental impact and long term sustainability. The fragmentation of responsibility for different sources and functions and the fact that stakeholders are not consulted, much less invoved, in decision-making compounds the problem.

Therefore, while civil engineers and hydrologists will continue to play a major role in articulating the technical possibilities and constraints of exploiting the water potential, integrated water resources planning will in addition require the expertise of several other disciplines: geologists, agronomists, public health engineers, environmentalists, legal experts, economists, and social scientists. These specialists should not only work as a team in the planning process but also work closely with those responsible for planning land use, transportation, urban settlements, and infrastructure. The creation of such an institutional structure at different levels in the hierarchy of natural hydrological units (river basins, sub-basins, and watersheds) is a major challenge. There is a parallel need for similar institutional arrangements for the management of international rivers, lakes, and seas in a manner that reduces the threat of conflicts and environmental degradation of common resources.

Bringing together the professional expertise of civil engineers and hydrologists with the expertise mentioned earlier to work as an interactive team is necessary but not sufficient for integrated planning. It must also provide mechanisms for mediating between conflicting interests to arrive at acceptable and implementable solutions. This cannot be done if planning, financing, and management—even if integrated at the technical level—vests entirely with the government as the latter tends to be arbitrary, opaque, and open to corruption and political interference. Effective monitoring and fair enforcement of rules—even if they are consistent and clear—is difficult and often not possible. Stakeholders have neither the interest nor the opportunity to participate in these processes.

There is a school of opinion, located mainly in international organizations and aid agencies, that advocates an arrangement whereby the State defines 'property' rights in water, legislates the basis of determination of such rights, and outlines broad principles and mechanisms for regulation/adjudication, leaving the development of resources wholly to the stakeholders and/or the private enterprise. The allocation of water between uses and users are to be determined by the trading of water rights through the market mechanism.

Being a fluid, common pool resource, commoditization of water and allocation through markets is inherently difficult except over a very limited domain. It is extremely difficult, indeed impossible, to define clear and enforceable property rights. The market is insensitive to issues concerning equitable sharing of water between different segments of a basin and within a particular system and also to issues concerning environmental impact and long-term sustainability. Moreover, a regime of fixed and permanent property rights impedes the flexibility to adjust allocation in the light of changing conditions of supply and demand. Public management through representative, democratic, autonomous organizations, with full participation by stakeholder representatives at all levels and phases, is essential to address these problems.

All this calls for far-reaching changes in the way water planning institutions are organized and function. There is a growing recognition that such changes are needed though the commitment and the will to bring them about is very uneven. Vested interests in the existing arrangements will have to be overcome, making the task at hand difficult. But based on specific instances cited in several papers, it is apparent that it is not impossible. Though many problems remain, substantial progress has been made in Europe in creating institutional frameworks and mechanisms for integrated water resource planning. A conscious effort to ensure that planning is a truly inter-disciplinary effort, active involvement and participation of all major stakeholders in making decisions, and strong enforcement mechanisms are distinguishing features of the cases cited. Some of these developments are of relatively recent origin, having been pressured and facilitated by an increasing awareness among the professionals and the public of growing water scarcity, deteriorating water quality, and pollution. They show that stakeholder participation in management leads to a better understanding of different viewpoints and constraints and leads, through a process of informal negotiation, to arrive at generally acceptable and enforceable solutions. The process is helped by the mediation and support of NGOs and experts.

In India the necessity for integrated planning and management of water with stakeholder participation is now widely recognized. This finds expression in the Central government's National Water Policy Statement of 1987 and, much more strongly and explicitly, in the revised version adopted in 2002. The operationalization of the policy is, however, largely left to the state governments whose commitment to this concept and willingness to implement it is uneven and so far rather weak. Some states, for example Tamil Nadu, have long been promoting 'farmers participation' in irrigation management by setting up water users' associations at the tertiary level. More recently, some states (notably Andhra Pradesh) have enacted legislation for formally including elected representatives of farmers in the management committees at all levels of irrigation systems. Some have brought together government agencies dealing with different sources and uses of water under a unified water resources organization. These are promising beginnings but only just that. They are incomplete, halting, and fall far short of integrated planning and management.

Thus in Andhra Pradesh, which has legislated perhaps the most radical change, not all stakeholders seem to find a place in the management committees even at the system level. (Practically all states, including Andhra Pradesh, are far from enthusiastic about setting up empowered and autonomous basin level organizations.) How much autonomy and power the system management boards will have to make and change allocation rules, to levy and collect water rates, and to enforce compliance are left vague. In reality, the effective power in all these respects still vests with the government and its bureaucracy. It is not even clear whether these committees are used to consult the user representatives and ascertain their views.

Entrusting the responsibility and the resources to the tertiary level users' associations for maintenance and repair has, by all accounts, made a significant difference to the quality and cost of works. There is little hard evidence to show that it has made a significant difference to the distribution of water and the efficiency of its use. Also this mechanism is not very effective unless the relevant information is freely and fully available to the public. Even then, stakeholders may not be in a position to articulate technically sound alternative solutions and their implications for various interests.

Changes in organizational structure by themselves will not be sufficient. Overcoming the narrow perspectives of specialists and getting them to work together and with stakeholders as a team is far more important and far more difficult. The process will be facilitated, perhaps even stimulated, by public action in civil society to secure fuller, freer access to information on the functioning of the water management organizations at all levels and intensifying the as yet nascent and scattered efforts of NGOs to mobilize the capacity to tackle complex technical issues.

The success of institutional restructuring also requires major changes in the legal framework and approaches to conflict resolution. They must address the basis for and the nature and content of 'water rights' both at an individual level and between different communities sharing the common water resource pool in a basin, aquifer, or watershed. The current concepts of 'rights' of different individuals and groups affected, beneficially or adversely, by water projects (both *inter se* and in relation to government) are quite vague, often conflicting, and not appropriate for integrated participatory management for efficient, equitable, and sustainable development. Similarly, the present mechanisms of monitoring and enforcement, depending wholly on the state bureaucracy, are costly and quite ineffective.

Decentralization of authority to autonomous self-managed organizations combined with built-in incentives (including, especially, proper pricing of water) for compliance should be given a much larger role. Similarly with dispute settlement procedures, instead of relying wholly or mainly on the bureaucracy and adversarial proceedings in courts and tribunals, alternative mechanisms to facilitate negotiated mutually acceptable compromises through negotiations between disputants directly or through mediation need to be explored. The Rotterdam Port case provides a good example of such alternative mechanisms.

FUTURE DIRECTIONS

In sum, there is a strong convergence of opinion that efficient, prudent, and equitable use of water resources calls for an integrated approach covering all sources and uses from the watershed through to the basin. This requires not only a multi-disciplinary team at the technical level but also consultation and involvement of stakeholders both in planning and in operations. Representatives of stakeholders must effectively participate in all phases and at all levels of management. The State should provide the basic enabling/guiding legal framework defining the nature of water rights and mechanisms for regulating prices and set up dispute settlement mechanisms. Management must, however, be left to autonomous organizations empowered to levy and collect water rates and regulate water allocation and enabled to enforce rules and be financially self-reliant. For this purpose it is important to create strong incentives for efficient operation by insisting on stakeholder contribution to the set-up costs, and on each project or system being financially self-reliant.

It is desirable that, as far as possible, rules for allocation of costs and benefits among different stakeholders, their respective rights and obligations *inter se* and relation to the system, and conflicts arising therefrom be settled within these organizations through a process of open discussion and negotiation between these interests, with provision for periodic renegotiation in response to the experience.

Pending these major institutional reforms, NGO initiatives can be very important for creating awareness by spreading essential water literacy in local communities and encouraging participatory exploration of local problems and remedies. It is worthwhile to explore the possibility of NGO initiatives to foster a serious and sustained dialogue among stakeholders of selected conflict-ridden systems to better understand the nature of their problems, and explore feasible solutions and their implications. This would give a better idea of the potential and problems of solving problems through negotiated social contracts. These NGOs could be a vehicle for the stimulation of freedom of information regarding rules, changes in rules, actual water deliveries in specific systems, as well as for arrangements to generate relevant, reliable, and timely information on water resource development and use.

For the institutional strengthening of the actors in water management and planning it is essential to enhance research on the following:

(a) the existing legal framework, and its underlying concepts; interpretations thereof, reflected in case law and tribunals awards; and implementation of judicial decisions;
(b) the elements of a framework of laws and institutions to facilitate and promote negotiated settlement among claimants for water;
(c) the structure and functioning of existing irrigation and water management institutions in selected systems, the perceptions of concerned interests (different user groups, managers, those adversely affected) regarding defects and their attitudes to alternative ways of organization;
(d) a critical assessment of the experiments with user participation and their lessons;
(e) collate and evaluate experiences of participatory, integrated management in other countries and their lessons for design of better institutional arrangements in India.

It would be worth considering the possibilities of using the IWT type of mechanisms for public discussion of specific water related disputes and, where a large and powerful stakeholder has been adversely affected and has legal claims for redress from offending stakeholders, to explore the scope for applying the strategy followed by the Rotterdam Port Authority. Also useful would be a close study of the legal, institutional, and operational aspects of participatory planning and management of water along the lines of the Water Boards of Netherlands and the French River Basin Planning Authorities. These suggest promising avenues for action research in the Indian context.

CHAPTER 1

Managing Conflict in the Bhavani River Basin

PAUL P. APPASAMY

INTRODUCTION

River basins in the arid and semi-arid regions of India are increasingly prone to competition between various users. Both surface water and groundwater are plagued by pollution, overuse, and other external problems, i.e. the use by one sector/user has begun to impinge on the rights of others. The traditional systems of water rights are breaking down and need to give way to a more rational allocation of water resources. In this context, conflicts over access to and use of water between uses and users, and between polluters and those adversely affected by pollution, are endemic and pervasive. All these myriad conflicts are manifest in the Bhavani river basin of Tamil Nadu.

WATER UTILIZATION IN THE BHAVANI BASIN

The Bhavani river, a tributary of the Cauvery, has its source in the Western Ghats, the mountain range which divides peninsular India into two parts. The eastern plateau is to a large extent semi-arid because the Western Ghats block the south-west monsoon. Bhavani, like other rivers which flow eastwards from the Western Ghats to the sea, provides badly needed water to the parched plains in the rain-shadow regions. The river passes through the Nilgiris and flows through Coimbatore and Erode districts before joining the Cauvery, the inter-State river which ultimately flows into the Bay of Bengal.

The water resources of the basin are intensively, and almost fully, exploited. In the pre-British period, the local kings constructed two *anicut* (water diversion) systems. The Kodiveri *anicut*, built several hundred years ago, diverted the waters of the river along two canals—the Arakankottai (54.4 km.) and the Thadapalli (144 km.)—irrigating about 9,900 ha.[1] Part of the command area is also used to grow commercial crops like sugar cane. A second *anicut* is located just above the confluence of the Bhavani with the Cauvery. It feeds the Kalingarayan canal (roughly 90 km. long) commanding about 7,300 ha. The farmers in the command of these two canals (together known as the Old System) are entitled to grow two paddy crops as well as commercial crops like sugar cane. They have a recognized right to the release of water (about 726 mcm a year) on a priority basis.

A new large stone masonry and earthen dam was constructed during the period 1948-56 at the confluence of the Bhavani and the Moyar rivers to extend irrigation in the basin. This project, known as the Lower Bhavani Project (LBP) and also as the New System, consists of the Bhavanisagar reservoir (with a storage capacity of about 940 mcm), a main canal of about 197 km., and branch canals and distributaries of 1,120 km. in length. It was originally designed to provide about 870 mcm a year to some 81,000 ha. for growing relatively less water-intensive, 'irrigated dry crops' like cotton, groundnuts and millets.

The Bhavanisagar reservoir is also utilized to culture fish. Fishermen in the area are licensed to harvest the fish and sell to the Tamil Nadu Fisheries Development Corporation, which in turn markets the catch through private traders. Illegal harvesting is done by unlicensed fishermen in the area, resulting in conflicts with the licensed fishermen. Pollution of the reservoir by industrial effluents is also a concern because of the reduction in yield and the disappearance of certain species of fish. When pollution levels are high, there have also been fish kills.

Traditionally, shallow wells using manual or animal-powered lifts were widely used both for domestic use and for agriculture. During the last 50 years the number of wells and, more importantly, the number of wells using energized pumps has increased manifold. Currently, there are some 51,266 wells in the basin, irrigating a net area of 45,554 ha. and a gross area of 62,723 ha. (MIDS, 1997). More than half of the wells are in the command area of the LBP system. Water levels in the command area remained stable at a depth of about 10 m. in winter and 30 m. in the summer months. Farmers in the canal command conjunctively use well water and surface water to raise different crops.[2]

Electric or diesel power is used to pump water through cement concrete or PVC pipes from the river to the field. The high initial investment is the major constraint. Unauthorized river pumping attracts penalties of up to 20 times the normal water rates. Direct pumping of water from the river, and from canals, has been objected to by farmers in the Old System and in the LBP. A 1985-6 estimate indicated that about 2.8 per cent of the net area in the basin was irrigated through river pumping (MIDS, 1997). The majority of farmers who directly pump water from the river or canals raise commercial crops like sugar cane, banana, and turmeric. Farmers who cultivate in the catchment area of the Bhavanisagar contribute to the siltation of the dam (Sundaresan, 1993). (The quantum of groundwater extracted in the basin is substantial but its magnitude is not known.)

The bulk of the water drawn through the Old and the New Systems as well as groundwater is used for irrigation. As we shall see presently, the usage of water for agriculture has increased much beyond the available supplies, resulting in reduced deliveries to the Cauvery delta and increased competition within the basin. Domestic and industrial uses, though relatively small (currently 5-10 per cent of the total use), are growing rapidly and are contributing to an increase in the intensity of competition and conflict between uses and users for available water resources.

Domestic Use

There are five towns which draw about 20 mld from the river for drinking water. About 60 per cent of the population in these towns are covered by protected water supply. None of the towns has an underground drainage system. Waste water combined with storm water flows through open drains and reaches the river (MIDS, 1997). There are also some 325 villages in the basin with a total population of about a million which use river water through various schemes implemented by the Tamil Nadu Water Supply and Drainage Board. The towns and villages are scattered throughout the basin—both upstream and downstream of the Bhavanisagar dam. There are concerns about the quality of water in these towns and villages on account of industrial and bacteriological contamination.

A substantial amount of the basin's water is also used to meet domestic and industrial needs in the neighbouring Noyyal basin. A reservoir in the upstream of the river, at Pillur near the base of the Nilgiris, was essentially meant for the generation of hydroelectric power. About 131.5 million litres per day (mld) of water is drawn from this reservoir to meet the needs of the Coimbatore urban area in the Noyyal basin. Another scheme has been commissioned to lift about 50 mld of river water over the ridge to the town of Tiruppur, also in the Noyyal basin. The rapid growth of the hosiery industry in Tiruppur has led to an escalating demand for water. There is a proposal to tap 185 mld of water at a point near the confluence of the Bhavani and the Cauvery. The advantage of this location is that it would be downstream of most of the agricultural activity in the basin, and would not, in a strict sense, be competitive in use. However, if there had to be some assurance that a continuous supply would have to be provided, a competitive situation might emerge.

Industrial Users

Industrial usage is smaller than domestic consumption, probably no more than 70-80 mld. Industrial demand is growing and putting some pressure on supply for other uses. But its impact on the basin's water situation is not so much in terms of quantity as on water quality.

Industries are located both upstream and downstream of the Bhavanisagar dam. The largest unit is a viscose rayon plant drawing about 50 mld, and this plant returns the waste water to the river. When the unit was set-up in the 1960s, the management entered into an agreement with the Public Works Department (the agency responsible for managing the river) to return the water after use so that the river flow would remain unaffected. It soon became apparent that the pulp and rayon plant was discharging wastes with a high organic content and sulphur-bearing compounds. Since the plant was located *upstream* of the dam, the effluents from the plant collected at the dam, particularly before the onset of the monsoon. There is a cluster of six bleaching and dyeing units *upstream* of the dam which take about 2.6 mld of water and discharge about 1.7 mld of effluents. Also, seven units, which intend to start operations, will discharge an additional 3.6 mld after treatment. They plan to divert the treated effluents on to land to avoid contamination of the river (MIDS, 1997).

During the summer months (April-May) when there is low flow in the river, water quality deterioration was observed. Fish kills have also occurred due to oxygen depletion. After the monsoon when the reservoir fills up, the water is released into the LBP main canal. On some occasions, coloured water was seen to flow through the irrigation canals to the fields. Farmers downstream claim that the effluents have adverse effects on the productivity of their lands. A detailed assessment of the impact on crop productivity has yet to be carried out (Sundaresan, 1993).

Downstream of the dam there are two large sugar mills which discharge about 2.35 mld of effluents after partial treatment for the cultivation of sugar cane and horticultural crops. There are concerns about the long-term impact of the effluents on the land and on the quality of groundwater in the nearby wells. Land disposal has become the preferred method of effluent disposal in many semi-arid regions of India where water is urgently needed for irrigation. Since effluent standards are much less stringent for land disposal compared to the discharge into rivers and streams, industries have chosen this option. In theory, if the treated effluents could be used as an irrigation source, it would be a Pareto-superior option (i.e. everyone would be better off). Unfortunately, the *continuous* application of treated effluents on to a finite area of land can cause long-term impact due to the build-up of salts on the surface and sub-surface layers of the soil, as well as contamination of the groundwater.

There are also a large number of dyeing units (66 units) and tanneries (62 units) located near the confluence with the Cauvery (near the town of Bhavani) and along the Kalingarayan channel. They draw water from the river/canal and discharge their effluents either directly or indirectly into drains or gutters. Some common effluent treatment plants (CETP) have been proposed for some of these units, since it would be too expensive for each unit to have individual treatment plants (MIDS, 1997).

Altogether, as the harnessing of the basin's water resources has reached, if not exceeded, its potential, competition and conflict over access to and use of water within and between sectors has increased. The emergence and spread of water pollution has added another, and increasingly important, dimension to these conflicts. In the following pages, we discuss the sources and nature of these conflicts, the manner in which they have been tackled, and the adequacy and effectiveness of these efforts.

CONFLICTS IN THE BASIN

Conflicts Within the Agricultural Sector

As the largest water-using sector in the Bhavani basin involving thousands of cultivators, the potential for conflict is greatest in the agriculture sector. There are substantial conflicts between:

- basin farmers and delta farmers in the Cauvery delta;
- farmers in the Old System and farmers in the LBP system;
- farmers in different segments of the LBP system; and
- farmers in the command areas of the system and those who pump water from the river and the canals.

BASIN FARMERS AND DELTA FARMERS

When the LBP project was planned in the first half of the twentieth century, farmers in the Cauvery delta expressed concern that the impounding of water in the Bhavanisagar dam would result in reduced flow into the Cauvery, and therefore into the delta. These apprehensions were taken into consideration in the following way by the then government:

> During the first half of February all flows in the river up to 1000 cubic feet per second (cusecs) will be passed down for existing irrigation and only any excess over 1000 cusecs will be impounded in the reservoir. Similarly during other periods, only excess over the limits fixed above will be impounded. (*Report on the Lower Bhavani Project—1946 Scheme*)

In other words, the riparian rights of the farmers in the Cauvery delta take precedence over the rights of the basin farmers.

The sharing of the waters of the Cauvery (which originates in Karnataka) is itself in dispute between the two riparian states of Karnataka and Tamil Nadu. A special tribunal has ruled that adequate water for the delta farmers requires the release of

TABLE 1: ISSUES FROM THE BHAVANI TO THE CAUVERY DELTA (1954-5 to 1991-2)

Period	1954-5 to 1963-4	1964-5 to 1973-4	1974-5 to 1983-4	1984-5 to 1991-2	Entire period 1954-5 to 1991-2
Issues to the Cauvery delta (mcm)	797	129	333	43	295
Nil issues (No. of years/ Total years)	0/10	4/10	3/10	6/8	13/38

SOURCE: Sivanappan (1995a)

5,880 mcm by Karnataka. The Bhavani waters are currently fully utilized in the basin with little or no surplus, as shown in Table 1. In the years when the delta receives an inadequate supply from Karnataka, the augmentation has to be provided by the Bhavani project, since the delta farmers have riparian rights. However, farmers in the basin have protested against the drawing of water from the Bhavanisagar dam for the benefit of farmers in the delta (Raju, 1997). That the latter have prevailed over the delta farmers is evident from the fact that the issues to the Cauvery delta have declined progressively since the commissioning of the LBP. In 13 out of 38 years there have been no issues, and in the final period (1984-5 to 1991-2) there were only two years when there were issues to the delta (Table 1).

THE OLD SYSTEM AND THE NEW SYSTEM

When the LBP project was implemented, the riparian rights of the Old System farmers were protected. Due to assured supply of water, farmers in the Old System are profligate in their use of water. It has been estimated that the crop water requirements plus the conveyance losses would require about 459 mcm per year whereas the issues to the Old System have averaged about 726 mcm per year (Sivanappan, 1995a). This suggests that about a 50 per cent saving in the quantum of water used is possible. The yearly issues to the LBP canal, on the other hand, have averaged 872 mcm (Table 2). The farmers' organizations in the LBP feel that the proportion of water to the LBP system could be increased without

TABLE 2: ISSUES TO THE OLD SYSTEM AND THE LBP (1954-5 to 1991-2)

Period	1954-5 to 1963-4	1964-5 to 1973-4	1974-5 to 1983-4	1984-5 to 1991-2	Entire Period 1954-5 to 1991-2
Issues to the Old System (mcm/year)	857	667	714	649	726
Issues to the LBP (mcm/year)	1,037	871	948	574	872

SOURCE: Sivanappan (1995a)

seriously affecting productivity in the Old System. The contention of the LBP farmers has some merit in that the command area of the LBP extends over 81,000 ha. whereas the Old System irrigates only 17,000 ha. (Vadivel, 1997). Furthermore, the return flow from the LBP augments the flow in the river and the canals of the system (Sivanappan, 1995b).

The LBP farmers for their part have not observed the original cropping pattern restrictions and have shifted to wet crops like paddy. The 1964 Committee on Plan Projects report observed:

> In the Lower Bhavani Project, the envisaged pattern of dry crop farming of over 200,000 acres (100,000 acres of cotton and 100,000 acres of groundnut and millets) has got vitiated by paddy farming, needing almost double the quantity of water likely to be available in normal years . . . thus denying [the] benefit of irrigation to some sectors of the command. (Government of India, 1964)

The project envisioned the growing of paddy in low-lying valleys to the extent of 10,000 acres. But within two years the area had shot up from 5,060 ha. to 25,060 ha. in 1957-8. In 1959 the command was divided into two zones and the ban on paddy cultivation was removed. By 1962-3, 52,630 ha. were under paddy cultivation. In 1969-70, a modified zonal system of year-to-year rotation with crop restriction was introduced. Farmers in the first zone were allowed to raise paddy only for the first crop in the first year and a dry crop in the second year, while farmers in the second zone did the reverse. The modified zonal system requires the release of about 1,033 mcm in any given year (689 mcm for paddy + 344 mcm for the dry crop), but in practice the average issue to the LBP has been only 861 mcm. The Committee on Plan Projects had warned back in 1964 that the project working should not be planned on an availability of more than 831 mcm.

Thus, the desire of the LBP farmers to utilize a greater proportion of the inflow into the dam is not consistent with the original objectives of the project. Improved management on their part could result in water saving and increased availability. For example, Palanisami (1995a) compares the areas in the LBP with and without the farmers' organizations (in the head, the middle, and the tail reaches) and demonstrates that the farmers' organizations have played a major role in ensuring more efficient allocation, with a consequent reduction in demand by about 2 per cent. Conflict resolution thus requires concerted action by both groups of farmers (i.e. in the Old System and in the LBP) to reduce their usage. Farmers' organizations and the on farm development (OFD) works can help improve water utilization in both systems.

CONFLICTS AMONG FARMERS WITHIN THE LBP COMMAND

The command area of the LBP system has been divided into the head, the middle, and the tail reaches, which in turn are subdivided into 10, 13, and 15 blocks respectively. Water users' associations exist at the sluice level (Irrigation Committee), at the branch level (Branch Committee), and at the distributary level (Council). Time schedules for water distribution have been developed and are strictly implemented, particularly in the tail end. Committees in the tail reaches of the scheme are more active than in the head reaches (Blomqvist, 1996).

A study by the Tamil Nadu Agricultural University shows that the *formation of committees has reduced the conflicts about water among the farmers* (Krishnamurthy, 1993, quoted by Blomqvist). Blomqvist argues that the water users' associations at the lower levels have been more effective in improving water management and reducing conflict among the farmers (within the LBP), while at the higher level they have worked on protecting the interests of the LBP farmers as a group *vis-a-vis* other interest groups such as the farmers in the Old System, farmers engaged in river pumping, or industries discharging pollutants.

FARMERS PUMPING WATER AND FARMERS IN THE CANAL COMMANDS

The possibility of raising commercial crops like sugar cane, banana, and turmeric has led farmers outside the command areas to pump water directly from the river. River pumping is done either by individual farmers or by groups using diesel or electric pump sets. In some cases, such pumping has been authorized, while unauthorized pumping attracts penalties levied by the government.

Farmers' organizations in both the Old System and the LBP system have strenuously objected to the unauthorized pumping from the canals and rivers.

> While the capacity of channels, sluices and cross-masonry works are fixed according to the requirement, the illegal tapping of water upsets the whole irrigation potential and the sufferers are the ayacutdars (i.e. farmers in the command areas) at the end. (Raju, 1997)

The farmers' organizations feel that the government has failed to regulate the illegal extraction for fear of political repercussions. The acute shortage of water in the non-command areas of the basin has made it economically attractive for farmers to raise commercial crops even after the payment of penalties and the added cost of diesel for the pump sets (Palanisami, 1995b). The farmers who have been 'regularized' do not have to incur the penalty, and may also apply for subsidized electricity supply. Thus, 'regularization', which is often a political decision, could substantially increase the profit margin of the concerned farmers. The farmers in the command areas can do little except voice their protest in different forums.

INTRA-SECTORAL CONFLICTS IN OTHER SECTORS

Intra-sectoral conflict is not as much in evidence in the other sectors. Reference has been made to the conflict between the licensed and the unlicensed fishermen. The Tamil Nadu Fisheries Development Corporation has granted fishing rights to a small number of fishermen because the availability of fish in the Bhavanisagar reservoir is limited. All the licensed fishermen live in two hamlets of the Bhavanisagar Municipality. Unlicensed fishermen from upstream areas violate the fishing regulations and also threaten the legal fishermen (Karunakaran, 1997).

There is no evidence of intra-sectoral conflict in the drinking water sector. Sometimes micro-level problems such as sharing of water from stand-posts in urban areas may lead to altercations. In rural areas, certain caste groups such as the Scheduled Castes are not allowed to draw water from the same water source. However, such conflicts, if they exist, have not emerged in any significant way in this basin.

Similarly, there is no evidence of conflict *among* industrial users either with regard to the extraction of water or the discharge of effluents. In the neighbouring Noyyal basin, there are differences between the dyeing units in respect of effluent treatment. Small units wish to have CETPs, while the large units prefer individual facilities. The Pollution Control Board, which previously insisted on CETPs for all units, has now agreed to both common and individual treatment plants.

Inter-Sectoral Conflicts

As long as the water resources of the Bhavani were predominantly used for agriculture, there was little potential for *inter-sectoral* conflict. However, urbanization and industrialization have led to increasing demand in the non-agricultural sectors both within and outside the basin.

INTER-BASIN CONFLICTS

The neighbouring Noyyal basin is a water-deficit basin and has diversified its economic base. Urban areas such as Coimbatore (population: 1 million) and Tiruppur (population: 300,000) are well known for their textile industries. Several hundred dyeing and bleaching units, which use a large amount of water, are located in Tiruppur.

The demand for both drinking and industrial purposes has far outstripped the local supply. The surface water resources of the Bhavani are the nearest source for both cities. Inter-basin diversion schemes have become a source of conflict among the existing users in the Bhavani basin. Already some 182 mld are being diverted from the Bhavani to the Noyyal basin to meet non-agricultural requirements in the Coimbatore and Tiruppur areas. The government has approved the drawing of an additional 185 mld of water

(100 mld for industries, 50 mld for drinking water in Tiruppur and 35 mld for wayside villages from the confluence of the Bhavani and the Cauvery). Farmers' organizations have protested the diversion of water outside the basin, and raised the prospect that the waste water generated will adversely affect agriculture in the Noyyal basin (Sengodu, 1997).

INDUSTRIAL POLLUTION

Industries within the Bhavani basin take water from the river/canals and discharge highly polluted effluents into the water or on the land. The discharge of untreated effluents by three industries (namely, a viscose rayon plant, dyeing units, and tanneries) has resulted in conflicts with other users in the basin.

The viscose rayon plant, set up in 1961 to produce wood pulp, viscose fibre, and rayon, takes in about 50 mld of water from the Bhavani and discharges 41 mld as effluent. Despite having effluent treatment facilities, the plant discharges organic wastes as well as sulphur-bearing compounds which accumulate in the Bhavanisagar reservoir when the dam is closed. A study conducted during 1971-81 indicated that there were severe pollution problems during periods of low flow.

> On March 30, 1974 when water inflow was extremely low, the pH of water was recorded below 4.0, dissolved oxygen 1.0 ppm, with saturation percentage 14.0 at Sirumugai. Dead plankton and fishes were observed on the surface of the water. The discharge of acidic effluents and the subsequent depletion of pH and oxygen were the cause of mortality of plankton and fishes. Similar oxygen depletion (1.5 ppm), decrease in pH (5) and accumulation of carbon dioxide (126 ppm) were noted in June 1976 when the inflow in the reservoir was of a low order. (Natarajan et al., 1981)

Data on species-wise landing in the Bhavanisagar reservoir collected by the TNFDC from 1971-2 to 1996-7 indicate the disappearance of certain species of fish. Fishermen also complain that the catch has been reduced on account of the effluents. The fish catch data show a decline during the 1980s (1982-92) compared to the earlier decade. The decline could be due partly to lower water availability in the reservoir during the period 1982-92 when rainfall was low. There has been a substantial increase in the availability in the last few years.

Farmers in both the Old System and the New System have objected to the discharge and accumulation of effluents in the reservoir. When water is released, coloured water was seen to flow to the field. There was also concern about the potential impact on drinking water in downstream towns and villages. NGOs such as the Bhavani River Protection Council took the matter to the Madras High Court. The plant was temporarily closed till the pollution problem could be sorted out. However, the closing of the plant raised other issues such as loss of employment. Thus, the pollution caused by the plant resulted in inter-sectoral conflict between industry and at least three other sectors—fisheries, agriculture, and drinking water.

The Pollution Control Board approved plans for several dyeing units to be located in Mettupalayam, *upstream* of the water supply intake for the town of Tiruppur. The Tiruppur Consumer Protection Council called for a citizens' protest (*bandh*) on 5 November 1994. Subsequently, the Collector of Coimbatore District requested the Pollution Board to locate the units at least 5 km. away from the river and to ensure the proper treatment of the effluent.

A large number of tanneries and dyeing units near the town of Bhavani draw water from and discharge waste water into the river/canals. Many of these units are in violation of the requirement that industries be located at least 1 km. away from any watercourse. The farmers' organizations have protested against the discharge of waste water into the river/canals (Sengodu, 1997; Raju, 1997).

These three cases illustrate instances of inter-sectoral conflicts that have arisen on account of industrial pollution. Effluents have the potential of affecting drinking water sources, fisheries, and agriculture. NGOs have also pointed out the potential damage that can be caused to aquatic ecosystems and biodiversity.[3] More detailed assessments have to be carried out in the basin to determine the impact on different ecosystems, agriculture, and health.

MANAGEMENT OF CONFLICTS

Water conflicts in the Bhavani river basin have risen and increased on account of the following features of development and the use of this resource:

- The construction of the Bhavanisagar dam has increased the area under cultivation more than four-fold, but it has also reduced the availability of water to the existing riparians in the basin and further downstream in the Cauvery delta.

- The change in the cropping pattern from dry crops like groundnut and millet to wet crops like paddy and sugar cane had caused conflicts within the Lower Bhavani Command, and these were resolved by the Irrigation Department by using a zonal system of water allocation.
- Unauthorized lifting of water directly from the river and from irrigation canals to irrigate non-command areas of the basin continues to cause conflicts with users in the command areas.
- The discharge of coloured effluents by the pulp and viscose rayon plant upstream of the Bhavanisagar dam has affected the ecology of the reservoir, particularly during periods of low flow. Other downstream industries, including distilleries/sugar, tannery, and textile industries, cause significant pollution problems in the basin, affecting both surface and groundwater.
- Over-exploitation of groundwater in the non-command areas has led to the lowering of the water table, thus affecting other users of groundwater.
- Diversion of water for urban uses in the neighbouring basin limits the surface water availability in the Bhavani basin.

Responses to these conflicts are varied and the results mixed. In some cases the aggrieved parties have chosen not to press their claims and in effect have accepted the loss or erosion of their entitlements. In some other cases, pressure from one or the other segment of users has induced the government to intervene by changing the allocation rules, imposing penalties on violators of the rules, and creating new institutions. In yet other cases, when the government did not act, the aggrieved parties took the matter to the courts. These initiatives have met with but limited success. Broadly speaking, conflicts in the basin have been avoided or have been reduced, but are seldom resolved, when *institutions* have had the flexibility to respond, and when the decision-making process has provided for the participation of key stakeholders.

Take for instance the conflict between the farmers in the Old System and the farmers in the LBP. When the project was implemented, the earlier users in the Old System who had riparian rights were guaranteed that their water allocation would not be affected. The LBP farmers, though critical of the excessive use of water in the Old System, have accepted the *status quo* and have not seriously questioned the justness of the riparian claims of their cousins in the Old System. More efficient use in the Old System would have released a substantial amount of water for the LBP command area as well as for other new users. But the existing system of the water law provides no disincentives for wasteful use of water in systems where allocation is guaranteed. On the contrary, it actually promotes the inefficient use of water. The law does not provide for adapting rights and entitlements to the changing patterns of water demand and technology of water use. The conflict between the two systems could be avoided if the legal institutions governing water allocation were more flexible and responsive to change.

Conflicts within the LBP, as seen earlier, arose when farmers deliberately changed the cropping pattern in the command areas of the LBP to wet crops like paddy and the water demand in the LBP area subsequently shot up. The behaviour of the LBP farmers in deciding to change the cropping pattern (to wet crops) illustrates the need to plan public projects taking into account the economic rationale of farmers. While a large public infrastructure project may provide badly needed water in a dry region such as Coimbatore-Erode, it cannot be assumed that farmers will faithfully observe the rules laid down by the administrative agencies. The violation was swift and dramatic; it was the government that had to modify the rules of the game to accommodate the needs of the farmers.

The initial attempt at zoning and removing the ban on paddy only aggravated the problem. Despite the subsequent introduction of a more complex zoning system for the equitable sharing of water *within* the LBP command, there are still conflicts between the head, the middle, and the tail reaches within the command. Water users' associations (WUA) have been effective to some extent in resolving some of these conflicts. This has also led to more efficient allocation. Thus, the *responsiveness* of the administrative machinery to the need to evolve a different water allocation system, viz., the zonal system, as well as the creation of WUAs by the farmers has led to more orderly and efficient allocation within the LBP command.

Nevertheless, it must be stressed that the driving force behind this move was the substantial expansion in irrigated agriculture through a large public project. If the LBP system had not been built, there would have been a considerable surplus in the Bhavani river flowing down to the Cauvery delta. Conflict emerged on account of the perceived or actual scarcity only *after* the construction of the LBP. The *post facto* behaviour of the farmers (in switching to wet crops) aggravated the scarcity and caused further conflicts among the different groups of users. *In contrast to the antiquated legal system, the administrative and social institutions have responded well to create a new set of rules for improved allocation.*

Water-lifting Devices

Water lifting from the irrigation canals to upland areas outside the canal command is yet another case where legal institutions have not been able to accommodate physical change (in this case, the technology of pump sets). Water lifting by bullocks was allowed as early as the 1930s in the Kalingarayan canal of the Old System. Generally, such lifting did not affect the supply to the downstream farmers. However, the introduction of diesel and later electric pump sets resulted in a sea change in respect to water lifting; the increased direct pumping from the river to raise water-intensive crops like banana has caused negative externalities for the downstream users.

Water lifting/pumping was deemed illegal and penalties were assessed by the administrative agencies. The courts have also upheld the government contention that water lifting through diesel and electric pump sets is illegal. While there have been some modifications permitting certain types of devices and regularizing others, violations of the rules regarding water lifting continue.

The 'riparian' rights of the existing users are protected to the detriment of new users in the non-command areas. Unlike the publicly subsidized farmers in the command area, these users have invested in pumps and in fuel and have often paid penalties because they perceive that the *net* benefit from irrigated crop production is still substantial. There is also an incentive for such farmers to use water as efficiently as possible. Yet the legal system punishes the efficient farmer in favour of those who are heavily subsidized by public provision. In this ironic situation, both the irrigated area and production have increased, and a group of farmers is economically better off as a result of the 'illegal' pumping. The claim is that pumping is affecting water availability to the 'legal' users. The conflict simmers on because the present legal system provides no remedy to accommodate the new users. The new users for their part use the political rather than the legal process for their continuance.

Urban Demand

Like most developing countries, India is becoming progressively more urbanized. Tamil Nadu is the third most urbanized State in the Union in terms of the percentage of the population living in urban areas. Urban demand for water for domestic and industrial use is growing. While the Bhavani sub-basin still remains predominantly rural, the neighbouring Noyyal sub-basin is urbanizing rapidly. The Coimbatore-Tiruppur-Erode corridor, home of the textile industry, is one of the fastest growing regions in Tamil Nadu and perhaps in south India. The nearest source of water to meet the urban demand is the Bhavani river. However, if the water resources of the basin are committed to a large extent to irrigation, the urban demand can be met only by reallocation.

Currently, the agency responsible for allocation is the Water Resources Organization (of the Public Works Department). Allocation is done administratively without consulting the users. Since the national and State water policies assign the highest priority to drinking water, requests by the water supply agency (Tamil Nadu Water Supply and Drainage Board) are rarely turned down. Unlike irrigation, drinking water has to be provided throughout the year regardless of the flow in the river. In this particular case, the reallocation was handled administratively through negotiations between three public agencies—the Public Works Department, the Water Supply Board, and the Electricity Board. Since the water is taken out of a hydroelectric reservoir, there was also some loss of generating capacity for which the Water Supply Board compensated the Electricity Board.

There could also be negative externalities on downstream agriculture, particularly during years of low rainfall. At the present time, the urban demand is less than 6 mcm compared to the agricultural demand of over 2,000 mcm. If the urban demand rises further, the pressures for reallocation will increase. Farmers have protested against this relatively small reallocation but have not gone to court as yet. It is conceivable that farmers, particularly those in the Old System, may be willing to give up a part of their share if they were suitably compensated. As urban water demand increases, the pressure for reallocation between agriculture and domestic use is bound to increase. But there are no well-defined rules or institutional mechanisms (such as water banks or water markets) or procedures to address the problems of reallocation and resolve the inevitable conflicts involved in the process.

In respect of water pollution, the passage of the Water Pollution Control Act and the Environment Act has put legal pressure on the industry to treat its effluents before discharge. In the long-term, process change may also be needed to meet effluent standards. Forcing the industry to act has required pressure by legislators, and ultimately by the courts. The Tamil Nadu Pollution Control Board (TNPCB) did not take the downstream impact seriously until an environmental group took the matter to court. Thus, *there was a failure on the part of the state in so far as the administrative agencies were concerned.* Nor was the industry interested in negotiating with farmers or with the citizens' groups in any 'Coasian' sense, leading ultimately to judicial proceedings which are still underway. Despite the large number of affected parties downstream, the existence of farmers' organizations and environmental organizations reduced the 'transaction costs' of pursuing a judicial resolution of the conflict.

However, the solutions to the environmental problems caused by clusters of small polluting industries such as tanneries and dyeing units require a different institutional approach. The TNPCB has recommended a collective solution—the construction of CETPs for clusters of units. Although these units have caused severe pollution problems, the level of conflict is not as high as it is in the case of the larger industries. Pollution caused by domestic sewage is evident near municipalities, but it has not yet become a source of conflict.

Thus, the response to pollution problems has been the enforcement of court decisions in the case of the large plants, and jointly financed CETPs for clusters of small units. Water quality/environmental problems are location specific, and the solutions necessarily have to focus on control at the point of discharge. The extent of pollution by non-point sources such as agricultural run-off is not known. Drainage (from irrigation) augments the flow, but may lead to the deterioration of water quality. There is no institutional mechanism to address problems of this nature. All these problems are exacerbated by archaic systems of water rights and by government policies such as restrictions on cropping pattern and differential pricing. Pollution control policies were relatively ineffective until intervention by the courts was sought by the affected parties. Non-point source pollution by agriculture, silviculture, etc., is poorly understood for lack of data, and has not emerged as yet as an area of potential conflict.

In summary, the conflicts that have emerged in the Bhavani basin have been managed wherever social institutions had the flexibility to respond. Existing systems of property rights to water have been like an albatross around the necks of users and management, frustrating negotiated settlements or the working of a market or any other rational solution. It is patently clear that 'riparian rights' is a regressive concept which must give way to more rational approaches to water management that are both efficient and equitable. On the other hand, where *social institutions* have been flexible (e.g. WUAs), conflicts have been resolved through equitable allocation.

The legal system has been relatively more successful in the case of pollution caused by large industries. It may also work in the case of small units if sufficient pressure is exerted by NGOs (evident in the case of the closure of some dyeing units in the neighbouring Noyyal basin). However, non-point sources of pollution such as agricultural run-off may require other approaches because of the diffused nature of the pollution and the large number of polluters.

The pollution/water quality conflict illustrates the need for all stakeholders (including the environmental groups) to *participate* in water quality management. The amendments to air and water quality legislation have now made it possible for citizens' groups by themselves to take polluters to court. Previously, only the State Pollution Control Boards could do so. Thus, the legal system has in a broad sense provided for wider citizens' participation. It is the administrative system which does not provide for consultation with stakeholders. Under pressure from the legislature, a technical committee of the Pollution Board met once with stakeholders, but could not take the process forward because participatory planning has not been the norm for administrative agencies. This must change in the future if water management is to be sustainable.

TOWARDS A NEW PARADIGM

The increasing pressure from both the agricultural and the non-agricultural sectors on the scarce surface and groundwater resources of the basin has the potential to generate conflict, and pollution places further strains on the limited sources, thus diminishing effective availability. Pollution also imposes negative externalities on downstream users in terms of health, lowered productivity, ecological damage, etc. In this context, the existing administrative and legal system is able to cope with the situation only to a limited extent. *Conflicts emerge when the social institutions do not have the flexibility to accommodate change and do not provide for the participation of stakeholders.* The solution, therefore, is to develop a process of planning and consultation that can constantly monitor and deal with change. Such a process must have the following elements.

Water Accounting

When water resources are becoming scarce (relative to demand), allocation or reallocation is not possible unless one can be precise about availability. Short-term changes (e.g. during a drought) require continuous monitoring of availability and use. For example, a 5 per cent saving in the agricultural sector would be sufficient to meet the entire non-agricultural demand. But, reallocation is not possible without accurate data being available on a continuing basis. The return flow (drainage) is so large that it is tapped and reused in one of the old canal systems. In other words, part of the irrigation water is used more than once in the basin. Industrial and domestic waste water is partially returned. Thus, the physical accounting must provide reliable and up to date information on the sources and uses of water, and also its re-use. It must also provide data on water quality, since quality affects downstream use. Remote sensing has made it possible to study water availability and use at the basin level, but this must be checked with ground-level data as well.

Basin Planning

A regional approach at the basin level must be adopted if the conflicts that are emerging at the local level are to be resolved. Such an approach requires the following initiatives:

1. Accurate data must be collected on water quantity and quality and on land use in order to carry out a water accounting exercise for preparing a basin plan. Geographical information systems as well as water system planning models now make it possible to work out the implications of alternative scenarios and to help arrive at an informed, generally acceptable plan of extraction and use.
2. Planning at the basin level requires the involvement of all the stakeholders—farmers, industrialists, workers, municipalities, NGOs, and government representatives. Of crucial importance in planning are the land-based economic activities which affect the quantity and quality of water in the basin.
3. Appraisal of the existing systems of water rights as well as the laws governing pollution would be helpful in resolving the conflicts among users. However, mediation rather than litigation may be a preferred way of conflict resolution in the long run. Forums or other institutional structures may have to be set-up for this purpose.
4. Environmental data (wherever available) must also be fed into the plan. Increasingly, allocations are being made to protect the environment (for endangered species, for maintaining water quality, etc.) on the basis of 'in-stream' uses.
5. Lastly, the costs and benefits of various strategies must be an integral part of the exercise if one is to ensure efficient allocation of resources.

Forum for Negotiation

The basin plan is not an academic plan. The planning exercise is a continuous one, which must have the capacity to meet new challenges and deal with changes. The legal system in particular may have to be modified to provide for changes in use. Thus, when new users create new demands on a resource which is limited in supply, solutions must be found based on the principles of equity. Similarly, polluters cannot indiscriminately impose damages on downstream users, including the natural environment. Economic instruments, mediation, and other mechanisms may have to be used. Various models of 'consultation' which have worked in other countries can be tried in river basins, such as the Bhavani.

Enforcement of the Rules

The administrative system must be sufficiently flexible to accommodate new users, changes in technology, etc. The users must have a 'voice' in the design of new rules of allocation and environmental management. Once the rules are promulgated, enforcement should be a shared responsibility between the agencies and the users to ensure that flexibility does not result in abuse or anarchy. Also, rules cannot be changed too frequently and without a clear rationale. Thus, due care must be given to develop institutional arrangements to formulate, monitor, and enforce the rules for allocation and for the control of pollution. These institutional arrangements must provide for extensive consultation with all the stakeholders in the basin.

Conflict resolution in the water sector is not a new concept. However, rapidly escalating demand combined with environmental concerns have sharpened the dimensions of the conflicts in various river basins. The Bhavani basin serves as an interesting case study to examine the nature of the conflicts as well as the mechanisms for the resolution of such conflicts. The experiences from such case studies should be helpful in determining the institutional structures that are needed, and to gradually put them in place in order to utilize water resources in a sustainable, equitable, and efficient manner.

NOTES

1. Traditionally, British units have been used in the agricultural sector. All figures in this paper are in metric units and where necessary have been converted as follows: 1 TMC = thousand million cubic feet = 28.7 million cubic metres. Metric units are commonly used in the industrial and drinking water sector. 1 mld = million litres per day = 100 cubic metres per day.
2. Conflicts about groundwater use are discussed in S. Janakarajan's paper in this volume.
3. In this paper, the damage to the ecosystem and biodiversity has not been taken up as an area of conflict. The discussion has been limited to the 'use values' of water in the economic sectors, because conflicts generally occur between different user groups. Intrinsic values like biodiversity and ecological values have become increasingly important in water resources management. Although academic institutions and

NGOs have raised these concerns in the Bhavani basin, they are not well articulated for lack of reliable data.

ACKNOWLEDGEMENTS

This paper is based to a large extent on two research projects on the Bhavani basin funded by the Swedish International Development Authority (1993-5) and by the Indira Gandhi Institute for Development Research, Mumbai and UNDP under the Capacity 21 Programme (1996-7).

REFERENCES

Aiyaswamy, P.K., 'Institutional Building: A Case of Water Users Association in Bhavani River Basin', *Proceedings of the Project Advisory Committee Meeting on Sharing Common Water Resources of the Bhavani Basin*, Chennai: Madras Institute of Development Studies, 1995.

Appasamy, Paul P., 'Sharing Common Water Resources of the Bhavani River, Tamil Nadu, India', in Jan Lundqvist and Torkil Jonch-Clausen (eds.), *Putting Dublin/Agenda 21 into Practice: Lessons and New Approaches in Water and Land Management*, VIIIth IWRA World Congress, Cairo, 1994.

Blomqvist, Anna, 'Food and Fashion: Water Management and Collective Action Among Irrigation Farmers and Textile Industrialists in South India', *Linkoping Studies in Arts and Science*, 148, Linkoping: Linkoping University, 1996.

Government of India, Committee on Plan Projects, 'Report on Optimum Utilisation or Irrigation Potential—Lower Bhavani Project (Madras State)', New Delhi, 1964.

Government of Madras, 'Report on the Lower Bhavani Project—1946 Scheme', Madras, 1948.

Karunakaran, R., 'Fisheries in Bhavani River Basin', Report (unpublished), Chennai: Madras Institute of Development Studies, 1997.

Lee, Donna J. and Ariel Dinar, 'Integrated Models of River Basin Planning, Development and Management', *Water International*, vol. 21, 1996, pp. 213-22.

MIDS, 'Physical Accounting of Water: Lower Bhavani Basin', Interim Report on Natural Resources Accounting of Water Resources, Chennai: Madras Institute of Development Studies, 1997.

Natarajan, A.V. et al., 'All India Coordinated Research Project on Ecology and Fisheries of Fresh Water Reservoirs', Final Report (1971-1981), Bhavanisagar Sub Centre, 1981.

Palanisami, K., 'Water Allocations and Role of Farmers Organisations: Results from Preliminary Studies in LBP, Tamilnadu', *Proceedings of the Project Advisory Committee Meeting on Sharing Common Water Resources of the Bhavani Basin*, Chennai: Madras Institute of Development Studies, 1995.

______, 'River Pumping in Bhavani Basin: Economics and Constraints', *Proceedings of the Project Advisory Committee Meeting on Sharing Common Water Resources of the Bhavani Basin*, Chennai: Madras Institute of Development Studies, 1995.

Raju, A.N., 'Bhavani River Basin', paper presented at the Workshop on Natural Resources Accounting for Water Resources of the Bhavani Basin, Coonoor: UPASI, May 1997.

Saravanan, P., Personal Communication, 1997.

Sengodu, K.P., 'LBP Irrigation—A View', papers presented at the Workshop on Natural Resources Accounting for Water Resources of the Bhavani Basin, Coonoor: UPASI, May 1997.

Sivanappan, R.K., 'Water Management in Bhavani Basin: Policy Issues', *Proceedings of the Project Advisory Committee Meeting on Sharing Common Water Resources of the Bhavani Basin*, Chennai: Madras Institute of Development Studies, 1995a.

———, 'Regenerated and Return Flow in Old and New Canals of Bhavani River', *Proceedings of the Project Advisory Committee Meeting on Sharing Common Water Resources of the Bhavani Basin*, Chennai: Madras Institute of Development Studies, 1995b.

Sreenivasan, A. and Sounder Raj, 'Effects of Certain Wastes on the Water Quality and Fisheries of Rivers Cauvery and Bhavani', *Environmental Health*, vol. 9, 1967, pp. 13-21.

Sundaresan, 'Environmental Appraisal of the Bhavani Basin', Tamil Nadu Water Resources Consolidation Project, 1993.

Vadivel, K., 'Irrigation Under Bhavani River: A Study', paper presented at the Workshop on Natural Resources Accounting for Water Resources of the Bhavani Basin, Coonoor: UPASI, May 1997.

CHAPTER 2

Competing Demands for Water in the Sabarmati Basin: Present and Potential Conflicts

VISHWA BALLABH AND KATAR SINGH

INTRODUCTION

Water is essential for all forms of life and is therefore invaluable. Although a renewable natural resource, water is exhaustible and has multiple uses and hence multiple demands. In many arid, semi-arid, and hard-rock regions of India, water resources have almost been completely exhausted and now water is not easily available even for drinking purposes. This chapter attempts to study and analyse the competing demands for surface and groundwater and water-related conflicts in the Sabarmati basin lying within the administrative jurisdiction of the State of Gujarat in western India. The study area is characterized by widespread industrialization, intensification of irrigated agriculture, and a rapidly growing population in Ahmedabad and Gandhinagar cities situated within the basin of the Sabarmati river.

The demand for water for domestic, agricultural, and industrial uses in the basin significantly exceeds the quantity of water available. This has led to conflicts among various uses as well as users of water. Growing industrialization has led to the pollution of the river Sabarmati and its tributaries which in turn has adversely affected the health of the rural people and the crop yields in the affected areas. The nature of conflicts and the problem of pollution are explored through case studies of three villages and one city, Ahmedabad, located in the basin. It is argued in this chapter that the present pattern of use of water is unsustainable and that in future more conflicts will crop up. It is suggested in this chapter that there is a need for evolving an appropriate institutional structure for regulating the use of scarce water resources in the basin as also to resolve various types of conflicts among water users.

A CONCEPTUAL FRAMEWORK

A universally accepted paradigm of human behaviour is characterized by the notion of rationality. However, it is also widely recognized that the 'rational' behaviour of an individual is tempered by the institutional environment in which the individual lives and works.

The institutional environment or structure comprises, *inter alia*, socio-cultural, economic, legal, and political norms defining the activities and behaviour of individuals which are permitted or prohibited in a particular action arena. These norms interact with physical and technological factors and the nature of their community in shaping their shared understanding of their world. Given the premise of rational individual behaviour and the scarcity of water, competition and conflicts among water uses are inevitable. But in order to understand the conflicts and the ways in which they are handled, all the above-mentioned variables and their interactions need to be considered.

The physical and technological aspects of resources within a water basin help determine their availability and economically feasible uses. Knowledge of various physical factors such as rainfall, soil texture, nature and extent of groundwater aquifers, and the overall hydrological cycle is important for efficient water management in a basin. Technological factors are equally important because they either constrain or facilitate the capacity of individuals and society to capture the resource and use it. For example, the traditional lift system, which depends upon human or animal power, can tap a significant amount of water only from a depth of roughly 50 feet (Dhawan, 1983: 9). In contrast, modern water-lifting devices using submersible electric motor pump sets can lift water from depths of over 1,000 feet or even more.

Supplying or selling water could be a means of livelihood for some people and a precious commercial commodity/resource for some others. The needs of a community in conjunction with its attributes and attitude towards this resource determine the pattern of its use, management, and conservation. Institutionalized rules associated with the capture, distribution, and use of water form the third set of factors that affect water use and management. These rules also determine who is eligible to make decisions in different arenas (Ostrom, 1990). Institutionalized rules are functional rules which are actually used, enforced, and monitored. These rules facilitate, constrain, or forbid individuals to make choices. Since supplying a new set of rules is a 'public good', it suffers from under-supplies and individuals prefer to opt a 'free ride' instead of engaging in the making, enforcing, and monitoring of rules.

The three sets of factors briefly described in the preceding paragraphs, i.e. physical and technological factors, attributes of a community and institutional factors—interact with the physical and socio-cultural world and create a wide variety of situations. Individuals operating in these situations take various kinds of actions to meet their needs and interests. Through this interactive process, water is allocated among its competing uses. If the needs are not met satisfactorily, competition and conflicts among various uses and users are inevitable.

A PROFILE OF THE SABARMATI BASIN

The river Sabarmati originates in the Aravalli hills in Rajasthan and passes through the state for some 48 km. before entering the state of Gujarat. After traversing a further 300 km. it empties into the Gulf of Cambay. The total drainage area of the basin is about 21,674 sq. km. of which 17,550 sq. km. (80 per cent) is in Gujarat, and the remaining 20 per cent is in Rajasthan. The entire area of the districts of Sabarkantha and Gandhinagar and about 75 per cent of the area of Kheda district fall in the Sabarmati basin. In addition, parts of Mehsana, Ahmedabad, and Banaskantha districts also fall in the Sabarmati basin (Table 1).

The basin is divided into three sub-basins on the basis of watershed sub-systems. The area covering the upper part of the catchment of the main river up to the Dharoi dam (constructed about 165 km. upstream of Ahmedabad) is designated as the Dharoi sub-basin. The second sub-basin includes the catchment

TABLE 1: DISTRICT-WISE AREA OF THE SABARMATI BASIN

S. No.	District	Total geographical area (km²)	Area within the basin (km²)	Percentage area within the basin
1	Sabarkantha	7,390	7,250	98.1
2	Kheda	7,194	5,439	75.6
3	Ahmedabad	8,707	2,378	27.3
4	Mehsana	9,027	975	10.8
5	Gandhinagar	651	648	99.8
6	Banaskantha	12,703	860	6.8

SOURCE: GOG, *Integrated Plan of Sabarmati River Basin*, 1996.

area between the Dharoi dam and the confluence of the river Khari. Since the river Hatmati is one of the main tributaries of the river Sabarmati in this sub-basin, it is named the Hatmati sub-basin. The third is the Watrak sub-basin, named after the river Watrak, another main tributary of the Sabarmati. Both the Hatmati and the Watrak rivers originate in different parts of Rajasthan and meet the Sabarmati on its left after traversing about 105 km. and 248 km. respectively from their origins (GOG, 1996). The salient features of the three sub-basins of the Sabarmati are shown in Table 2.

The upper sub-basin once had dense forests but the area is now thinned and degraded. The groundwater formation takes place under confined to semi-confined aquifers in the upper reaches followed by unconfined aquifers in the alluvial plains of the lower reaches in the Sabarmati basin. The geological configuration, type of aquifers, and topography along with precipitation determine the total water yield and the amount available for utilization through surface and groundwater. These factors also affect the type of technology used to capture water, particularly groundwater. The lower portions of the Hatmati and the Watrak sub-basins, being located downstream of Ahmedabad, need to be considered separately because they are affected by the severe pollution from urban and industrial wastes from Ahmedabad.

Demand for Water in the Sabarmati Basin

The Gujarat State Water Resource Planning Group (GSWRPG) has estimated the total water requirement for the Sabarmati basin as a whole and for the three sub-basins separately, taking into

TABLE 2: GENERAL FEATURES OF THE SABARMATI SUB-BASIN

Particulars	Dharoi Sub-basin	Hathmati Sub-basin	Watrak Sub-basin
Catchment area (sq. km.)	2,640	5,[illegible]73	9,337
Geology	hilly and rocky	rocks followed by alluvial plain	alluvial plain
Physiography	undulating dissected hills, sharp hills to hill slopes	gently sloping pediments to gently sloping alluvial plain	gently sloping to nearly levelled alluvial plain
Run-off	very high	high to low	low
Water-holding capacity	poor	good	good
Groundwater formation	confined and semi-confined aquifers	semi-confined to unconfined aquifers	unconfined acquifers
Irrigability	good	good	not irrigable in some areas due to salinity in soil
Forests	traditionally well forested, now degraded	poor and degraded forest	no forest
Main communities	dominated by Bhil tribe, interposed with new settlements	caste-based agrarian communities dominated by Patidars and Chowdharies	caste-based agrarian communities dominated by Patidars

SOURCE: Compiled from GOG, *Integrated Plan of Sabarmati River Basin*, 1996.

consideration such factors as the size of rural and urban populations, livestock population, industrial growth, and demand for water for agriculture. According to their estimate, in 1995 the total water requirement in the basin was about 3,912 mcm per year. Of this, the domestic water requirement amounted to only 365 mcm per year (9 per cent). Most (90 per cent) of the water is used in agriculture for irrigation purposes (Table 3).

There are reasons to believe that the quantity of water actually used is higher than the figures shown in the table. For instance, the domestic water requirement was estimated based on the norms

TABLE 3: ESTIMATES OF WATER REQUIREMENT FOR DIFFERENT USES IN THE SABARMATI BASIN

(mcm/year)

S. No.	Sub-basin	Domestic uses	Livestock	Industrial uses	Irrigation	Total
1	Dharoi	8.8	0.88	2.00	163.79	175.4
2	Hatmati	209.9	21.00	26.00	1,268.55	1,525.4
3	Watrak	79.7	8.00	1.75	2,055.85	2,145.3
	Total	298.4	29.9	29.75	3,488.1	3,912.1

SOURCE: GOG, *Integrated Plan of Sabarmati River Basin*, 1996.

prescribed by the government of Gujarat (140 litres per capita per day [lpcd] for urban areas and 40 lpcd for rural areas), which are well below the actual quantities of water used (GOG, 1996: 174). Many villages within the Sabarmati basin have access to as much water as is available in the urban areas. The rapid expansion of economic activity and the resulting improvement in living standards in these villages has induced a rising demand for modern sanitation facilities and in turn an increase in domestic requirements of water. Rural domestic water use has also increased due to a decline in the real private cost of collection of water following the establishment of public water supply systems either by the village panchayats or by the Gujarat Water Supply and Sewerage Board (GWSSB). Recent surveys (Prabhakar et al., 1997 and Ballabh and Shunmugam, 1998) suggest that the actual water use in rural areas is much higher than 40 lpcd.

Urban consumption is also under-estimated. The norm of 140 lpcd for domestic use is unrealistic, especially for large cities like Ahmedabad. A recent study estimated the demand for water in Ahmedabad city to be 75 lpcd for huts; 150 lpcd for chawls; and 260 lpcd for houses, flats, and bungalows (TCE, 1997). The Ahmedabad Municipal Corporation (AMC) is currently supplying 152 lpcd. And consumption is growing rapidly; it has been estimated that by the year 2010 full satisfaction of the needs of urban people would require 270 lpcd.

The GSWRPG has also under-estimated the industrial water requirement. Its figures are based on the area of industrial estates developed by the Gujarat Industrial Development Corporation (GIDC) and the size of the industries established there. For the industries outside of GIDC estates the requirement is taken to be 50 per cent of the water consumed by the GIDC estate industries. This estimate does not take into account the industrial waste-dissolving function of water.

The estimated use for livestock (90 mcm per annum) is nearly thrice the GSWRPG figure (30 mcm) (Prabhakar et al., 1997). The figure for agricultural use excludes the quantities of water supplied from the basin to areas outside of it. The Dharoi dam irrigates 45,000 ha. in Mehsana district and another 33,600 ha. outside the basin are irrigated by the Sabarmati waters downstream of Ahmedabad.

For these reasons, the actual consumption of water in the basin seems considerably higher than the official estimates. But even the official estimate is about 33 per cent higher than the estimated 2,910 mcm of potential created for surface and groundwater utilization in the basin (GOG, 1996). The additional water seems to have been extracted from mining groundwater.

Supply of Water in the Sabarmati Basin

Water flows in the Sabarmati and its numerous tributaries have been harnessed by two large, eight medium, and 569 minor surface irrigation projects. All are currently in operation. Practically all the surface irrigation potential has been fully developed and there is no further scope for new surface irrigation projects in the basin (GOG, 1996). Since the Sabarmati is a seasonal river, stream flow almost stops after December-January. This imposes a serious constraint on the surface irrigation systems.

That supplies are inadequate to meet the demand is reflected in the fact that the systems serve no more than a quarter of their designed command area. In 1995-6, the total culturable command area of all the surface irrigation systems taken together was 321,821 ha. and the irrigation command area (ICA) was 222,337 ha. However, the area irrigated by the surface irrigation systems between 1993-4 and 1995-6 averaged only 55,333 ha. or less than 25 per cent of the ICA. The surface irrigation system is also unable to provide sufficient water required for optimum crop yields. For example, in the years 1994-5 and 1995-6, the Dharoi canal system provided only two irrigations to crops in its command which is significantly less than the requirement.

The Dharoi reservoir was constructed not only to provide irrigation but also as a source of drinking water to Ahmedabad and Gandhinagar cities and to supply water required by the two thermal power stations located in these cities. Water used by the power stations is recycled, after treatment, for domestic use. The original plan envisaged that a total of approximately 250 mcm of water per day would be supplied to these two cities which agreed to share the cost of construction. Actual water allocation for these two cities averaged 175 mcm during 1980s and 1990s. As a proportion of water available in the reservoir, the supply earmarked for the two cities ranged from a minimum of 7 per cent to 100 per cent.

In the drought years of 1986-7 and 1987-8, almost all the water stored in the Dharoi reservoir was allocated to these two cities (Table 4). The State government has also made provision for carrying over and reserving water to the next year and storing it until the next monsoon begins. Such reservation often leads to wastage of water (Shah, 1997). An additional 63 mcm has been reserved for meeting the domestic water needs of 559 villages and eight towns of Mehsana district affected by flourides. All these facts suggest that drinking water receives a higher priority over irrigation, with cities and towns receiving a higher priority over rural areas. Consequently, the supplies meant for agriculture are highly variable from year to year, and even during a year.

In view of this, farmers make alternate arrangements for irrigation using groundwater. Groundwater irrigation is the mainstay for most of the farmers in the basin. Exploitation of groundwater has spread rapidly. Besides an increase in the number of wells, deepening and finding new and dependable locations for wells—all of which are costly—have become common phenomena, particularly in the Hatmati sub-basin. Farmers do not exclusively depend on surface irrigation alone. The conjunctive use of water is *de facto* practised extensively. But even in surface irrigation commands (e.g. Dharoi) the water tables have gone down substantially, signalling a tendency for consumption to outstrip supply.

According to official statistics, the total area under well irrigation (406,700 ha.) is about twice the planned command of surface systems and more than seven times these are actually served by them. Dug wells are the major source for irrigating 251,000 ha.

TABLE 4: YEAR-WISE TOTAL INFLOW (WATER STORED) AND ALLOCATIONS TO AHMEDABAD AND GANDHINAGAR CITIES FOR DRINKING PURPOSES

Sr. No.	Year	Inflow in mcm (water stored)	Allocation to A'bad and Gandhinagar in mcm	Percentage of total inflow
1	1976-7	2,096	251	11
2	1977-8	2,583	200	7
3	1978-9	578	189	32
4	1979-80	366	116	31
5	1980-1	1,256	175	13
6	1981-2	429	195	45
7	1982-3	359	205	57
8	1983-4	1,286	183	14
9	1984-5	832	274	32
10	1985-6	361	133	36
11	1986-7	135	127	93
12	1987-8	57	58	100
13	1988-9	773	123	15
14	1989-90	351	160	45
15	1990-1	1,408	147	10
16	1991-2	640	194	30
17	1992-3	2,169	198	9
18	1993-4	1,147	193	16
19	1994-5	2,428	170	7
20	1995-6	299	252	84
21	1996-7	361	124	36

NOTE: Rounded off to the nearest full number.
SOURCE: GOG, Department of Irrigation, Dharoi Subdivision, Dharoi; Mehsana, 1997.

But tube wells have also become important and now serve 123,000 ha.

COMPETITION AND CONFLICTS: SOME CASE STUDIES

Given that available supplies in the basin are short of requirements—and all evidence points to the widening of the gap—competition for capturing scarce supplies and the resulting conflicts is naturally acute. This is reflected in: (a) a growing—and now widespread—tendency to pilfer water meant for urban areas by agriculturists en route; (b) the rapid growth of well and tube well irrigation, leading to the widespread and progressive lowering of the groundwater table in the basin; and (c) the growing problem of water pollution arising from the discharge of untreated industrial effluents and urban wastes. The nature of these conflicts, however, varies, as will be evident from the following illustrative examples.

Increasing Upstream Competition for Water: Lambadia Group Panchayat

The Lambadia Group Panchayat in the upstream Dharoi sub-basin consists of a cluster of seven villages. The river Sabarmati passes through this cluster. All the villages, except Lambadia, are inhabited by the Bhil tribe who settled here two or three generations ago. Domestic water requirement was met through the water stored in pits dug in local rivulets and small streams draining into the Sabarmati river. In more recent times, several other sources have become available. These include open dug wells, hand pumps, and (in one village) community drinking water supply schemes. Although availability of drinking water is not a problem in the villages, a few of the hamlets do not have sources located in their vicinity and, therefore, women have to walk 1-2 km. to fetch water.

The water table is 20-5 feet lower than the level that existed in the late 1970s. Villagers attribute this to the excessive pumping of groundwater for irrigation and the development of surface irrigation systems on the upstream tributaries, compounded by deforestation in the upper reaches, causing soil erosion and loss of topsoil. These developments together have reduced the inflow of water into the Sabarmati. Earlier, there used to be enough water in the Sabarmati until the end of April but it dries up by the end of February or by the first fortnight of March.

Till about 1970, the villagers grew only rain-fed crops during the monsoon. In winter they used to migrate to other parts of Gujarat and Mumbai in search of employment. When diesel pumps became a dominant source of irrigation in other parts of Gujarat replacing the traditional human- and bullock-driven systems, the Patidars from other areas moved in here. Since there was a ban on the purchase of land from tribals, the Patidars started renting out diesel pumps and pipelines to landowning tribal farmers for irrigating their crops on a share basis. The owners of diesel pump sets received 25 per cent of the crop share, or 50 per cent if they also met the costs of diesel and maintenance. Several factors, such as the annual renewal of contract, a high degree of honesty among the tribals, the contiguity of homesteads and agricultural fields, helped diesel pump set owners to monitor and regulate the behaviour of the water buyers.

Gradually, over time, many tribals were able to generate surplus income from increased agricultural production. Coupled with government-supported credit and subsidy, this enabled them to acquire diesel pump sets of their own. This has helped to reduce the number of absentee water sellers in the area. About 20-5 per cent of the land is still irrigated using the rented diesel pump sets. With some of the tribals entering the water-selling business, currently about 25 per cent of the cultivated area is irrigated.

Mining Water for Profit: Dashotar Village

Dashotar village, located in Sabarkantha district, falls in the Hatmati sub-basin of the Sabarmati basin. It is a relatively large village with a population of 3,288. Though the Scheduled Castes and Baxi Panch castes are numerically predominant, the agrarian scene is dominated by some hundred households belonging to the Patidar community. The main crops grown are cotton, maize, jowar, and castor during the monsoon season and wheat and lucerne during the winter season. During summer, farmers who own tube wells cultivate fodder crops. Because of the flourishing dairy business, a significant proportion of land is allocated to fodder crops in all the three seasons.

Traditionally, domestic requirement of water was met through the step well locally known as *vav*. The construction of the *vav* is done in such a manner that evaporation losses are minimized. Water is collected mostly by women using the steps for going down and climbing up. Two such wells are still in use in the village. Built some 300-400 years ago, one well is for the affluent upper castes and another for the lower caste communities. A few old women from Scheduled Caste families report that they were not allowed to collect water from the *vav* located in the middle of the village.

As the population grew, the water demand for domestic use also rose, even as the traditional sources were declining. The villagers responded by digging more wells and converting one of the *vav* into a deep tube well. When that failed, another tube well was constructed and water carried by a pipeline. The panchayat charges a water tax for these services which, however, is not sufficient for meeting even the operational costs. Recently the village has also been brought under a government water supply scheme for a group of villages, including Dashotar.

Overall, the water supply for domestic use at the present time does not appear to be a problem. A rapid survey of five to six nearby villages, however, revealed that it is becoming a serious problem in other parts of the area; villages which are dependent only on their panchayat are now facing an acute shortage of water and the quality of the water is also not good.

Water use in agriculture in the village has undergone tremendous changes between 1960 and 1995. Traditionally, water was lifted from open wells in leather buckets using bullock power from a depth of no more than 50-60 feet. This system was known as *kos*. According to old villagers, there were more than 200 *kos* in the village until the early 1960s when diesel pump sets were introduced to lift water from open dug wells. When the yield of the wells started declining, farmers tried to capture water through making investment in vertical and horizontal bores in these open wells. As a result, the village now has more than 50-60 dug-cum-bore wells. With the coming of electricity, in the early 1970s, almost all the farmers in the village shifted to electric pumps. Subsized credit and cheap power brought newer possibilities; many farmers went for deep tube wells, but the scope was limited because of a hard pan at about 80-90 feet from the ground level. Success rates in drilling were as low as 20-5 per cent. At present there are 20-5 deep tube wells and over 150 dug-cum-bore wells in the village.

The village is also served by a distributary of the Left Bank Canal of the Dharoi reservoir project. The village is situated near the tail end of the distributary. The supply of canal water is erratic and unreliable. However, the village records show that about 500 ha. of land are irrigated by the canal and 150 ha. by tube wells. Many farmers lift water by diesel pump sets and have invested in underground pipelines (3-4 km.). A similar pattern was reported from other villages located on both the Right and the Left Bank Canals. As in other parts of Gujarat (Shah, 1993), water markets are well developed in Dashotar and the nearby villages.

The primary purpose of investing in tube wells, diesel pump sets, and dug-cum-bore wells, as indicated by the farmers, is to have assured access to water for irrigation. They sell water to others only when they have a surplus. Selling and buying of water, nevertheless, is widespread in the village. And despite tremendous growth in tube well irrigation, the water table has not been declining as it has been in other parts of north Gujarat such as Mehsana. Two factors have checked the decline in the groundwater table in the area. The hard-pan underground restricts access to the water aquifer, and the canal water supply, though erratic and uncertain, helps recharge the aquifer.

Water Scarcity and Pollution in Ahmedabad City

Ahmedabad city, situated on the banks of the Sabarmati, has a population of 2.9 million. Its total area is about 269 sq. km. of which the Ahmedabad Municipal Corporation (AMC) area covers 190.8 sq. km. It is one of the biggest and fastest growing industrial centres of the country. Its water supply comes partly from surface water released from the Dharoi dam some 165 km. upstream and partly from the pumping of groundwater. As mentioned earlier, the actual releases of water from the Dharoi reservoir for Ahmedabad city have been well below the original plan.

Besides, water released from Dharoi has to travel 165 km. before reaching Ahmedabad. In between the two points farmers have constructed a number of tube wells to irrigate their lands. The wells are located in fields adjacent to the river and linked through pipelines to the riverbed and to the fields. Farmers are not supposed to construct such wells without first securing government permission. But this is seldom done. Farmers claim a natural/riparian right over river water since they are located upstream and their forefathers settled in the area because of its proximity to the water source. Large-scale pumping of groundwater in this fashion in effect reduces the surface flow. The huge difference between the releases at Dharoi and the arrivals at Ahmedabad suggest that pilferage on the way is done on a large scale. But no action has been taken to tackle this problem.

A recent report (TCE, 1997) estimates the total quantum of water available to the AMC at 486 mld. About 47 per cent of this (227 mld) is derived from French wells and infiltration works located on the river. The balance 219 mld is accounted for by groundwater pumped by the 64 tube well stations owned and operated by the AMC as well as by a large (but unknown) number of private bore wells supplying water for domestic and industrial use (40 mld).

The demand has been growing rapidly. It has been largely met by the expansion of the groundwater supply through an increase in the number of wells and the deepening of the existing wells. As a result, the groundwater table has been falling; the water table is estimated to be falling 2-3 m. a year (Gupta, 1993). The depth of the AMC tube wells has gone down to 250 m. Water levels in the western, central, and eastern parts of the city have dropped from 12-15 m. in 1940 to 60-80 m. in 1984 and further to 90 m. in 1997. Many tube wells have been abandoned because of lowered yields and high costs. Yet digging/drilling of new wells by the government and the private sector continues unabated.

The ever-growing gap between the demand for and the supply of water has led to the decline in both the quantity and quality of the water supply to the residents of Ahmedabad city. Two coping strategies have emerged in response to these problems. First, the construction of private tube wells in the city has been going on at an alarmingly high rate. Second, to capture the maximum possible quantity of water during the limited period when the AMC supplies water, residents use electric motor pumps of 0.5 hp or more to draw more than their fair share of water. This adversely affects the supply at the tail end. In particular, the poor (including slum dwellers) who cannot afford to buy motor pumps suffer the most. The AMC's response to the water scarcity is to find newer sources of water, and it is currently involved in protracted negotiations with the government of Gujarat to provide water to Ahmedabad city from the Narmada project.

Pollution Impact on Downstream Villages

Industrial growth in the basin has not only cut into the water available for other uses, but it has also led to a deterioration in water quality. The seriousness of this problem can be judged by the fact that the industrial BOD (biochemical oxygen demand) load from Ahmedabad-Gandhinagar alone accounts for 30 per cent of the domestic BOD, compared to less than 10 per cent in Delhi, and many other cities (CPCB, 1989: 45). Water quality in the river, the French wells, and tube wells has deteriorated. The Sabarmati is rated as one of the most polluted rivers in India (Pangotra and Shukla, 1994).

Industries account for 80 per cent of the total waste water

generated in the city. Most of it gets diverted into the river. The quantity of dissolved solids (TDS) has been rapidly rising due to the intrusion of polluted water and the fly ash content of slurry from thermal power stations into the French wells. Water quality parameters of these wells are currently within desirable limits, but are deteriorating. TDS levels in bore wells, however, have already exceeded desirable limits (TCE, 1997).

Pollution of the Sabarmati has had serious consequences for agriculture and for people in areas downstream of the city. Water seepage on both sides of the riverbed extends up to 250 m. the total area adversely affected by pollution is estimated at 4,500 ha. Initially, in 1940, the AMC constructed a drain to carry sewage water. Subsequently, the drain was closed and treated sewage mixed with fresh water from the Dharoi was supplied through a canal, namely the Ftehwadi, for irrigation.

When sewage water is allowed to pass through the infiltration layer, it is automatically filtered and nitrogen, potassium, and heavy metals present in it are significantly reduced. But the raw sewage water also contains pathogens which may cause cholera and typhoid and which harbour salmonella bacteria, protozoa, eggs of worms, and all types of viruses (Gupta, 1993). Untreated or partially treated water could be used to provide limited irrigation only to selected crops such as jute, trees, pastures, and other crops which are not consumed by human beings. The treated water can be used to irrigate all types of crops. However, many researchers claim that irrigation with sewage and polluted water adversely affects the size of grains, their storage, life taste, and cooking characteristics. Further, polluted water also causes the wilting and drying of the crops irrigated with it and attracts several pests and engenders diseases. This has been confirmed by a study of Gyaspur village in Ahmedabad district and a survey of seven villages conducted by the Institute for Studies and Transformation (IST), Ahmedabad.

In the Fatal Grip of Sabarmati Polluted Water: Gyaspur Village

Located on the downstream side of the Sabarmati, the village of Gyaspur is surrounded by the river on three sides. It is about 12 km. away from the main Ahmedabad city. According to the

1991 census, the population of the village was 5,291 and the number of households was 1,015 comprising Thakores (400 households), Prajapathis (65), Bharwads (72), and others (478). The total geographical area of the village is 826 ha., of which about 31 per cent is cultivated. The main crops grown in the village are paddy, jowar, fodder, and pulses during the rainy season and wheat during the winter season.

The villagers fear that they may have to leave the village because their cultivable lands are increasingly becoming unproductive. Besides contaminating the drinking water, the pollutants discharged into the river are leading to crop failures and an abnormal rise in the incidence of different diseases. Since the late 1970s drying up of crops and pest infestations have become very common in this village. A sharp rise in insect-borne diseases, skin diseases characterized by irritation, black patches, and dry skin, and cases of miscarriages have also been reported by villagers.

Approximately 45 per cent of the cultivated land is irrigated. The main sources of irrigation are some open wells, tube wells, and the Fatehwadi canal. In addition, there is a tube well owned by the AMC, which exclusively supplies drinking water to the villagers. There are 48 open wells, of which one has been abandoned and three are dysfunctional. There are five tube wells which exclusively provide irrigation water. All of them are privately owned by the farmers. Besides, diesel pump sets (40-50) of 7.5-10 HP are used for drawing water directly from the Sabarmati river and the Fatehwadi canal.

Originally, the village used water from the AMC sewage drain for irrigation. As per the farmers' perceptions, the practice in the beginning actually increased the fertility of agricultural lands in the village. The increase in yield was, however, short lived and the adverse consequences of this practice were soon apparent. In 1973-4, the supply was stopped. After the construction of the Dharoi dam, the Fatehwadi canal was excavated to supply to mix of sewage water and fresh water. The sewage water is treated at a plant located near the village before it is mixed with the fresh water.

The wilting and subsequent failure of the growing crops in the agricultural lands adjacent to the treatment plant is a very common phenomenon in the village. All over the village, pest infestations and outbreaks of diseases are on the increase and crop yields have gone down by as much as 50 per cent. The taste of the food grains has also deteriorated and so has their keeping quality; the storage life of the grains which in the past used to be 12 months is now reduced to less than three months. Cultivation of high-value cash crops like potato, papaya, sugar cane, watermelon, sweet melon, and guava is now almost impossible. In a study conducted by the IST, Ahmedabad in seven downstream villages of the Sabarmati basin, 50 per cent of the respondents reported crop failures; 73 per cent reported yield losses; 69 per cent reported change in the quality of grains, and 93 per cent reported pest infestation in crops due to polluted water (IST, 1996).

The impact of polluted water is not limited only to the crop lands and livestock but also extends to human beings. The polluted water stagnates in ditches and ponds in the village and has become a good medium for the breeding of mosquitoes. The incidence of malaria is very high in this village. Further, dangerous disease-causing pathogens are carried to the downstream villages also. Besides, due to high degrees of soil moisture created by the water, a lot of shrubs and trees have come up on the village common lands. The busy growth of the shrubs and trees provides a safe hideout to thieves and other anti-social elements, leading to an increasing number of thefts and looting, due to which the main path to Ahmedabad has fallen into disuse.

The area of the *kakus* plant is used for the disposal of dead animals. The carcasses pose serious health hazards. The dogs who live on the rotten remains of the dead animals catch the dreaded disease rabies. Many men, women, and children are bitten by these dogs every year and die of hydrophobia. The bad smell and pathogens generated from the putrefaction of the carcasses are spread throughout the surrounding areas, making them uninhabitable. Most of the children in this area suffer from swollen stomachs, jaundice, and polio. The Pirana sewage plant also emits a bad odour and pollutes this village. Older people and children suffer from serious respiratory problems, vomiting, and headaches. According to a local doctor, the incidence of diseases is highest during the monsoon season and this, in his opinion, is due to the pollution created by the two treatment plants.

Until 1960, the open well water was safe to use for all domestic purposes and no problems were encountered. But now the groundwater aquifer is also contaminated due to the percolation of polluted water. Water stored in vessels for two-three days causes the appearance of dark and white spots on the containers. Water drawn from 200-m.-deep bore wells appears normal at first glance but food prepared using this water develops a pinkish layer on top within a few hours after cooking.

There is poor coordination between the pollution-monitoring and the controlling agencies. Also, the powers of such agencies are restricted by legal complexities. Added to this is the double standard used by the organizations involved in the supply of drinking water to rural areas (Shunmugam and Ballabh, 1997, p. 10). The data supplied by two organizations—the AMC and the GWSSB—from an analysis of a sample of drinking water from a well in Gyaspur village, located downstream of Ahmedabad, prove this point (Ballabh and Shunmugam, 1998). The AMC concludes that the water is potable whereas the GWSSB's conclusion is that it is not.

CONFLICTS OVER WATER: A SYNTHESIS

The analyses and discussion presented in the previous sections are synthesized and the present and potential conflicts in water use and management in the Sabarmati basin are identified here. The synthesis is done with reference in the conceptual framework developed at the beginning of the chapter.

We begin with the assertion, based upon the empirical evidence presented in this chapter, that the total quantity of water used in the basin is more than the quantity supposedly developed from both surface water and groundwater sources and that the demand for water has been rapidly increasing over time in both urban and rural areas. A conservative estimate suggests that the demand for water in the basin would increase to 4,343 mcm/per year in the 2001 and 5,782 mcm/year in 2021 (GOG, 1996). It is also projected that the domestic and industrial demands would increase more rapidly than the demand for agricultural uses. This means that the present use pattern is not sustainable and that water is becoming scarcer by the day and that the stage is being set for al-round competition for capturing the rapidly depleting water sources in the basin. We have identified the following conflicts in the use and management of water resources in the basin.

Conflicts between Upstream and Downstream Users

The rapid increase in the use of water in agriculture in the upstream area, as typified by the Lambadia case study, is causing scarcity in the downstream area. In the absence of well-defined property rights in river/stream water, surface water sources in the basin are *de facto* open-access resources and therefore are being over-exploited; the riparian doctrine does not promote the socially optimum use of water.

Pump set-owing farmers not only irrigate their own land but also sell water to others. In fact, many absentee owners of pump sets rent them to the landowners and appropriate quite a big chunk of the surplus generated by irrigated agriculture. Downstream people attribute the declining water table to the excessive demands made by these upstream farmers. The upstream of the Sabarmati basin has a hard-rock geological formation and confined aquifers. Therefore, the overdraft of water from such aquifers is difficult to correct through increased recharge. Although the tension between upstream and downstream users in the Dharoi sub-basin of the Sabarmati has not yet reached alarming levels, indications are that these issues will become important in the near future.

A similar situation obtains in the alluvial plains of the basin. The extraction of water from unconfined groundwater aquifers affects the movement of surface water. For example, in the late 1960s and the early 1970s, the flow in the Sabarmati declined significantly due to the excessive withdrawals by the AMC (Moench, 1995). The decline in the river flow has been made good by increased releases from the Dharoi reservoir since 1976 (Patel, Sharma and Ramanathan, 1979). However, it appears that the increased withdrawals in the recent past for domestic and industrial purposes have adversely affected the availability of water in the vicinity of Ahmedabad and downstream villages.

Conflicts between Agriculture and Domestic Uses

The excessive withdrawal of groundwater for agriculture as exemplified by the Dashotar case study in the Hatmati sub-basin is leading to a rapid decline in the water table in the area. This has had an adverse impact on the availability of potable water for domestic use in the rural areas. Farmers make substantial investment in informal water companies and in transporting water to distant places for irrigation. They consider water lying beneath their land as their private property and therefore do not hesitate to mine as much groundwater as they can (Moench, 1995; Singh, 1997).

Issues concerning the regulation of groundwater withdrawal have been discussed by many scholars (see, for example, Moench, 1993; Dhawan, 1989; and Shah, 1993). These studies point out that direct and indirect regulations have not helped in restricting farmers' investment in the tube wells. The Sabarmati sub-basin is no exception to these general observations. In fact, these regulatory measures are counter productive and have generated a rent-seeking attitude on the part of the enforcing authority (Dhawan, 1989; Shah, 1993). Overall, the impact of regulatory mechanisms on the extraction of groundwater is almost nil; the rich and the elite continue to benefit at the cost of the poor.

The government's response to the shortage of rural drinking water has been to bring more and more villages within the domain of state-sponsored regional rural drinking water supply schemes. The schemes have mitigated the problem to some extent by supplementing the local water supply but they suffer from erratic power supply, bureaucratic inefficiency, and inappropriate pricing (Prabhakar et al., 1997). More and more the provision of water supply through long-distance pipelines also means the diversion of water from other uses to drinking purposes.

Conflicts in Allocation of Surface Water

The government of Gujarat has clearly defined allocation priorities for surface reservoir water. Domestic use has the first charge on the water supplies available. This is followed by agriculture, industry, and other purposes (GOG, 1996). The main objective of the public irrigation systems is to allocate adequate water on time and equitably across various user groups and areas within the command. If water is not sufficient to protect crops in the whole of a command area, then water allocation to the farmers is decreased proportionately and the crops grown in the tail end are left out (GOG, 1990; see also Kolavalli et al., 1994, for a detailed discussion). For example, in years of scarcity the Right Bank Canal has received priority over the Left Bank Canal from the Dharoi reservoir.

As between urban people and rural people, the former, being more articulate, organized, and politically powerful, manage to appropriate more than their fair share of fresh water while the latter have to bear with whatever is left (Prabhakar et al., 1997). This is evident from the fact that drinking water allocation to

Ahmedabad and Gandhinagar cities receives priority over allocation to the rural areas from the Dharoi reservoir. At the same time, since the government is unable to check large-scale unauthorized pumping by cultivators along the riverbanks, the intent of this policy is frustrated. Effective supply to Ahmedabad and Gandhinagar is much less than intended.

Industrial Pollution and Rural People: Conflicts between Unequals

The magnitude of industrial pollution in the basin has been increasing at a rapid rate and it is the rural people who suffer most from the pollution. Several public interest litigation cases are pending in the Gujarat High Court. In a landmark judgement, the Gujarat High Court declared 'The polluter has to pay', and directed that no expenditure is to be incurred for the treatment of industrial discharge from the public exchequer. The litigation was about the pollution caused by chemical industries located near Ahmedabad. The High Court issued directions to the state government to close down the industries classified as the specified industries causing damage to water bodies. The Supreme Court of India in 1980 invoked Article 21 of the Constitution which guarantees the right to life and hence to water and environment and thereby restores the people's natural right to water quality.

A discussion with some of the political leaders in the basin indicated that while everyone is suffering and even willing to cooperate in the legal battles, people are disenchanted with the legal and bureaucratic hassles and delays. One of the reasons for the tardy progress in settling such cases is that the people who suffer from industrial pollution are poor, dispersed, and unorganized, even as those who pollute are economically more powerful and better organized. The state policy also favours the polluters on the ground of industrial growth and employment generation. In fact, during the 1990 and 1994 assembly elections in Gujarat, villagers in the affected area tried to seek a promise from the electoral candidates for the redressal of their grievances. The then chief minister, Chimanbhai Patel, had promised that the norms of pollution control would be strictly enforced and that none of the industries would be allowed to bypass these regulations. However, political expediency and the powerful industrial lobby stalled all efforts to control the pollution, and the people of this area continue to suffer.

In the Sabarmati basin, Khari, a minor tributary of the Sabarmati, carries heavy polluted loads from the Vatva and Odhav industrial estates of the GIDC before its confluence with the Sabarmati downstream of Ahmedabad city. It has also been reported that the directions of the Gujarat Pollution Control Board for setting up of common effluent treatment plants in these industrial estates have not been heeded by the GIDC (Shunmugam and Ballabh, 1997, p. 21). The AMC raises protests against the industries upstream of Ahmedabad when it comes to the pollution of river water on which it mainly depends for the drinking water supply to the city. However, the AMC itself is responsible for the pollution of both surface water and groundwater in the downstream villages due to the letting out of untreated and partially treated municipal sewage into the river. This shows the double standards of organizations involved as both service providers and as water pollution control agencies (Ballabh and Shunmugam, 1998).

The response of NGOs to the problem of water scarcity and the resulting conflicts in the Sabarmati basin has been multifaceted. Some NGOs are trying to involve farmers in the management of surface water and groundwater. Others are engaged in collecting and disseminating information relating to the consequences of mismanagement of water in the basin. These experiments and efforts are of critical significance in expanding our understanding of the art of the possible, and some of the NGOs claim that they have identified technologies and institutions that could mitigate the conflicts at the micro level. These are, however, too small in scale to cope with the conflicts and the real dimensions of the problem. The government, on the other hand, believes that the water scarcity problem in the Sabarmati basin could be solved only by transporting water to the basin from water-abundant areas. Such an approach is unlikely to resolve the conflicts associated with water use and users but it can perhaps reduce the intensity of the scarcity. Besides, the solution will not be efficient and just, and the poor and the weak would continue to remain marginalized.

In Gujarat, the Narmada project is considered to be the lifeline of the state, and is expected to solve all problems related to water (Moench, 1995). Some of the areas within the Sabarmati basin which fall in the command of the Narmada canal, including Ahmedabad and Gandhinagar cities, are likely to receive from the project a substantial amount of their domestic water requirements if the current protracted negotiations between the AMC and the government of Gujarat are successful. According to some informed sources, the Narmada waters can reach up to Ahmedabad even if the height of the Sardar Sarovar Dam is not increased.

If this happens, it would certainly solve the problems associated with the drinking water supply in Ahmedabad. However, it is unlikely to resolve the conflicts associated with the overdraft of water for agricultural and industrial purposes and the pollution caused by industrial discharge into the water bodies. In sum, the total quantity of water used in the basin markedly exceeds the total quantity available and the gap is going to widen over time. This has led to growing competition and conflicts among various uses and users of water. There is need for evolving an appropriate institutional structure for the judicious allocation of water and for regulating its use. Otherwise, there will be more competition, more pollution, and more conflicts in the future.

Efficient and effective management of water in a basin requires that the software and hardware of it are *in tandem* with each other, the software being the legal and institutional aspect of water management, and the hardware being the political-cum-organizational structure existing in the basin. The case studies demonstrated that this is hardly the case in the Sabarmati basin. There is complete anarchy in the use and management of both surface water and groundwater in the basin. The Irrigation Department has little control over even the surface water systems beyond damming and releasing water for drinking and irrigation purposes.

Defective pollution laws and their lax enforcement allow industry to inflict water pollution-related externalities on rural populations. The Irrigation Department is unable to check the illegal appropriation of water. The hierarchy of claims on water is determined more by default than by rational choice. Rural people and urban slum dwellers suffer most from this default option. Their vulnerability and powerlessness, their sparse distribution over a vast territory, and their lack of organization all combine to deprive them of the political power and influence needed to deal with the water problems they encounter.

CONCLUDING REMARKS

As in most of the arid, semi-arid, and hard-rock regions of India, the existing and the potential demand for water in the Sabarmati basin also far exceeds the present and potential supply of water if the present pattern of and trend in water use continues unabated. The widening gap between the demand for and supply of water in the basin is leading to increasing conflicts among water uses and users. The rapid growth of industries, increasing urbanization, and the intensification of irrigated agriculture in the basin over the last few years have further aggravated the water scarcity problems and the competition for and conflicts over water use. In the absence of a clear-cut policy for the allocation of water and the regulation of water use and of an appropriate institutional structure for promoting sustainable and equitable use of both surface water and groundwater in the basin, conflicts among water uses and users are bound to increase in the future. There is an urgent need to regulate and monitor the allocation and use of water and augment its availability in the basin through appropriate policies and institutional structures that are owned and controlled by associations of water users but which are managed professionally.

The growing industrialization and increasing urbanization in the basin have resulted in the pollution of water in the Sabarmati river and its tributaries and the consequent deterioration in the quality of both drinking water and irrigation water in the basin. People living in villages downstream the river are most adversely affected by the growing water and air pollution; their crop yields have declined drastically; and the incidence of malaria and water-borne diseases such as cholera and typhoid has increased recently. Being poor and unorganized, the victims of the pollution cannot even claim any compensation from the industries and the municipal corporations which are responsible for their miseries. The industrial lobby is so strong that even the judiciary and political leaders find it difficult to do anything to ameliorate the lot of the poor victims of the pollution. The NGOs' experiments, though valuable, are far too small in scale *vis-a-vis* the extent of the problem.

The Sabarmati water resources are already over-exploited and are severely polluted. Ensuring their sustainable use requires effective measures for more efficient management of water and for arresting the present trend in pollution. The problems of the Sabarmati basin can only be resolved by the pragmatic and integrated management of different categories of sources and uses at the basin level with the active participation of all stakeholders, i.e. different user groups, including the polluters who benefit at the cost of others.

NOTE

The authors are respectively Professor and Director, Institute of Rural Management, Anand 388001, India. Several individuals and organizations helped us with information and ideas. We are grateful to them all. The authors would also like to thank the IDPAD Seminar participants for their comments on the paper which were helpful in its revision. Particular thanks are due to Professor J.M. de Goede and Professor A. Vaidyanathan for their comments on an earlier draft of this paper and for providing valuable suggestions for its development.

REFERENCES

Ballabh, V. and V. Shunmugam, 'Distributed Governance in Basin Management—A Paradisiacal Denouement?', *Social Engineer*, vol. 7, no. 2, 1998, pp. 129-45.

CPCB, *Basin Sub-Basin—Sabarmati Basin Report*, New Delhi: Central Pollution Control Board, 1989.

Dhawan, B.D., 'Preventing Groundwater Over-Exploitation', paper presented at the Workshop on Efficiency and Equity in Groundwater Management, Anand: Institute of Rural Management, 1989.

______, *Development of Tubewell Irrigation in India*, New Delhi: Agricole, 1983.

GOG, *Plan for Operation and Maintenance (POM) for Five Medium Irrigation Systems: Fatehwadi Canal System*, Gandhinagar: Water Resources Department, Government of Gujarat, 1990.

______, *Integrated Plan of Sabarmati River Basin*, Gandhinagar: Gujarat State Water Resources Planning Group, Gandhinagar, Government of Gujarat, 1996.

Gupta, S.K., 'Water for Recharging of Aquifer in Ahmedabad, India', *Indian Journal of Earth Sciences*, vol. 20, no. 1, 1993.

IST, *Sabarmati Basin Pollution and Monitoring Project: Survey Report*, Ahmedabad: Institute for Studies and Transformation, 1996.

Kolavalli, S.S.H. Kalro, G. Naik and Nitin Shah, 'Joint Management in Medium Irrigation Systems in Gujarat: Two Cases', in M. Svendsen and Ashok Gulati (eds.), *Strategic Change in Indian Irrigation*, Washington D.C.: International Food Policy Research Institute, 1994.

Moench, Marcus, 'Maintaining the Desert Mirage: Political Economy and Questions of Groundwater Management in Gujarat', paper presented at the International Conference on the Political Economy of Water in South Asia: Rural and Urban Actions and Interactions, Chennai: Madras Institute of Development Studies, 1995.

Ostrom, Elinor, *Governing the Commons: The Evolution of Institutions for Collective Action*, USA: Cambridge University Press, 1990.

Patel, A.P., A.R. Sharma, Ramanathan, 'Integrated Approach to the Management of Water Supply for the City of Ahmedabad and Its Metropolitan Area', *Current Trends in Arid Zond Hydrology: Today and Tomorrow*, Delhi, 1979.

Prabhakar, R., Vina Vani, Ketan Tatu and S. Patel, *Case Studies on the Status, Dynamics and Sustainability of Fresh Water Resources in Gujarat*, Anand: Institute of Rural Management, 1997.

Pangotra, Prem and P.R. Shukla, *Management Plan for River Sabarmati*, Ahmedabad: Centre for Environment Planning and Technology (CEPT), 1994.

Shah, Anil C., 'Dealing with Scarcity Through Integrated Planning: How to Proceed?' Keynote address, IDPAD Seminar on Managing Water Scarcity: Experiences and Prospects, held at Amersfoort, The Netherlands, 13-17 February 1997.

Shah, Tushaar, *Ground Water Markets and Irrigation Development: Political Economy and Practical Policy*, Mumbai: Oxford University Press, 1993.

Shunmugam, V. and V. Ballabh, 'Decision Making for Sustainable Basin Management—A Case Study of Pollution Management in the Sabarmati Basin', paper presented at the National Workshop on 'Environment Database for the Indian Economy', Mumbai: Indira Gandhi Institute for Development Research, 15-16 October 1997.

Singh, Katar, 'Property Rights and Tenures in Natural Resources', in John M. Kerr, D.K. Marothia, K. Singh, C. Ramaswamy, William R. Bentley (eds.), *Natural Resource Economics, Theory and Application in India*, New Delhi: Oxford & IBH, 1997.

TCE, *Water Supply Project Identification Report: Ahmedabad City Water Supply Project*, Mumbai: Tata Consulting Engineers, 1997.

CHAPTER 3

Urban Water Supply and Management: Delhi

ASHIS BANERJEE AND SURESH K. ROHILLA

The state of Delhi, covering an area of 1,483 km^2, currently (2001) has a population of 13.8 million. It is predominantly urban; the bulk (87-8 per cent) of its population lives in Delhi city and New Delhi which constitute the third largest and the fastest growing urban centre in the country. The remainder live in rural areas of Delhi.

Physiographically, a quartzite ridge running through the middle of the state is a prominent feature. On either side of it are a number of micro watersheds through which rainfall is drained. On the west of the ridge, the water drains into the Najafgarh Jheel; on the east, it drains into the Yamuna. The gently sloping land east of the ridge consists of alluvial soil, including the nearly closed alluvial Chattarpur basin, with swamps being common in the flood plains of the river.

The climate is of the semi-arid type, with an average annual rainfall of around 600 mm, subject to wide year-to-year variations and marked by high seasonal concentration. The lowest recorded rainfall (in 1929) was 263 mm and the highest (in 1975) 1,253 mm. Most of the precipitation (over 80 per cent) occurs during the south-west monsoon between June and September.

SOURCES OF WATER SUPPLY

Water supply for the domestic needs of the population and for agriculture, industry, and commercial activity is obtained partly from surface sources and partly from groundwater. Of the estimated

TABLE 1: PRESENT RAW WATER AVAILABILITY SOURCE WISE (ANNUAL)

Water Sources	Supply to Delhi (*in mcm annual*)
Yamuna	348.4
Ganga	165.9
Bhakra Storage	331.8
Groundwater (Government wells)	82.9
Groundwater (private tube wells)*	165.4

NOTE: *Field estimates based on the estimated number of private tube wells.

total annual availability of 1,100 mcm, 840 mcm is obtained from surface sources (348 mcm from the Yamuna, 332 mcm from the Bhakra system, and 166 mcm from the Ganga) under inter-state agreements between Delhi and the neighbouring states of Haryana, U.P., and Punjab. The entire supply from these sources is channelled through the public water supply system and accounts for the bulk (over 90 per cent) of its total supply. Taken together with water drawn from government-owned wells/tube wells, the public system accounts for nearly 85 per cent of the state's total water supply. Only about one-sixth of the total supply is from the private sector, almost entirely through wells/tube wells (Table 1).

Taking public and private sectors together, groundwater contributed nearly a fourth of the total supply. According to a field survey undertaken by INTACH in 1997, the state had approximately 32,280 private tube wells, which pumped an estimated 165.4 mcm a year. The distribution of these wells by region and user categories is given in Table 2. The survey reveals that the majority (two-thirds) of wells are owned and operated by households for their own domestic and agriculture use. About a third are owned by industrial establishments and about 8-9 per cent by commercial enterprises and a variety of institutions.

Water Requirements

The available supply is inadequate to meet the norms of consumption considered desirable for a predominantly urban and relatively prosperous region like Delhi. Of the state's total population (13.8 million in 2001) less than 3 per cent live in the villages; another 9 per cent are in semi-urban areas. The bulk of the Delhi population

TABLE 2: PRIVATE SOURCES OF WATER SUPPLY (GROUNDWATER) YEAR 1997

S. No.	Head	Total number of tube wells						Total Tube wells	Total in mgd	Extraction mld
		Alipur	Kanjhawala	Najafgarh	Mehrauli	Shahadra	City Block			
I	Households (for domestic and irrigation)	5,439 *	1,465 *	7,818 *	987 *	402 *	3,889 Δ	20,000	157.8 *	717.99
II	Industries							9,600	6.8 †	30.94
III	Others							2,680	37.6 #	171.08
1	Offices	18	23	24	43	15	314	437	6.99	31.80
2	Farm houses	4	2	3	364	2	0	375	6.00	27.3
3	Embassies High Commissions	0	0	0	0	0	115	115	1.84	8.37
4	Institutions							493	7.89	35.89
5	Buildings	0	0	2	10	1	327	340	5.44	24.75
6	Cinemas	1	2	1	2	3	68	77	1.23	5.59
7	Clubs	0	0	0	5	3	46	54	0.86	3.91
8	Hospitals Nursing homes	6	27	9	25	34	119	220	3.52	16.01
9	Hostels Dharamshalas	0	2	0	17	4	87	110	1.76	8.00
10	Hotels	0	1	0	2 0	127	130	2.08	9.46	
11	Apartments	21	140	11	20	103	34	329	5.26	23.93
	Total							32,280	272.2	1,238.51

NOTES. * calculated @ 2,000 gallons/hr and 4 hrs/day x 180 days
† calculated @ 2,000 gallons/hr and 4 hrs/day of 10 per cent of industries
# calculated @ 2,000 gallons/hr and 8 hrs/day
Δ calculated @ 2,000 gallons/hr and 12 hrs/day (1 gallon = 4.55 litre)

SOURCE: Gross Estimates.

lives in a highly urbanized and metropolitan environment. Apart from domestic requirements, substantial quantities are needed for industrial, commercial, and institutional use, as well as for agriculture. Estimates based on 'norms' of con-sumption for different uses at the current (2001) level of population are set out in Table 3.

In making these estimates, the norms for domestic consumption in rural areas (70 lpcd) and for semi-urban areas (135 lpcd) are in line with those adopted by DDA's Master Plan for Delhi (MPD 2001). The 'norms' assumed for urban areas in the state—which are relatively more prosperous than the average city—is considerably higher (225 lpcd). This is in line with recommendation of the Central Public Health and Environmental Organization of the Ministry of Urban Development for cities with more than 100,000 people. Altogether the total requirement for domestic use work out to 4,538 mld.

Agricultural requirements are estimated on the basis of the extrapolations of past trends in irrigated areas and allowing for changes in area and the water consumption per hectare of different crops. The estimate for 2001 (219 mcm equivalent to 500 mld) is about 25 per cent below the estimated consumption a decade earlier (270 mcm), largely reflecting the progressive decline in irrigated cropping in the state.

Inadequacy of Supply

Based on the above-mentioned assumptions, Delhi's total water requirement in 2001 works out to 5,350 mld (Table 3), which is considerably less than the gross availability— 3,874 mld. One must also reckon with the fact that a considerable amount of the public

TABLE 3: TOTAL WATER DEMAND, NCT DELHI (NON-IRRIGATION + IRRIGATION) YEARS 1991-2001

Sector	Demand			
	1991		2001	
	mgd	mcm (annual)	mgd	mcm (annual)
Non-irrigation	434.82	721.5	997.49	1,655
Irrigation	220.24	270.3	178.44	219
Total	655.06	991.8	1,175.93	1,874

system supply is lost in conveyance and distribution (due to wastage, losses, fire hydrants, and free supply to economically weaker sections of society). Distribution losses of water drawn from private wells are likely to be much less. Actual consumption levels are, therefore, likely to be well below the desirable norms.

There are clear indications that the city's water supply system is unable to cope with the demand and is under considerable stress. The supply in the public system is irregular and inadequate. Though 3 out of every 5 households have access to piped water, half the citizens are reported to be without potable water. The quantum supplied varies widely from 30 lpcd in rural areas to as much as 500 lpcd in New Delhi and the Cantonment. Households have constructed tanks to store municipal supplies and also set up bore wells and tube wells. A significant percentage of the population depends to some degree on groundwater, the dependence being as high as 65-70 per cent in the periphery of the city.

The public water supply system has not been able to maintain its facilities properly, and nor has it been able to add significantly to its supply sources. Moreover, water is priced at uneconomic levels. Delhi's water rates are among the lowest among the nation's cities (Table 4). Apart from wastage and other losses in distribution, a sizeable volume is in fact supplied free. This is reflected in the fact that less than 60 per cent of the water 'produced' is in fact billed.

TABLE 4: COMPARISON OF WATER CHARGES BILLED FOR DIFFERENT CITIES IN INDIA WITH DIFFERENT LEVELS OF CONSUMPTION

	Level of Consumption		
	20 kl (in Rs.)	40 kl (in Rs.)	60 kl (in Rs.)
Delhi	9.1	27.3	37.1
Chandigarh	14.0	49.0	79.0
Jaipur	21.0	45.0	69.0
Bangalore	8.0	21.5	49.0
Ahmedabad	30.0	60.0	90.0
Hyderabad	19.0	57.0	65.0
Baroda	18.0	36.0	54.0

SOURCES: Data from DWS & SDU. Data from A.S. Dhami, 'Water management of mega-cities—charges for water through metering of flat rate system', in Indian Water Works Association, 26th Annual Convention, 1994 and Data from the Public Health Engineering Deptt., Jaipur, Rajasthan.

In 1993-4, according to data obtained from DWS & SDU, 2,270 mld of water was produced; of this only 1,286 mld was billed. Of the quantity which is billed, only three-fourths is actually collected. In effect, barely 43 per cent of the water provided by the public system yields any revenue. The combined effect of low rates and poor recovery is that the system is not in a position to ensure the proper maintenance of the distribution network and to undertake large-scale investment for augmenting supply.

The position in all these respects is getting worse. Delhi's population and economic activity has been growing rapidly; the population has grown by 45 per cent during 1991-2001 and its SDP by nearly 70-80 per cent. Even as the demand for water has increased, the quantum available to the public system from surface sources has remained more or less constant. As a result, the city's dependence on groundwater has increased considerably. The groundwater scenario—in terms of rate of extraction and utilization by various user categories—has been constantly changing. While it is difficult to assess the trends in the total volume extracted or its use for different purposes, there is evidence of growing pressure on groundwater resources.

Progressive deepening of tube wells and falling water table levels reflect this pressure. Data from GWB's national hydrograph stations indicate the declining trend in water levels over the past decade in most parts of Delhi. Except in the flood plains of the Yamuna (where a rise in the water table of up to 2 m. is observed), water tables have fallen by 2 m. to 8 m.) The maximum fall (over 8 m. being recorded in parts of the Mehrauli, Najafgarh, and Kanjhawala blocks. Yet another indication is the failure and abandonment of several tube wells due to the lowering of the water table, decrease in discharge, and poor water quality.

The Central Ground Water Board has already sounded an alarm over the sustainability of groundwater exploitation at current levels. Already the current rate of exploitation is estimated to be about 90 per cent of the annual rate of replenishment. In some parts of the city, extraction exceeds replenishment (see Table 5).

There is also a serious problem of water quality. Groundwater in the western parts of the state (Najafgarh and Kanjhawala) is brackish to saline at shallow depths. Flouride concentration is higher than the mandatory limit (1.5 mg/litre) in several localities (Kapashera, Western Colony, Azadpur, Punjabi Bagh, Kingsway

TABLE 5: GROUNDWATER POTENTIAL AND EXPLOITATION IN NCT DELHI

Block	Total area (sq. km.)	Total replenish-able ground water resources (mcm)	Committed net draft year 1991 (mcm)	Present withdrawal for domestic purposes (mcm)	Total ground-water with-drawal (mcm)	Balance provision for domestic, industrial, and other uses (mcm)
Najafgarh	253	27.24	32.00	8.19	40.19	nil
Kanjhawala	275	44.92	34.00	4.05	38.05	6.87
Alipur	259	60.70	30.00	6.70	36.70	24.00
Mehrauli	149	18.75	18.00	10.25	28.25	nil
Shahadra	79	17.52	4.00	2.90	6.90	10.62
City	470	122.50	0.00	110.50	110.50	12.00
Total	1,485	291.64	118.00	142.59	260.59	53.49

SOURCE: 'Development and Augmentation of Groundwater Resources, 1996', CGWB report (unpublished).

Camp, and Bijwasan). High nitrate concentrations have been recorded in rural areas (Naraina, Azadpur, Moti Bagh, Lawrence Road, Nagal Dewat, NCERT and IIT Campuses). Studies by environmentalists have found considerable concentration of heavy metals in some areas (Moti Bagh, Kingsway Camp, Meera Bagh, Keshavpur, and Neera Nagar) and cyanide in Mall Road and Timarpur.

Future Plans

Requirements will continue to rise with increases in population and economic activity. Plans to meet the growing demand rely heavily and almost exclusively being on reservoirs across the Ganga and its tributaries, which are under construction or are planned. City planners expect the Tehri Dam on the Ganga to provide 725 mld; Kishua across the river Tons 1,680 mld; and Renuka across the river Giri 1,248 mld, making a total of around 3,900 mld. This is about 1.7 times the volume currently available from existing surface sources.

The prospects of realizing additional supplies from these sources are, however, not bright. Tehri's progress is excruciatingly slow as it is bogged down in environmental and ethical controversies.

Whether it will be completed on target (2004) is in serious doubt. Work on the Renuka has yet to commence, while the project report on Kishua has not been finalized. Whether they can steer clear of environmental imbroglios and be implemented in the present severely constrained financial situation of Delhi and the other concerned governments remain open questions. Even if they are taken up, they will take a couple of decades to complete and their waters begin to flow to Delhi. Recent assessments by the CWC also suggest that no raw water supplies from these dams can be expected in the near future.

Whether and to what extent these projects will contribute to increased water supply are thus in serious doubt. At the same time, falling groundwater levels and the deterioration in water quality suggest that even the present levels of supply are not sustainable. In that event, and given the continued rapid growth of demand for water, people may be induced to further increase the rate of groundwater exploitation. Since the rate of extraction is already close to the rate of replenishment, there is a real risk of exhausting this resource leading to the collapse of the eco system. The unfolding scenario is thus grim. A rethinking of the strategy and priorities for expanding and improving the water supply system consistent with sustainability and environmental health is imperative.

In this context, efforts to reduce transmission losses, stricter demand management to induce conservation, and recycling are obviously important. These solutions to the problem are recognized and there are welcome signs of their beginning to be addressed. Equally important, however, is to explore alternative, feasible, and cost-effective strategies to augment supplies from local sources.

Alternative Strategy

For this purpose the Irrigation and Flood Control Department of the Delhi Government commissioned INTACH to study the possibilities of urban and regional rain-water harvesting and prepare a blueprint to exploit them. The results of this study, summarized below, highlight several concrete ways—utilizing natural drainage channels; revival and improvement of natural storages; converting abandoned quarries/man-made storages for rain water; construction of check dams and revival of traditional water-harvesting systems, and roof-top harvesting—in which local rainfall could be harnessed more fully and effectively. If even 25 per cent of the state's rainfall (910 mcm) can be trapped and stored—and this seems feasible—the water requirements of the region will improve substantially. The INTACH plan also proposes a novel way for large-scale treatment of sewage and its distribution for non-potable uses. These proposals together have the potential to substantially augment the capacity of the state to meet its water requirements at a significantly lower cost than alternatives currently under consideration.

Rain-water Harvesting

Rainfall in Delhi is mainly concentrated in the three months of the monsoon season. Hard urban surfaces and urbanization of watersheds has led to the loss of this rain water through rapid and increased runoff. If a fraction of this rain water could be trapped, the need for ecologically controversial dams and canals can be reconsidered, river-water flow can be increased, groundwater can be conserved, capital resources can be utilized in a cost-effective manner, inter-state river disputes can be reduced, a host of ecological advantages garnered, and the urbanscape drastically altered for the better. The rain over the NCT, Delhi (area 1,483 sq. km.) is equal to 202,000 million gallons annually, equivalent to 2,500 mld. If even 25 per cent (a feasible figure) of the rain water is trapped or stored, it would substantially improve the water regime in Delhi at comparatively lower costs than by the conventional methods presently envisaged.

The rain-water harvesting components in the NCT of Delhi are as follows:

Natural Drainage Channels in the Urbanscape

A large number of the natural drainage channels wind their way across the city and the region on their way to the river Yamuna. A few are in good condition whereas many have become carriers of dirty water, with certain stretches being no better than open lavatories. It is possible to construct pipes in the centre of the channels to carry the sewage water flowing through them. Together with measures for checking other undesirable activities along with the provision of alternatives, the natural drainage channels could be restored to their original function for the continuous on-channel recharge of groundwater, the transfer of water from these channels to various storages, and also the storage of water in the channels. The channels, once filled with water, could thus constitute urban waterways. Their cleanliness could be maintained with the use of the root zone system, mosquito weed, fisheries, and aeration. The channels would be shaped, landscaped, and deepened where necessary. Their banks would be the green lungs of the city. Their use for urban water sports, recreation, and certain commercial uses would be examined and integrated into the system as revenue-generating components. The Najafgarh Drain and the Barapullah Nallah are the main examples of this component.

Man-made Storages: Abandoned Quarries

The Delhi Ridge is an extension of the Aravali Ranges. Quartzite is the host rock of this system and is very well exposed. Bajri/kankar occurs as a direct derivative and China clay as an associated mineral formed by the alteration of the original pigmatites. The Ridge area has been extensively quarried for supplying construction material to the city for a long time. The topography of the ridge is dominated by well-exposed boulders, and low ridges and gentle valleys which provide drainage patterns. The area is now devoid of perennial streams and rivers. However a few *nallahs* flow in times of heavy rains through the small valleys over clay/quartzite sand formations.

The following are the results of extreme quarrying and mining:

- 8 sq. km. of land have been rendered waste due to mining.
- The total volume of voids caused by mining is 29.16 mcm.
- In Bhatti (south Delhi) alone about 300 old pits are scattered over an area of about 2.6 sq. km.
- In two very large pits, measuring more than 0.5 sq. km. each, groundwater has been exposed at a depth of 45 m.
- Tajpur quarry, near Badarpur, can store about 4.0 mcm of water.

It is worth mentioning that due to the environmental insensitivity of the development authorities these quarries have been identified as sanitary landfill sites to dump the enormous quantities of solid

TABLE 6: MAJOR QUARRY PITS AT BHATTI

Pit No.	Area (in ha.)	Catchment Area (in ha.)	Average Annual Rainfall in Catchment Area (in kl)	Expected Charge Received (in kl)
2	80	150	900,000	606,000
11	15	80	480,000	207,000
12	46	180	1080,000	517,000
10	3.5	154	900,000	291,000
Total	144.5	564	3,360,000	1,621,000

SOURCE: GM, DSMDC.

waste generated in the Delhi area. Such use of quarries is a short-term measure, as the quarries will rapidly get exhausted, and furthermore this would contaminate the fresh water aquifers.

On the persuasion of INTACH, the Delhi Government has initiated a pilot project of rain water harvesting in the Bhatti quarry pits covering some 144 ha. Their catchment (564 ha.) has an estimated average annual rainfall of 3.4 mcm, of which roughly half is the expected charge.

Besides the Bhatti mines and the Tajpur quarry pits, there exist a large number of quarries which have been fully worked to the water table or which have been abandoned for other reasons. It is proposed to identify and modify these quarry pits for storing rain water from local catchments or for storing surplus water transferred to them from drainage channels. This would help augment the water supply, check the decline in groundwater, and suitably reclaim these abandoned quarries. Such treatment would improve the micro climate and make the area suitable for plantation and wildlife habitats.

Historical Water Bodies

A similar treatment would be accorded to the various *baolis* (step wells) and *hauzes* (tanks) so as to revive them and use them for storage purposes. The revival of water bodies constructed in the past could play a significant role in the augmentation of the water supply system in Delhi. For example, the Hauz Khas tank with a surface spread of 16 ha. and a storage capacity of 0.2 mcm could

TABLE 7: HISTORICAL WATER BODIES

Name	Location
Tughlaqabad Jheel	Tughlaqabad Fort
Satpula Dam	Khirki (Press Enclave)
Anang Dam	Mehrauli
Mughal Tank	Narela
Hauz Khas	Hauz Khas
Red Fort Moat	Red Fort

be easily revived and would help in recharging groundwater. Seven *baolis* (step wells) in Tughlaqabad Fort, if maintained properly, could serve local needs at least partially.

Natural Storages

These basically appear in the form of lakes, major depressions, and wetlands. With rapid urbanization these water bodies face the threat of being encroached (built upon) and silted up. They could be identified on a regional basis and developed along with their watershed areas. Depending on their proximity to the water channels they could also be used for the storage of surplus waters from natural channels. Their capacities would be augmented by desiltation. The following lakes have been identified during the preliminary phase for detailed work:

TABLE 8: MAJOR LAKES IN NCT DELHI

Name	Location
Najafgarh Jheel	Southern Haryana-Delhi border, near Gummanhera village
Nallah Mundela	Near Shamspur Khalsa village, in Najafgarh Block
Bhalaswa Lake	North Delhi, near Bhalaswa village
Jahangirpuri Marshes	Near Jahangirpuri, North Delhi
Rithala Lake	Near Rithala village, North Delhi
Naini Lake	Model Town, North Delhi
District Park Lake	Paschim Vihar, West Delhi
Bhagwati Talab	Near Madipur village, West Delhi
Wazirpur Lake	Ashok Vihar, North Delhi
Kabir Basti Lake	On Rani Jhansi Road
Shastri Park Lakes	Near Shastri Park and Seelampur, East Delhi
Chaukhandi Lake	Near Chaukhandi village, West Delhi

(contd.)

TABLE 8 (contd.)

Raghubir Nagar Lake	Near Raghubir Nagar and Tagore Garden, West Delhi
Prasad Nagar Park Lake	Near West Patel Nagar
Shantivan and Vijay Ghat Lakes	Shantivan and Vijay Ghat, Ring Road
Seelampur Lake	Seelampur, East Delhi
Jhilmil Colony Lake	Jhilmil Colony, East Delhi
Harinagar Lake	Harinagar, West Delhi
Mayapuri Lake	Near Mayapuri Flyover, West Delhi
Mahavir Jayanti Park Lake	Near Mahavir Nagar Park
Dhaula Kuan Jheel	Dhaula Kuan
Dwarka Lake	Near railway line, Dwarka
Moradabad Pahari Lakes	Near Moradabad Pahari village, South Delhi
R.K. Puram Lake	R.K. Puram
J.N.U Lakes	J.N.U. South Delhi
IIT Lake	IIT, South Delhi
Shyam Nagar Lake	Shyam Nagar
Hauz Shamsi	Vasant Kunj, South Delhi
Tughlaqabad Fort Lakes	Tughlaqabad Fort, South Delhi
Sainik Farms Lakes	Sainik Farms, South Delhi

Water Harvesting in the Aravalis

The terrain is not only studded with quarries but it also permits the construction of a large number of check dams. Some old check dams (*bandhs*) are in existence but these have fallen into disuse and disrepair as tube wells have now become the main source of irrigation water. The *Delhi Gazetteer* of 1976 states:

> A major portion of the bunds are in the Mehrauli block. Along with steep slopes in the terrain here, there are also flat grounds which can be bounded by the embankment and where rain water can be stored. Total area benefitted by these bunds is 368 hectares (910 acres). These bunds have been constructed over nallahs coming from a hilly tract on the southern boundary of the territory flowing through Mehrauli block. The rain water coming into the nallahs is checked by these bunds and spreads over large tracts upstream. The water is drained out through regular outlets after the monsoon. The retention of water results in soakage of soil which helps in the raising of rabi crops, controlling soil erosion and raising groundwater levels.

There are a number of other minor bandhs and fresh water sites that need to be identified, revived, and maintained, and their catchments and watershed areas need to be protected. This would definitely result in a rise in the water table as well as make available a significant quantity of surface water.

TABLE 9: BANDHS IN NCT DELHI

Name of Bandh	Area Benefited (in ha.)	Present Condition
Dera Bandh	375	in good condition but, squatters have come up on both sides of the bandh
Asola Bandh	250	in good condition
Gadaipur Bandh	125	in good condition
Sultanpur Bandh	250	in good condition
Rajokri Bandh		aquired for defence works
Bijwasan Bandh	375	in good condition
Deoli Bandh	50	in good condition
Fatehpur Bandh	850	in good condition

SOURCE: *Delhi Gazetteer*, 1976.

Village Ponds or Johads

Historically, these ponds were the focus of community life. People's participation during the process of building *johads* and their regular maintenance were a sustained and collective effort by the community. No village was complete without a *johad* surrounded by lots of trees and platforms for social gatherings and meetings. *Johads* were never built in isolation, and were a well-planned system connecting local streams and rivulets and their tributaries; and they served to moderate floods. With the introduction of piped water supply and the promise of the same to all villages and with the rapid course of urbanization, these *johads* have been undervalued, and are threatened with destruction and abandonment. Many *johads* have silted up or have been deliberately filled and their land used for urban purposes. Elsewhere the ponds have been used for sewage disposal and their catchments built upon. The hydrological cycle has thus been interrupted.

A survey of existing village ponds being conducted by INTACH reveals that due to lack of foresight and the hunger for land the catchment areas of these ponds are being destroyed (filled, encroached, and built upon). The fertile pond beds are not even being used for the useful purposes of agriculture, but have been converted into residential colonies, government offices, community centres (*barat ghars*), bus stations, or urban parks, and thus these treasures are lost for ever.

There are 113 urban villages that have been engulfed by urban

TABLE 10: NUMBER OF VILLAGE PONDS IN DELHI, 1971–96

Year	No. of ponds
1971	350
1981	300
1996	290 *

NOTE: *Primary survey.
SOURCE: *Census Handbook, Village and Town Directory* (Delhi) for 1971, 1981.

agglomerations, and due to the hunger for land the village ponds of these have been filled and built upon. Several villages are on the verge of being engulfed in a similar manner under various urban extension programmes of the Delhi Development Authority. For example, there is a big village pond in the Narela sub-city project and 4-5 village ponds in Palam village in the Dwarka sub-city project of the DDA. A village pond in Narela is now being used as a bus terminal; another big pond in Najafgarh has been proposed by PWD as a site for a bus station; and the village pond land in village Begampur (Rohini Phase III) has been sub-divided into plots and distributed to the landless.

On the other hand, the committee report on recent (1995) floods by the Irrigation and Flood Control Department, Delhi Administration has asked for Rs. 50 million for constructing new ponds. If all the village ponds in the existing rural settlements and some of those on the verge of being engulfed are nurtured and if their catchments protected it would result in the recharging of groundwater, improving the local environment, bringing about sustainability, and encouraging people's participation in managing water resources.

Roof-top Harvesting

The rain that falls on the hard roof-tops in the metropolitan cities is also an unutilized source of water, and water goes waste when it falls into urban drains and ultimately into the river Yamuna. It can be used locally at the campus level through underground storage or for artificial groundwater recharge using appropriate technologies. Roof-top harvesting is proposed at the campus level and the identified campuses along with their estimated roof areas are categorized below. An assessment of the building type and the associated rain volume harvested have also been calculated and are shown in Table 11.

From Table 11 we find that the approximate total roof area available in Delhi is around 1,711.77 ha. Average rain falling over Delhi is 611 mm. annually, of which 495 mm. precipitates during the monsoons. The runoff from the roof-tops that can be harvested amounts to 8.5 mcm annually.

Reservoirs in the Yamuna Flood Plain

The recent inter-state agreement on the sharing of the Yamuna waters has resulted in an annual allocation of 724 mcm of the

TABLE 11: POTENTIAL FOR ROOF-TOP WATER HARVESTING IN DELHI (ESTIMATES)

	Roof Area (ha.)	Water Volume Harvested (in cubic metres)
Apartments	80.15	396,742.0
Auditoria	5.35	26,482.5
Cinema Halls	18.22	90,189.0
Buildings	67.45	333,877.5
Clubs	3.7	18,315.0
Embassies	57.55	284,872.5
Hospitals	24.0	118,800.0
Hostels	1.9	9,405.0
Hotels	15.2	75,240.0
Industries	609.1	3,015,045.0
Institutions	29.15	144,292.5
ISBTs	2.25	11,137.5
Museums	4.5	22,275.0
Offices	50.0	247,500.0
Stadium	50.0[1]	247,500.0
Schools	39.4[2]	195,030.0
Other educational institutions	99.6	493,020.0
Total	139.0	688,050.0
Runways	18.0[3]	89,100.0
Cantonment		2,654,437.5
At the rate of 12.5 per cent of the total area	536.25	
Grand Total	1,711.77	8,473,261.5

NOTES: *Based on the rainfall during the three month of the monsoon season 495 mm.

[1]25 × 2.5 [2]197 × 0.2 [3]1.2 km. × 50 m. × 3

Yamuna waters to Delhi. Of this allocation 580 mcm flows during the monsoon season. Nearly half of it (282 mcm) flows away unutilized for lack of storage and reservoir space. (Source: 'Development & Augmentation of Groundwater Resources in NCT Delhi 1996', A CGWB Report, unpublished.) It is proposed to locate reservoirs on the flood plain itself at suitable locations and in the form of wetlands to store a substantial portion of these waters. These waters would be directly utilized and would also enable the enhanced performance of a large number of shallow tube-wells to be located in the vicinity.

The flood plain reservoir site proposed by INTACH would be created by scooping the soil to slightly above the sub-soil water level and using the excavated material to pile up bandhs up to above the Highest Flood Level (HFL) mark. Scouring protection works may have to be undertaken and gates installed for the inflow of flood waters. Minor reservoir sites would be identified at a later stage. The reservoirs would have a natural shape like the wetlands. In addition, the possibility of the regular dredging of the main channel to increase the on-channel pondage upstream of the three existing barrages would also be examined in the subsequent phase of the work.

Well Fields on the Flood Plains

A scheme for the development and management of the unconfined aquifer group period underlying the Yamuna flood plains involving dewatering during pre-monsoon period and refilling during the monsoon period is another major proposal. This has been mooted by CGWB and dovetails into the proposal for flood plain reservoirs. Shallow tubewells (30 m. deep) could be constructed at favourable

TABLE 12: PROPOSED LOCATION OF MAJOR FLOOD PLAIN RESERVOIRS

S. No.	Location	Bank
1.	Adjacent to Alipur bandh, opposite Burari	Left bank
2.	From Sonia Vihar to Old Railway Bridge	Left bank
3.	Adjacent to Noida bandh, from Shakarpur to New Railway Bridge	Left bank
4.	Adjacent to Noida bandh, from Mayur Vihar to Nizamuddin	Left bank
5.	Adjacent to Okhla	Right bank

locations in the 97 sq. km. of the flood plains. It is proposed to pump these tube wells during the months of May and June. Up to 300 tube wells pumping a total of 270 mlpd for 60 days would be possible. The dewatered aquifer zones are proposed to be refilled during the following monsoon period, adapting appropriate artificial recharge practices. It is proposed to carry out a similar exercise along the major storm water channels such as the Najafgarh Drain, Barapullah, Mungeshpur, and Nallah Mundela.

TREATMENT AND RECYCLING OF RAW SEWAGE

Eco-parks

Presently, 1,700 mld of domestic sewage is generated, and against this the sewage treatment capacity is 1,270 mld. Thus, while 430 mld flows into the river untreated, a further 1,000 mld receives only primary treatment and is discharged into the river at unacceptable levels of Biological Oxygen Demand (BOD) and Suspended Particulate Matter (SPM). By 2001 the sewage generated is likely to be 4,098 mld provided there is requisite water supply and the proposed treatment capacity at that point is 3,540 mld.

Industrial waste waters up to 320 mld are generated in addition to the above, and these are only partially treated at the sources. Thus, the total waste water generated in 2001 would be 4,400 mld approximately.

The conventional Sewage Treatment Plants (STP) are costly, highly power consumptive, and of course require land allocation. After treatment the water quality level would be 20 mg/l BOD and 30 mg/l SPM. As against the above, eco-park or biological sewage treatment parks use the Root Zone System (RZS) for treatment of sewage water. The RZS is an alternative technology for sewage treatment which is practised in over 500 international installations and has also been tested successfully in Delhi. The level of BOD after the suggested RZS would be around 5 mg/l, which is also the prescribed international standard for bathing water. This system is cost-effective in that it requires less capital investment (Re. 0.30/litre), has lower operating and maintenance costs, requires much less power, and has a much longer lifespan as compared to the existing and proposed Sewage Treatment Plants of the DWS & SDU in Delhi.

The operations involved in an eco-park are as follows:

1. Sewage inlet into a 'settlement tank' to remove metal and other heavy particles.
2. Inlet from 'settlement tank' into 'technical tank' containing lime and other chemicals. Aeration of sewage with the help of electromechanical equipment discharging micro-fine bubbles of air.
3. Inlet from 'technical tank' into the 'tree and shrub plantation area', which would also contain carnivorous aquatic plants such as 'ulticularia', which eats mosquito larvae, thus preventing mosquito growth.
4. Introduction of sewage into fishery ponds or irrigation channels before disposal into rivers or streams.

The maintenance of quality and quantity will be done at the entry and exit points. Large areas are required to treat the sewage and those are available at the existing and proposed sites of the STPs, the requirement being 1 ha./million litres of sewage.

The treated waters would then be distributed in different zones for horticultural and irrigation purposes as well as in a system of dual water supply for house gardens, water closets, and certain industrial uses, especially power plants. Thus, by recycling waste waters an additional resource can be generated amounting to 2,270 mlpd. The balance 50 per cent released into the river would improve the quality of the river water.

CONCLUSION

In the last couple of years, since INTACH started promoting alternative ideas and brought the concept of rain- and flood-water harvesting into focus, perceptions have changed from cynicism and scepticism to a realization that these ideas are not only feasible but it is also imperative that these alternatives be put into practice. Citizens' groups from various parts of Delhi, especially those who experience major shortages in water supply, have become articulate in demanding that the solutions suggested by INTACH be implemented. They have protested against any other use of such lands as have been identified for potential water harvesting.

One of the major reservations in official circles was that, while it might be possible to implement these ideas in remote and rural areas, the urban metropolitan situation does not permit water harvesting because of the heavy demographic pressure on land, high land values, encroachments, and possibilities of water contamination. In fact, these reservations were also made to serve a reckless drive towards land reclamation on flood plains, urban densification, and landfills through garbage dumping in potential harvesting sites such as exhausted quarries.

The crucial lesson learnt from the last two years of lobbying, studying, and intensive interactions with a wide variety of government and non-government agencies is that, while it may take some time for policy thinking to change, once adequate homework is done and claims are given a scientific basis acceptability will gradually grow. The involvement of various departments like the Irrigation and Flood Control Department, the Delhi Water Supply and Sewage Disposal Undertaking, the Delhi Development Authority, the Forest Department, the Lt. Governor's Office, the Chief Minister's Office, the Central Ground Water Board, and the Ministry of Water Resources means that once the proposals are finally submitted to the government there will already exist an awareness that close coordination will be required between these agencies and institutions in order to implement the proposals.

However, the final lessons will only be learnt after the proposed system has worked for a few years. The rates of aquifer recharge will vary from place to place; certain water storage sites may show up faults which will render them less efficient than may have been anticipated at this stage. Certain legal difficulties may also arise with respect to the jurisdiction of different agencies. Nevertheless, since most of the land in question is in the ownership of the government and since the usefulness of water harvesting is increasingly being appreciated, there is a very strong case for its implementation. The biggest point in its favour is that as compared to all other means of augmenting water supply it is so inexpensive that even if optimal results are not obtained immediately it would go a long way toward mitigating the ongoing crisis.

Besides, with decentralizing water availability and the gradual recharging of aquifers the moisture content of the soil can be expected to increase and this will in turn enable a wider spread of greenery. If the Departments of Forest and Horticulture and related agencies take proper advantage of the changed scenario, a positive effect on the micro climate is likely to take place. Then the growing pollution of Delhi is also likely to be mitigated. Some experts have recently argued that there is a threat of desertification in Delhi. Such a threat will also be contained through water harvesting. Finally, the success of water harvesting in an urban area like Delhi will have a spread effect on the rest of the country because of the high visibility enjoyed by the metropolis capital city.

REFERENCES

Civic Guide, Municipal Corporation of Delhi, New Delhi, 1996.

Delhi Vision–2001: Civic Infrastructure and Environment, Proceedings of Seminar Organized by the PHD Chamber of Commerce and Industry and the Government of the NCT of Delhi, 1996.

'Development and Augmentation of Groundwater Resources of NCT Delhi' (unpublished report, 1996), Central Ground Water Board, Ministry of Water Resources, Government of India, 1996.

Manual of Water Supply and Treatment (3rd edn), prepared by an Expert Committee, constituted by Government of India, Central Public Health and Environmental Engineering Organization, Ministry of Urban Development, New Delhi, 1991.

Rana, K.S., S.C. Sud and A.K. Mohinta, 'Overview of Reservoirs Upstream of the National Capital Region', Central Water Commission, Ministry of Water Resources, Government of India (unpublished). A paper presented at a seminar organized by INTACH on Water Resources of NCR, New Delhi, held in November 1997.

Spatio-Economic Development Record, vol. 2, no. 6, Nov.-Dec. 1995 (Spl. Issue of INTACH Initiative on Harvesting Water).

Zerah, Marie Helene, 'Towards a more efficient management of water supply in Delhi: Issues of water pricing and cost-benefit analysis', paper presented at a seminar organized by INTACH on Delhi's Water: Problems and Alternatives, New Delhi, July 1995.

CHAPTER 4

Water Use in the Punjab Region: Conflicts and Frameworks for Resolution

KANCHAN CHOPRA AND SEEMA BATHLA

INTRODUCTION

When viewed from a global perspective, India does not feature as a region characterized by water scarcity. However, inter-regional variations in water availability and use are of crucial significance, in particular in a country of its subcontinental dimensions. The water situation in the north-western part of the country is, in some sense, atypical as of now. It could, however, turn out to be a pointer to the kinds of situations that may arise in other river valleys characterized by surface water irrigation systems. The region has a well-developed canal and groundwater irrigation network which has sustained a mechanized agriculture that produces a large food-grain surplus for the rest of the country.[1] Over the last decade or so, two kinds of water-related issues have emerged as significant: the first relates to the fall in the groundwater table in parts of the state; the second deals with waterlogging and salinity in certain other pockets. Both issues reflect faulty water management and are the consequence of the existence of an intensive agriculture

*Professor and Research Associate respectively, Institute of Economic Growth, New Delhi, India. This paper is a revised version of a paper presented at the IDPAD Seminar held in Amersfoort, The Netherlands in October 1997. The authors wish to thank the participants at the seminar and editors of the present volume for comments and suggestions. The usual disclaimer applies.

driven by market forces. Further, another distributional conflict emanating from the functioning of surface water distribution mechanisms has existed in the region for some time now.

This chapter examines two kinds of water conflicts existing in the region:

I. The first relates to surface water as an instance of an interpersonal distribution problem.
II. The second focuses on declining groundwater tables (essentially the consequence of a trade-off between present and future production).

We shall discuss first the issue of declining water tables and examine different estimates of this emerging scarcity. The magnitude of scarcity likely to arise over a longer period of time is also assessed on the basis of estimates obtained from a twenty-year simulation model to answer the question. This is followed by a discussion on the nature of the conflict between production in the present and in the future inherent in the situation and also spells out the nature of the more long-standing distributional conflict over surface water to which it may be related. The concluding section outlines possible solutions: by way of technology, interventions through market instruments, and through quantitative restrictions, administered centrally or locally.

IS THERE LIKELY TO BE AN EXTENDED WATER SCARCITY IN THE REGION?

The nature of water scarcity in this region has been studied over the last decade or so by groundwater experts as well as by economists. One of the major points of contention has been with respect to the nature of the fall in the groundwater table observed in different parts of the state at regular intervals.

The picture is as follows. With little or no augmentation of the canal system, surface water supply has remained at an almost constant level since the early 1980s.[2] This fact, together with the rapid rise in the number of tube wells in the state in the 1970s and 1980s, has meant that the burden of increased demand has been borne by groundwater sources. For the state as a whole, annual net recharge and draft almost balance each other. Predictions of impending water scarcity are based on the picture that emerges in certain pockets, in particular as these are seen to be expanding over time. It is seen that six of the twelve districts of the state show an annual draft higher than 100 per cent of the recharge. Estimates of the number of 'dark blocks'[3] in the region indicate a rising trend, though in years of high rainfall their number goes down. Estimates of rates of fall in the groundwater table vary widely, from 1.07 m. per year to something like 14 cm. per year. This variation is the outcome of differences in methodologies used, time periods covered, as well as levels of aggregation adopted by the study. Results from some studies are listed in Table 1.

It can be inferred that:

1. Studies referring to one point of time and to a larger region of investigation give larger estimates of decline. Compare, for instance, the rates of decline given in the studies of Malik and Faeth, and Prihar et al. with those given by Dhawan and Baldev Singh.
2. Econometric analysis for smaller regions with longer time periods tends to give lower estimates of rates of decline in the water tables. This is due to the crucial role of rainfall and other hydrological parameters which are of a cyclical nature and the impact of which can be captured in such studies.
3. The selection of the initial and the terminal years may crucially impact the estimates obtained.[4]

TABLE 1: WATER TABLES IN PUNJAB: EVIDENCE

Author	Region and methodology	Time period of study	Annual rate of decline		
			Average	Pre-monsoon	Post-monsoon
Prihar et al. (1993)	Punjab: Withdrawal normative requirement	1989-90	1.07 m.	–	–
Malik and Faeth (1993)	Ludhiana: Fixed recharge	1989-90	70 cm.	–	–
Dhawan (1995)	Ludhiana: Depth to table in observation wells	1973-93	–	8 cm.	10 cm.
	Patiala: Depth to table in observation wells	1973-87	–	19 cm.	26 cm.
Singh (1992)	Patiala: Depth to table in observation wells	1979-87	–	10 cm.	14 cm.
		1973-87	–	24 cm.	35 cm.

4. Pre-monsoon and post-monsoon estimates may vary. This is significant from the viewpoint of the nature of agricultural demand which is seasonal in nature.

A TWENTY-YEAR SIMULATION MODEL FOR A DISTRICT IN CENTRAL PUNJAB

It is of some interest in this context to examine the results obtained from a recent study based on a twenty-year simulation model for a district in central Punjab.[5] A simulation model based on water availability and use patterns is set up for Ludhiana, an agriculturally developed district which is beginning to experience shortages of water in relation to total requirement. The geographical area of the district is 386 thousand hectares. It had an average rainfall of 634.17 mm. over the period 1973-92. The population is 2.47 million, with a density of population of 640 persons per sq. km., substantially higher than the rest of the state. Growth rate of population, inclusive of immigration, is 2.93 per cent per annum. The economy of the district continues to be agriculture-based, despite the emergence of a strong industrial sector. According to the Census of 1991, 36.3 per cent of the main workers are cultivators and agricultural workers. About 86 per cent of the land is used for agriculture, with the cropping intensity increasing from 174 in 1980-1 to 193 in 1992-3. The average growth rate in the ratio of gross to net sown area over a span of 13 years is 0.82 per cent. Nearly all the present gross cropped area of 595,000 acres in the district is irrigated.

As in other parts of the state, there has been a shift in the allocation of area across crops. The rice-wheat rotation accounts for 80 per cent of the gross cropped area, the annual growth rate of area under paddy having been 7 per cent over the 12 years from 1980-1 to 1992-3. Yield per hectare of wheat and paddy is 38 and 37.6 quintals per hectare respectively. This has meant an increased demand for water. In 1989-90, the water availability in Ludhiana was 67.3 thousand hectare metres (ham.) from surface water and 141.5 thousand ham. from groundwater sources. The net annual draft from the latter source was estimated to be 197.5 thousand ham.

To study the water regime and its utilization within various sectors over time, an integrated resource stock-cum-use model of the ecological-economic type is developed.[6]

The merit of the model lies in its adoption of a comprehensive methodology incorporating:

(a) On the supply side, all ecological parameters such as rainfall variations from year to year and estimates of groundwater recharge from all resources, including that from irrigated fields. Rainfall predicted on the basis of the observations for the last twenty years is the 'driver' of the model. In determining groundwater supply, the stock of groundwater available every year, termed as the initial level of groundwater or the base recharge, is also accounted for.

(b) On the demand side, the scenario of agricultural and non-agricultural demand is predicted separately. The normative demand for groundwater as distinguished by use for agriculture (irrigation) and non-agricultural uses like drinking water by human and livestock population and industrial water utilization is estimated for the base year. The demand for water is then projected for non-agricultural uses for rural and urban areas separately, and for agriculture uses in the rural areas, with the former linked to population growth rates and the latter as a function of economic variables, specifically growth in the intensity of cropping.

(c) Likely changes in the groundwater table over the twenty-year period from 1993 to 2013 are then estimated. Assuming there are no policy interventions, no change in the supply of water from major and minor irrigation works, no change in cropping pattern and other social and institutional structures, the existing pattern of resource use, termed the 'reference run', is allowed to operate in the future. Comparing the annual water recharge and its demand from different sectors, extrapolation over a period of 20 years (1993-2013) is carried out to determine the sustainability or otherwise of the observed demand patterns. Further, sensitivity analysis is carried out to examine the impact of variation (within specified ranges) in the values of various ecological parameters on the demand and supply of groundwater resource.

The twenty-year simulation model gives estimates of the level of the groundwater table and the average annual fall over the twenty years under the following situations:

I. *Reference Run* based on hydrological conditions of the district, a minimum parameter value of return flow and demand based on a minimum rate of growth of cropping intensity.

II. *Sensitivity Run* which analyses situations in which ecological and economic variables vary within a fixed range. Increases in cropping intensity (to a maximum of 250 per cent by the year 2013) determine demand for water in one variant (Senrun 1). In a second variant (Senrun 2), growth in demand is determined by crop profitability. Paddy, being relatively the most profitable crop, growth of area under paddy determines the demand for irrigation water.

Table 2 summarizes the situation with respect to the extent of groundwater exploitation in the district obtained from the three situations described above. For all the scenarios, the simulated results reveal that nearly 70 per cent of the aggregate demand for irrigation water from all sources is required to be fulfilled from sub-surface sources. The groundwater irrigation demand continues to be higher than the average water supply of 208 and 234.6 thousand ham. under the reference and sensitivity runs for all years (refer to Appendix Tables 1 and 2). This implies that scarcity of water as reflected in the water balance shows an increase over time. However, irrigation demand in the Senrun 2 situation, computed on the basis of area elasticity and profitability of paddy, may not increase much after 2004. The average annual fall in the water table estimated from this run is 10.5 cm. as against 15.1 cm. in the reference run and 15.9 cm. in Senrun 1.

Figure 1 and Appendix Table 3 give the variations over time in the depth to the groundwater table. The high rainfall years are predicted to have low water requirements from the ground. Depth to the water table in these years is not as low as in other years. Low rainfall years result in more extraction of water and a fall in the water table: the year 2009 for instance will apparently be a drought year, with very high demand for groundwater. The water table could fall to 4.69 m. in this year.

It seems that scarcity of groundwater relative to demand is likely to be a permanent feature of some parts of the north-west region of India. The strength of the present model lies in the fact that it arrives at this result after taking into account year-to-year variations in

TABLE 2: RESULTS OF THE SIMULATION MODEL: LUDHIANA (1993-2013)

	Water balance (thousand ham.)			Change in depth to water table* (in metres)			Average annual fall (in cm.)
	1993	2003	2013	1993	2003	2013	
Reference Run	-82.86	-161.48	-257.82	1.43	2.79	4.45	15.1
Senrun 1	-80.09	-171.66	-303.41	1.10	2.58	4.27	15.9
Senrun 2	-80.09	-188.64	-234.43	1.10	2.57	3.19	10.5

NOTE: *The change is from an initial of depth to water table of 8 metres in 1992-3.

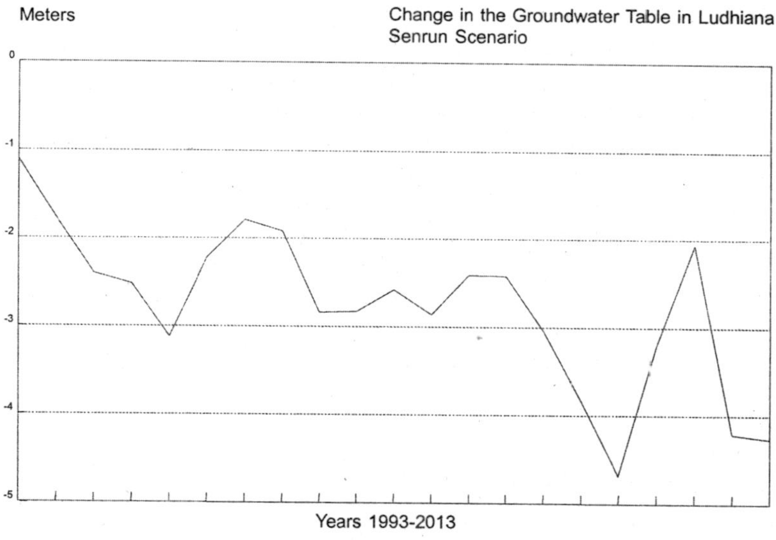

FIGURE 1: CHANGE IN GROUNDWATER TABLE: LUDHIANA (1993-2013)

rainfall, randomizing for variations in parameters such as recharge and estimating demand on the basis of alternative scenarios.

It must be added that the above model examines only the quantitative nature of the problem of water scarcity. With growing industrial demand for water, qualitative aspects of water supply are also likely to emerge as significant.

CONFLICTS OVER WATER USE: NEW OR LONG STANDING?

The operation of a number of factors has resulted in a situation where most commentators predict an emerging groundwater scarcity (viewed together with a near constant supply of surface water, this can be interpreted as an emerging overall scarcity) for the region. The earlier insistence on a 'finely balanced conjunctive use of surface and groundwater'[7] seems to have given place to a more pessimistic assessment of the situation.

This scarcity as it emerges can be seen as a clear case of a conflict arising out of the working of profit-maximizing behaviour when faced with a set of input and output prices in a specified institutional framework. Groundwater being a common pool resource, its extraction levels are determined by the prices of complementary inputs such as the power used for extraction or the price of the crop for which it is used. In the case of Punjab, increase in area devoted to paddy cultivation and the development of groundwater resources took place at around the same time. A large part of paddy cultivation depended and still depends on water from tube wells.[8] The particular conglomeration of input and output prices prevailing in the 1980s resulted in this new dimension of land use, along with its attendant ecological hazards.

It is also important to understand the nature of an older conflict which existed in the context of surface water and which led, in part, to the demand for subsidized credit for tube wells in the 1970s and 1980s. Effective access to canal water was never in line with the demand for it. The individual cultivator's access to canal water is determined by releases at three points: from the reservoir; from the main channel; and from the watercourse leading to the field. Eventually, the interaction of the factors determining the releases at each of these three levels determines the individual's access to water.

At the reservoir level, releases are determined by a mix of factors which take into account the requirements of the irrigation and power components of the major surface water project, the Bhakra which is a multipurpose project.[9] The 'reservoir factor' usually announced at the end of September determines the flow in the following agricultural season, and some experts maintain that there is a bias in favour of the requirements of power because of the higher revenue returns from it.

Further, the channels and courses that take water to the fields are both run on a rotational basis and the final availability of water to the farmer depends on the combined impact of the two kinds of rotation. This rotational system results in a decrease in the irrigated area as a proportion of the cultural command area towards the tail-end of the distribution system. Other inadequacies of the distributional mechanism include the presence of a large measure of uncertainty regarding the volume of water that the farmer is likely to receive. It is never certain that a particular farmer's turn to receive a large amount of water will coincide with the time when he needs it.

Excessive centralized control of the distribution mechanism and the absence of any channel for providing feedback with regard to its working result in the perpetuation of the drawbacks of the system.[10] This distributional inequity inherent in this system, which was the consequence of a typical centralized mode of functioning, resulted in a conflict between farmers located in the head reaches and those located at the tail-end of the distribution system. This situation was perhaps partly responsible for the overwhelming interest in access to tube wells and the consequent spurt in groundwater extraction.

POSSIBLE SOLUTIONS

In a sense, both kinds of conflict referred to above are of the distributional kind. One relates to inter-farm inequity arising out of the institutions for distribution; the other is in the nature of an inter-temporal conflict with the market throwing up solutions that give precedence to the present over the future. It is interesting that together they epitomize the dimensions of market failure and state failure in regulating the use of water.

Taking groundwater scarcity first, possible solutions suggested have been in the nature of technological solutions or price-based interventions. We shall also examine the possibility of newer kinds of market- or non-market-based solutions such as the emergence of users' cooperatives that lay down quantitative restrictions on the use of scarce water.

Technological Solutions

Solutions focused on evolving better technology in the face of groundwater shortage are aimed at improving efficiency of extraction. Some of these such as turbine pumps and submersible pumps focus on better methods of extraction of water.[11] They are targeted at the individual farmer and bypass completely the issue of an overall scarcity of water. Some of the solutions offered focused on end-use conservation of electricity by enhancing the efficiency of pumping equipment through retrofit measures.[12] While such measures have the benefit of reducing CO_2 emissions by about 33 per cent as well,[13] they do not seem to be directed towards conservation of water *per se*.

Technological measures that are aimed at increasing recharge address the issue of groundwater shortage more directly. Groundwater recharge may be increased by artificial recharge programmes, management of soil and water in the catchment, and by augmentation of surface water systems. In the Punjab region, artificial recharge of groundwater, though initiated as early as 1973, has not met with much success. A water-spreading technique is deemed to be the most suitable for the plains of central Punjab. For the foothills, it is suggested that a series of check dams and/or water-harvesting tanks across ephemeral streams would help in recharging groundwater aquifers.[14] While these schemes are certainly commendable, the impact they can have on the situation is somewhat limited.

Augmented supply of surface water within a scheme of conjunctive use would itself result in an augmentation of the groundwater table. Its impact would, however, depend on the magnitude of unutilized water still available in the state. An estimate based on discharge data for the years from 1986-7 to 1989-90 places the figure at 3.023 maf through rivers and 0.495 maf through drains and *nallahs*.[15] Plans are afoot for conserving 0.70 maf in a new surface water scheme. This may result in some augmentation of the groundwater tables as well.

Output and Input Price-based State Interventions

In a market-driven agriculture such as that of Punjab, water use is determined by the relative profitability of different kinds of land use, which in turn is dependent on relative input and output prices. The perception has existed for some time now that the introduction of paddy in the state is responsible for the increased and unsustainable demand for groundwater. The Johl Committee (1986) opined that a shift away from paddy was desirable. Sugar cane was suggested as the alternate crop. However, it is important to remember that, in view of the high recharge back to groundwater in its case, net water use by the paddy crop is only about a third of total use and the paddy-wheat combination may on balance be consuming less water than sugar cane.[16] In view of this, it is only substitution with maize that may be the appropriate policy from the viewpoint of prudent groundwater use. Relative profitability of the maize crop is very low, however, and intervention in output prices a difficult policy option. Development of alternative sources of supply for paddy procurement has been suggested as a *modus operandi* to wean people away from paddy.

It is also maintained by some that agricultural intensification as represented by increased cropping intensity is responsible for the stress on water resources in Punjab. One aspect of agricultural intensification is the emergence of water-intensive crops including paddy as the third crop.[17] In order to discourage this trend, the government has stopped giving support price to paddy that arrives before the month of August. The trend of increased agricultural intensification, however, is difficult to reverse, and policy can only aim at giving the right signals by appropriate pricing of complementary inputs. Time and again, commentators have suggested a more rational pricing of the energy used for extraction as the correct policy direction to adopt.

Table 3 presents the results of an exercise showing the impact of energy cost on area under paddy. Using cost of cultivation data for 1990-1 for estimating returns over cost and a value of 2.07 for elasticity of paddy area with respect to its profitability, the change in the relative area devoted to paddy with increased energy price is found.[18] The results indicate that with an increase in the flat rate to Rs. 55 per HP per month, the energy cost per hectare rises to Rs. 1,446 per hectare. There is a fall in the returns to paddy and a consequent decrease in the proportion of cropped area under paddy to 40.8 per cent. Under the unit price structure, the same effect is obtained by a rise in the unit rate to Re. 0.32 per KWh.

Are such increases in electricity tariff feasible? Conflicting signals seem to be emanating from the state. On the one hand, there is talk of a reform of the subsidy regime in agriculture and a possibility of higher power tariffs emerging in the medium term.[19]

TABLE 3: ALTERNATIVE ENERGY PRICE STRUCTURES AND AREA UNDER PADDY

Energy Price Structure and Rate	Percentage Increase	Ratio of Returns to Cost	Percentage Area to Paddy in Season
Flat Rate (Rs./HP/Month)			
1.50	–	1.161	57.3
2.55	10	1.00	40.8
3.60	20	0.99	39.8
4.65	30	0.976	38.3
Unit Rate (Rs./KWh)			
1.0.1089	–	1.161	57.3
2.0.12	10	1.092	50.2
3.0.13	20	1.087	49.7
4.0.32	194	1.001	40.9

On the other hand, the government of Punjab has made possible the free supply of electricity to farmers in the state. The solution to the inter-temporal conflict does not seem to be in sight.

Market-generated Solutions

The questions that seem pertinent to ask, in particular in a state characterized by commercialized agriculture, are: If indeed a water scarcity is emerging in certain regions, how is the market reacting to it? Are water markets emerging? Under what terms is water being sold or shared?

This is a little-researched area. An effort was made to collate information obtained from interviews with farmers in some of the water-scarce districts. Water markets do seem to have emerged in some areas. Scarcity or otherwise of water gets reflected in the differential prices at which water is sold and the extent of ease with which farmers are willing to part with water. Alternatively, the rates at which land is leased depend on the availability or otherwise of access to a source of water.

The evidence, though somewhat fragmentary, seems to indicate a pattern. There exists a premium varying from Rs. 2,500/4,000 to 6,000 on land with access to water in some areas. Further, electric tube well-irrigated land commands a premium of about Rs. 2,000 compared to diesel tube well-irrigated land. This

TABLE 4: LAND LEASE RATES AND ACCESS TO WATER IN VILLAGES IN PUNJAB, RS. PER YEAR PER ACRE (1997)

	Ludhiana		Muktsar	Moga
	Hiran	Tibba		
1. Without water			3,000-4,000	5,000-5,500
2. With water (electric tube well)	8,000	8,500	7,000-10,000	7,000-9,000
3. With water (diesel tube well)	5,500-6,000	6,500	–	–

SOURCE: Information obtained from village residents.

differential reflects the understanding that electricity is a subsidized—currently free—input.

Evidence also exists that water rates in situations of emerging water markets have been rising. They are found to range from Rs. 75 per acre to Rs. 120 per acre for wheat and paddy respectively, in situations where the canal rate in the village is Rs. 300 per acre for six months. Further, the existence of a perceived shortage finds expression in the disinclination to sell water during the paddy season.

What direction will people's incentives to solve these perceived shortages take? Will cooperation between users leading to voluntary limits on extraction set by people sharing water from the same aquifer evolve? Such cooperation has been known to emerge in similar situations in California in the USA, in Valencia in Spain, and in some other regions of the world. In the absence of a state-led intervention, this seems one possible institutional solution that could emerge. Whether or not participation evolves in the management of any resource depends, however, on a variety of factors.[20] It could take the form of water users' associations involving limited cooperation on specific issues.

CONCLUDING REMARKS

Water has been a crucial input in agriculture in the medium rainfall region of Punjab. Not much addition in surface water resources has been made in the last three decades or so. The groundwater situation seems to indicate that a scarcity is likely to emerge over the medium run, though its extent may vary from year to year, depending on rainfall. Conflicts over distribution in the context of surface water and over sustainable use in the context of groundwater resources are reviewed in this chapter.

Frameworks for suitable solutions seem to be difficult to work out. The alternatives examined are technological alternatives and state- and market-based interventions. The emergence of water markets as well as water- and land-linked markets is also reviewed. Future development is anticipated to be in the direction of limited cooperation between private agents in the appropriation and use of water.

NOTES

1. See, among others, Chopra (1990) for a documentation of the nature of agricultural development in Punjab.
2. Estimates are in the range of 1.44 million hectare metres (Khepar and Sondhi, 1992), 12.34 million acre feet (Gurcharan Singh, 1992), and a range varying from 12.125 to 13.88 million acre feet for the period from 1969-70 to 1984-5 (Chopra, 1990).
3. Dark blocks are areas where groundwater use is greater than 85 per cent of its utilizable recharge. Bathla (1997) estimates these dark blocks to be 69 out of a total of 118.
4. Between 1990 and 1997, for instance, the number of 'dark' blocks had decreased, mainly on account of the rainfall recharge effect.
5. See Appendix 1: Flow Chart of the Water Module. For more details, see Bathla, 1997.
6. See Braat and Leirop, 1987.
7. See Chopra (1990) for such a view.
8. This is due in part to the more reliable nature of access to water from private tube wells. See Chopra (1982 and 1990).
9. For a detailed exposition of the issues involved, see Minhas et al. (1972).
10. For a review of the working of this institution, see Malhotra et al. (1984) and Reidinger (1974).
11. See Kaushal et al. (1992) for a description of some of these.
12. Malik (1997) discusses one such set of measures which he claims are both privately and socially cost-effective.
13. This is due to a decreased demand and hence decreased generation of thermal power.
14. See Kaushal et al. (1992) for details.
15. This amounts to about 25 per cent of the 12 to 13 maf of surface water being currently used.

16. The PAU Expert Committee (Prihar, et al. 1990) estimated that 64 per cent of the water percolates back to the water table.
17. Based on reports from field information from PAU, 1997.
18. The proportionate decrease in the area under paddy to the total area is found using the following equation:
    Y = a + bX, where Y = proportionate area under paddy and X = total returns/total costs. Here a = -0.616 and b = 1.024. For details, see Bathla (1997).
19. See the Finance Ministry's white paper on subsidies (May 1997).
20. For an analysis of the factors in an adjoining area in Haryana, see Chopra et al. (1990).

REFERENCES

Bathla, Seema, 'Sustainability of Groundwater Resources in Punjab', M. Phil. thesis submitted to the Delhi School of Economics, University of Delhi (unpublished), 1997.

Braat, L.C. and W.F.J. Van Lierop (eds.), *Economic-Ecological Modelling*, Netherlands: Elsevier Science Publishers B.V., 1987.

Chopra, Kanchan, 'Alternative Sources of Irrigation and Land-use Pattern in Punjab', *Indian Journal of Agricultural Economics*, vol. 37, no. 2, 1982.

———, *Agricultural Development in Punjab: Issues in Resource Use and Sustainability*, New Delhi: Vikas Publishing House, 1990.

Chopra, Kanchan, G.K. Kadekodi and M.N. Murty, *Participatory Development: People and Common Property Resources*, New Delhi: Sage Publications, 1989.

Dhawan, B.D., *Groundwater Depletion, Land Degradation and Irrigated Agriculture in India*, New Delhi: Commonwealth Publishers, 1995.

———, 'Groundwater Depletion in Punjab', *Economic and Political Weekly*, vol. 28, no. 44, October 1993.

Government of India, *Census of India, General Population Tables, Series I*, New Delhi: Registrar General and Census Commissioner, India, 1991.

———, *Cost of Cultivation of Principal Crops in India*, New Delhi: Directorate of Economics and Statistics, Department of Agriculture and Cooperation, Ministry of Agriculture, 1991.

Jain, K.K., H.S. Bal and I.P. Singh, 'Economic Analysis of Irrigation in Punjab', *Indian Journal of Agricultural Economics*, vol. 44, no. 3, 1989.

Johl, S.S. et al., *Report of the Expert Committee on Diversification of Agricultural in Punjab*, Submitted to the Government of Punjab, Chandigarh, 1986.

Joshi, P.K. and N.K. Tyagi, 'Sustainability of the Existing Farming System in Punjab and Haryana: Some Issues in Groundwater Use', *Indian Journal of Agricultural Economics*, vol. 46, no. 3, July-Sept. 1991.

Kaushal, M.P., S.K. Shakya, S.K. Sondhi and B.N. Rao, 'Appropriate Technology for Management of Water Resources in Falling Water Table Area of Punjab', *Water Resource Day Proceedings,* vol. I, Ludhiana: Punjab Agricultural University, 1992.

Khepar, S.D. and S.K. Sondhi, 'Water Resources Development and Management Problems and Strategies for Irrigated Agriculture in Punjab', *Water Resource Day Proceedings,* vol. I, Ludhiana: Punjab Agricultural University, 1992.

Malhotra, Ashok, 'Availability, Assessment of Groundwater Resource in Punjab State', *Water Resource Day Proceedings,* vol. I, Ludhiana: Punjab Agricultural University.

Malhotra, S.P., S.K. Raheja, D. Seckler, 'Performance Monitoring in the Warabandi System of Irrigation Management in Niranjan Pant (ed.), *Productivity and Equity in Irrigation System*, Delhi: Ashish Publishing House, 1984.

Malik, R.P.S., 'Food Security and Environmental Stability', paper presented at the International Agricultural Economics Association, mimeographed, Delhi: Agricultural Economic Research Centre, 1997.

———, *Electricity Prices and Sustainable Use of Groundwater: Evaluation of Some Alternatives for North-West Indian Agriculture*, Delhi: Agricultural Economics Research Centre, 1993.

———, 'Public Policy and Sustainable Agricultural Development: A Natural Resource Accounting Approach', *Indian Journal of Agricultural Economics*, vol. 49, no. 4, Oct.-Dec. 1994.

Malik, R.P.S. and P. Faeth, 'Rice-Wheat Production in North-West India', in Paul Faeth (ed.), *Agricultural Policy and Sustainability: Case Studies from India, Chile, Philippines and the United States*, Washington, D.C.: World Resource Institute, 1993.

Minhas, B.S. et al., *Scheduling the Operations of the Bhakra Dam*, Calcutta: Statistical Publishing Society, 1972.

Narang, R.S. and H.S. Gulati, 'Optimum Cropping Pattern for Judicious Use of Water Resources', *Water Resource Day Proceedings,* vol. I, Ludhiana: Punjab Agricultural University, 1992.

Prihar, S.S. and S.S. Grewal, 'Improving Irrigation Water Use Efficiency in Rice-Wheat Cropping Sequence: Technical and Policy Issues', in J.S. Kanwar (ed.), *Seminar on Water Management: The Key to Developing Agriculture*, New Delhi: Agricole Publishing Academy, 1988.

Prihar, S.S., S.D. Khepar, Raghbir Singh, S.S. Grewal and S.K. Sondhi, *Water Resources of Punjab: A Critical Concern for the Future of*

Its Agriculture, 2nd edn, Ludhiana: Punjab Agricultural University, 1993.

Prihar, S.S. and V.K. Arora, 'Optimising Irrigation Schedules in Relation to Water Supplies', Lead Chapter, *National Symposium on Management of Irrigation System*, Karnal: CSSRI, 24-27 Feb. 1988.

Chand, Ramesh, 'Potentials and Prospects of Agricultural Diversification in Punjab', Draft Report, Delhi: Institute of Economic Growth, 1996.

Reidinger, R.B., 'Institutional Rationing of Canal Water in Northern India: Conflict between Traditional Patterns and Modern Needs', *Economic Development and Cultural Change*, vol. 23, no. 1, 1974.

Saleth, R.M., 'Factors Affecting Farmers' Decision to Buy Groundwater: Empirical Evidence from Indo-Gangetic Region', *Indian Journal of Agricultural Economics*, vol. 46, no. 3, July-Sept. 1991.

———, 'Efficacy of Power Tariff as an Instrument of Groundwater Management: Implications for Ecology, Efficiency and Equity', Delhi: Institute of Economic Growth, mimeo, 1993.

Sandhu, B.S., N.K. Narda and H.S. Uppal, 'On-Farm Efficient Management of Irrigation Water', *Water Resource Day Proceedings,* vol. I, Ludhiana: Punjab Agricultural University, 1992.

Sharma, S.K., *Methodology for Groundwater Resource Estimation*, New Delhi: Central Ground Water Board, no date.

Singh, Baldev, 'Groundwater Resources and Agricultural Development Strategy: Punjab Experience', *Indian Journal of Agricultural Economics*, vol. 47, no. 1, Jan.-Mar. 1992.

Singh, B. and G.S. Sekhon, 'Nitrate Pollution of Groundwater from Nitrogen Fertilizers and Animal Wastes in the Punjab, India', *Agriculture and Environment*, vol. 3, 1976, pp. 57-67.

Singh, Gurcharan, 'Utilisation of Water Resources of Punjab', *Water Resource Day Proceedings,* vol. 1, Ludhiana: Punjab Agricultural University, 1992.

Singla, T.N., 'Groundwater Recharge Programme: Present Status and Scope', *Water Resource Day Proceedings*, vol. 1, Ludhiana: Punjab Agricultural University, 1992.

APPENDICES

APPENDIX 1: FLOW CHART: WATER MODULE

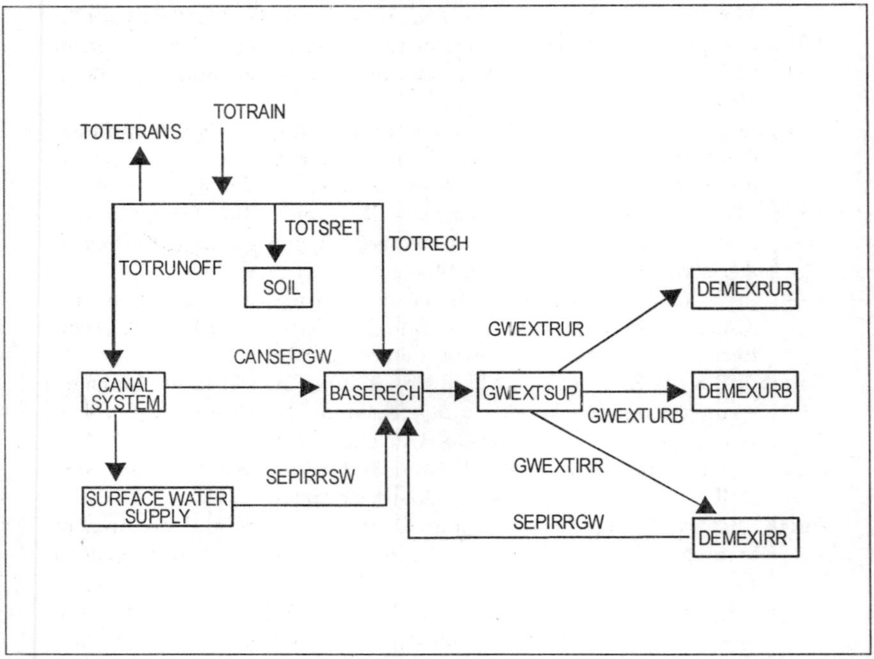

APPENDIX 2: TABLES

TABLE 1: COMPONENTS OF THE GROUNDWATER CYCLE IN LUDHIANA UNDER SENRUN 1 SCENARIO, THOUSAND HAM

Year	Volume of Rainfall	Soil Retention	Rain Recharge	Surface Run-off	Evapo-Transpiration
1993	242.41	138.76	44.92	17.80	40.93
1994	234.48	134.73	43.49	17.05	39.21
1995	190.12	112.92	35.50	12.64	29.07
1996	236.25	138.17	43.97	16.40	37.71
1997	166.74	96.59	30.97	11.87	27.31
1998	255.76	154.08	47.89	16.30	37.49
1999	303.11	179.42	56.55	20.34	46.79
2000	295.23	173.91	55.03	20.09	46.20
2001	207.78	129.04	39.16	12.00	27.59
2002	194.28	123.72	36.81	10.23	23.53
2003	265.41	159.33	49.66	17.10	39.33
2004	247.85	147.88	46.32	16.26	37.39
2005	295.25	179.76	55.41	18.21	41.87
2006	302.03	182.82	56.61	18.97	43.63
2007	261.65	151.40	48.59	18.68	42.97
2008	200.12	115.61	37.15	14.35	33.00
2009	135.91	81.75	25.44	8.70	20.02
2010	301.22	173.46	55.89	21.78	50.09
2011	446.52	253.61	82.62	33.42	76.87
2012	212.39	123.61	39.49	14.94	34.35
2013	216.95	129.08	40.52	14.35	33.00
Average	248.17	146.65	46.29	16.74	38.49

(contd.)

TABLE 1 (contd.)

Year	Canal Seepage	Return Seepage from Groundwater-Irrigated Fields	Return Seepage from Surface Water-Irrigated Fields	Base Recharge	Groundwater Supply
1993	72.702	100.97	25.37	120.32	263.32
1994	72.702	96.29	25.37	111.78	251.50
1995	72.702	88.88	25.37	97.41	233.04
1996	72.702	86.73	25.37	91.76	227.84
1997	72.702	81.95	25.37	87.08	216.13
1998	72.702	87.60	25.37	84.74	230.71
1999	72.702	92.05	25.37	87.59	242.23
2000	72.702	93.06	25.37	92.05	245.15
2001	72.702	87.03	25.37	93.06	230.29
2002	72.702	85.74	25.37	87.03	227.44
2003	72.702	87.97	25.37	85.74	233.48
2004	72.702	87.35	25.37	87.97	232.36
2005	72.702	90.52	25.37	87.35	240.83
2006	72.702	92.05	25.37	90.52	245.20
2007	72.702	89.29	25.37	92.05	238.72
2008	72.702	83.46	25.37	89.29	224.52
2009	72.702	76.35	25.37	83.79	207.31
2010	72.702	85.45	25.37	76.87	230.84
2011	72.702	96.12	25.37	84.01	258.64
2012	72.702	82.98	25.37	88.94	225.86
2013	72.702	81.09	25.37	82.98	221.58
Average	72.702	88.23	25.37	90.57	234.62

TABLE 2: TOTAL WATER DEMAND AS PER IRRIGATION, DRINKING AND INDUSTRIAL USES IN LUDHIANA UNDER SENRUN 1 SCENARIO, THOUSAND HAM

Year	Gross Irrigation Demand	Water Demand: Rural Areas	Water Demand: Urban Areas	Total Water Demand
1993	555.34	3.63	9.20	568.16
1994	558.89	3.66	9.43	571.98
1995	566.05	3.71	9.91	579.68
1996	573.32	3.77	10.42	587.51
1997	580.66	3.83	10.95	595.44
1998	588.11	3.89	11.51	603.51
1999	595.65	3.95	12.10	611.70
2000	603.29	4.01	12.72	620.02
2001	611.02	4.07	13.37	628.46
2002	618.86	4.14	14.05	637.04
2003	626.79	4.20	14.77	645.76
2004	634.83	4.27	15.52	654.62
2005	642.97	4.33	16.31	663.61
2006	651.21	4.40	17.14	672.76
2007	659.56	4.47	18.02	682.05
2008	668.02	4.54	18.94	691.50
2009	676.57	4.61	19.90	701.09
2010	685.26	4.68	20.92	710.86
2011	694.03	4.75	21.99	720.77
2012	702.94	4.83	23.12	730.88
2013	711.95	4.90	24.30	741.15
Average	628.83	4.22	15.46	648.50

TABLE 3: WATER DEMAND/EXTRACTION FROM GROUND IN LUDHIANA UNDER SENRUN 1 SCENARIO, THOUSAND HAM

Year	Ground-water Irrigation Demand	Ground-water Demand: Rural Areas	Ground-water Demand: Urban Areas	Total Ground-water Demand:	Change in Water Table (metres) Senrun 1	Change in Water Table (metres) Reference Run
1993	335.35	3.10	4.96	343.42	-1.10	-1.43
1994	359.14	3.11	4.99	367.24	-1.77	-1.85
1995	381.73	3.15	5.21	390.09	-2.40	-2.85
1996	387.67	3.19	5.41	396.27	-2.52	-2.52
1997	418.05	3.24	5.72	427.02	-3.12	-3.65
1998	367.00	3.29	5.98	376.26	-2.22	-2.48
1999	347.26	3.33	6.20	356.79	-1.79	-1.92
2000	360.31	3.38	6.52	370.21	-1.92	-2.07
2001	412.82	3.43	6.85	423.11	-2.84	-3.2
2002	409.67	3.48	7.20	420.36	-2.83	-3.54
2003	398.10	3.54	7.57	409.21	-2.58	-2.78
2004	417.48	3.59	7.96	429.03	-2.86	-3.08
2005	393.63	3.65	8.36	405.65	-2.41	-2.47
2006	398.72	3.71	8.79	411.22	-2.42	-2.44
2007	438.38	3.76	9.24	451.39	-3.04	-3.20
2008	482.52	3.82	9.71	496.05	-3.84	-4.15
2009	528.04	3.90	10.40	542.35	-4.69	-5.14
2010	443.84	3.96	10.86	458.66	-3.20	-3.2
2011	393.67	4.04	11.62	409.32	-2.06	-1.38
2012	513.46	4.08	12.00	529.54	-4.22	-4.40
2013	512.43	4.13	12.46	529.01	-4.27	-4.45
Average	414.25	3.57	8.00	425.82	-2.77	-2.96

CHAPTER 5

Conflict Over the Use of Groundwater: Some Evidence from Tamil Nadu

S. JANAKARAJAN

INTRODUCTION

The progressive decline in groundwater levels observed in many different parts of India indicates that the rate of exploitation of this resource is outstripping the rate at which it is being replenished. The proliferation in the number of wells and the trend toward deepening reflects both the growing scarcity of supplies relative to demand and the intensification of competition for access to this valuable resource. This is not surprising given that irrigated land is far more productive than rain-fed land, and that well irrigation—being more amenable to flexible management of water supply to match crop needs—makes an even larger difference to productivity. Even as scarcity has increased competition for groundwater as a source of irrigation, the growing demand from industries and urban areas as well as the pollution of water bodies have led to new problems and conflicts. This paper explores, on the basis of field studies conducted in the state of Tamil Nadu, the nature and extent of these problems and the ways in which users respond to the scarcity and conflicts thereby generated.

For this purpose, it is first necessary to appreciate the special nature of groundwater as a resource and the legal position regarding its ownership and use. Groundwater is, for the most part, derived from the seepage of surface run-off of rainwater and of irrigation from surface systems. Depending on the topography and the sub-surface geology, it accumulates in underground aquifers and is available for use by pumping. However, unlike in the case of land, the location and extent of aquifers cannot be defined precisely. Being liable to wide variations across space and time, aquifers are inherently difficult to map. For this reason property rights over groundwater are ill defined.

Under Indian law, the state 'owns' underground resources and can, if it so chooses, impose regulations on its exploitation. Private individuals and firms are entitled only to right of use. Under the Indian Easement Act of 1863, a person has the right to extract and use resources (including groundwater) under the land which he owns subject only to the restriction that his use should not adversely affect the similar right of others. In effect, the right to exploit groundwater is linked to landownership, thereby excluding those without any land from access to this resource. This leads to a tendency among inheritors of small landholdings with wells to partition rights to use wells along with the land. At the same time, it discourages the commercial exploitation of groundwater independent of landownership.

Those who have land can pump as much as they can without reducing the supply available to others. But the latter is difficult to monitor and in a situation where landownership is distributed among a large number of people, the costs of seeking a remedy through the courts are far too high. The chances of enforcing the verdict are also rated low on the basis of experience. The idea of treating groundwater as a common pool resource has not been seriously considered till recently. Under this regime, the growing demand for groundwater is met by more owners digging more wells; competitive deepening of wells in the face of declining water tables, and the emergence of informal water markets. Eventually, in many cases, this leads to overexploitation. All these tendencies are in evidence in Tamil Nadu, and each generates its own conflicts.

The chapter is divided into three parts. The first part, which follows the introduction, deals with the conflicts occurring in the rural areas. The second analyses the conflicts arising out of the increasing industrial/urban needs for groundwater. The last part presents a brief summary and outlines some policy suggestions for further discussion.

CONFLICTS IN THE USE OF GROUNDWATER IN THE RURAL AREAS

As indicated earlier, conflicts in the use of groundwater are the reflection of ambiguous property rights enjoyed by the well owners, resulting in (a) fragmentation of wells into different shares along with land and leading to conflicts between different sharers of the same well; (b) competitive deepening of wells and leading to conflicts between well owners who share a common aquifer; (c) trading in groundwater and leading to conflicts between water sellers and water purchasers; and (d) unregulated pumping, contributing to the drying up of the surface-water bodies and the depletion of groundwater.

FRAGMENTATION OF WELLS

The working of property rights in land, the right to exploit groundwater, and the operation of the law of inheritance result in the sub-division of both land and wells, especially among small holdings. A survey conducted during 1992 in the 27 villages of the Vaigai river basin (in southern Tamil Nadu) shows that, on an average, about one-third of the total of 1,100 sample wells are jointly-owned. The incidence of jointly-owned wells varies from zero to as high as 67 per cent (Janakarajan and Vaidyanathan, 1997). A similar survey conducted in the 26 villages of the Palar river basin in northern Tamil Nadu indicates that the incidence of jointly owned wells is about 20 per cent of the total of 1,064 sample wells (Rajagopal and Vaidyanathan, 1997). A third survey of 11 villages[1] in Tiruvannamalai-Sambuvarayar district shows that the proportion of jointly-owned wells is nearly half.

An analysis of the situation in eight of the 11 villages surveyed showed the following:

1. Nearly one-fourth of the farm households surveyed did not have any wells of their own, nor did they have a share in jointly-owned wells.
2. Nearly 60 per cent of the wells are under single ownership; a sizeable proportion of these belong to landowners with two or more wells.
3. There are 304 jointly-owned wells with 522 co-shares. About 40 per cent of the latter have a half share, roughly a fourth have a third, and one-sixth have a fourth share. The rest share 5 or more co-shares.
4. Besides landless agricultural labourers who are excluded from having any access to groundwater, as much as 70 per cent of farmers with less than half an acre, and 60 per cent of those

with less than one acre, are also excluded from having direct access to this precious resource.

5. The proportion of farmers owning wells and the average number of wells owned across size class increases as size class increases. Better access to land is thus associated with better access to groundwater.
6. There is a negative association between the size class and the incidence of the number of shares in a well. There is hardly any case of a shared well among farmers owning more than 15 acres. In contrast, well owners with less than one acre have on the average two co-shares. Joint overall farmers with small holdings have a relatively high incidence of shares below one-fourth. For instance, all the four farmers who own a very small fraction of < 0.1 share in a well, 11 out of 16 farmers whose share in a well is 0.1, and 21 out of 25 farmers whose share in a well is between 0.1 and 0.2 belong to farmers owning less than two acres.

Details of the management of jointly-owned wells have not yet been documented in all cases. But information obtained from interviews with knowledgeable villagers suggests that the incidence of conflicts is quite high and widespread. The practical difficulties involved in extracting water from a single well by, say, three or four sharers, are the most important sources of conflict.

The most common practice of joint well management seems to be to install a single pump set and run the motor in rotation for a fixed number of hours, the cost being shared equally among the sharers. A frequently encountered problem, however, is the lack of cooperation among the shareholders in observing the agreed cost sharing as well as the available water. The failure of joint well management is primarily due to financial reasons. Even if everyone shares in the initial costs of the installation of pump sets, many disputes occur in the using/sharing of water because of the erratic power supply. Many of these disputes are sought to be settled by the village panchayats (informal village courts). But this is not always effective; disputes keep cropping up again and again in the use of water during times of scarcity or due to the erratic power supply.

Another way of using the water from a jointly-owned well is for each shareholder to install individual pump sets (either electric or diesel operated) in that well. In this case, the joint well owners land in serious difficulties, as the available water gets drained more rapidly when a number of pumps are used. The problem is aggravated when the shareholders install high-powered motors in a competitive manner with a view to extracting more water. The incidence of such disputes increases if the cultivators of different castes share a well.

In many of these cases, one of the sharers with a better resource position buys up the shares of the others. There have been instances in three of the study villages in which the poor farmers sold their tiny parcels of land also along with their shares in a well. The most glaring case was the one in which a single farmer who shared a well with 15 others bought up all the water shares along with the land of all the other co-owners!

When shareholders have landholdings of different sizes, mobilizing resources for deepening is difficult, especially if a sharer apprehends that the benefits of deepening will not be shared in a fair manner. Conflicts arising from such situations are again referred to the village panchayats. It is reported that in such cases, the panchayats suggest he physical division of the disputed well into as many shares as needed, leaving it to the individual shareholders to dig and deepen their delineated parts as they desire. Such fragmented wells were found in significant numbers in all the villages surveyed. Although this is the widely adopted solution, it encourages competitive deepening even within a well and results in the emergence of wells within a dug well. Very soon, in such cases, the sharers whose resource position is weak find it difficult to survive. Eventually, the well is dominated by the one sharer endowed with better access to resources. There are also instances in which the wells were completely abandoned because there were too many sharers, resulting in far too many disputes for any solution to be effective.

CONFLICTS BETWEEN WELL OWNERS

The rapid expansion of groundwater irrigation has resulted in the steady decline of the water table in several parts of the country. This is widely interpreted as indicating that the pumping rates are much higher than the recharge rates and that the secular lowering of the water table has resulted in the mining of water (see, for instance, Bhatia, 1992; Rao, 1993; Moench, 1992; Vaidyanathan, 1996; Janakarajan, 1997). Typically, this is the result of the proliferation of wells and the phenomenon of competitive deepening. Both give rise to conflicts among farmers exploiting a given aquifer.

The most appropriate way of ascertaining the extent and the implications of the secular lowering of the water table is to monitor periodically the depth and the volume of water extracted from a representative sample of wells in different regions. But this kind of information is not available from any published source. Data from the observation wells of the Central and State Groundwater Boards give only the levels but not the volumes extracted relative to the recharge. In any case, the observation wells are not a representative sample. In a recent survey of wells selected from different segments of the Vaigai basin in Tamil Nadu, we ascertained from each sample well owner two important pieces of information, namely (a) the depth of the well when it was originally dug (original depth), and (b) the depth of the well at the time of the survey (current depth). The difference between the 'original depth' and the 'current depth' for any given well gives a measure of the extent to which the water table has declined over a period of time (Janakarajan and Vaidyanathan, 1997).

The weighted average depth of wells in both wet and dry lands has risen by about 50 per cent, from 30·1 feet at the time of the original commissioning to 46·7 feet currently. Moreover, the original depths of the sample wells dug at different points of time increased over time, implying that a newcomer today has to dig deeper than what his predecessor had to say, 10 years ago.

In the context of a falling water table, the deepening of one well affects others, leading to a stiff competition among the well owners. As a consequence, conflicts emerge in exploiting groundwater. An aggrieved party—whose well yield is reduced or which at the outer limit goes dry due to the deepening activity of his neighbour—is entitled to seek justice through the court of law. Property rights in groundwater are ambiguous and indeterminate. Also, for reasons noted earlier, the fact of interference is difficult to establish, litigation is expensive and time consuming, and it is even more difficult to ensure that the awards are enforced.

Since competitive deepening raises costs all round, cooperation among well owners to prevent it and at the same time to ensure an equitable sharing of the available supplies through collective arrangements for monitoring and enforcement would seem rational choices for all concerned. That this does not happen is a puzzle which must be explained. Perhaps vesting rights with communities rather than with individuals would help facilitate such a solution. The issues involved in this problem need much closer study.

The social and economic consequences of the progressive lowering of the water table are quite worrying. After all, well irrigation has become a gamble. Not all those who invest in wells are successful. Many fail and lose in the race of competitive deepening. They either sell their land or get themselves locked up in the debt trap. The direct consequence of all this is the emergence of a new dimension of inequality between those who have sustained the race of competitive deepening and those who have not. While the former have emerged as the potential water sellers, the latter are often reduced to the status of water purchasers. Competitive well deepening has emerged as a new source of differentiation among farmers (Janakarajan, 1997; Vaidyanathan, 1996).

CONFLICTS BETWEEN WATER SELLERS AND WATER PURCHASERS

Like joint well ownership, the emergence of the water market in rural areas facilitates a wider sharing of this scarce and valuable resource even by resource-poor farmers who cannot afford a well of their own. This phenomenon has spread but unevenly. The extent of trading in groundwater varies from place to place, depending upon the availability of groundwater and the need for it.

A detailed survey of 7 villages in the Vaigai basin shows an extensive prevalence of water sale in all the villages. Of the total of 893 well owners in all the villages put together, 76 (9 per cent) sold water to as many as 811 water purchasers, to a gross cropped area of 1,195 acres in the year 1990-1. In terms of the proportion to total gross cropped area, this may appear small, but what is important is to note that the number of water purchasers is several times that of the water sellers (Janakarajan and Vaidyanathan, 1997). A little more than three-fourths of the water purchasers are poor farmers whose holding size is less than one hectare. Moreover, a vast majority of the water purchasers belong to the socially deprived castes. In fact, the Scheduled Castes, the most deprived group in the social hierarchy, constitute 27.3 per cent of the water purchasers. A similar study in another location also shows the social polarization and the unequal bargaining capacity of buyers and sellers (Janakarajan, 1992).

A well owner, being himself a farmer, will sell only whatever surplus is left after meeting his own requirements. Since the availability of water from wells varies between and within seasons, the surplus available for sale will tend to be more volatile than the total supply. And this volatility is likely to be passed on almost entirely to the seller. This means that the seller has to contend with a highly precarious and uncertain supply. Second, there is an informal rule that a water purchaser should purchase water only from the nearest well owner. This rule is enforced basically with a view to avoiding conflicts which may otherwise arise, because water may have to pass through the field channels of the others to reach the buyer's plot. Conflicts between water sellers and water purchasers arise due to the violation of this rule (Janakarajan, 1992). However, if a concerned well owner agrees, water can be purchased from the next nearest well owner. Even if all those concerned agree, the distant water seller would insist that the water purchaser should have a hose pipe of the necessary length with a view to conserving water lost through seepage. A water purchaser, however, is reluctant to do so since there is no guarantee that a water seller would sell water regularly to him (Janakarajan and Vaidyanathan, 1997).

The unequal trading relationship results in the exploitation of the weaker agent, not only through the mechanism of price (paid for the water purchased) but also through several non-price measures. One common non-price mechanism used by the water sellers is to exploit the free or under-paid labour services of the water purchasers. The water purchasers cannot refuse as they are constrained by the threat of retaliation by the sellers stopping the water supply in the middle of a season (Janakarajan, 1992).[3]

In the Vaigai basin, we found several cases where the payments for water were required to be made through labour compensation and in several other cases through output (grains). In the process, the water market gets interlocked with the other agrarian markets such as labour, credit, and produce (Janakarajan, 1992 and 1997).[4] There are also recorded instances in the survey villages of Tiruvannamalai district in which the water purchasers have been forced to lease out their parcels of land in favour of the water sellers, on the terms dictated by the latter (Janakarajan, 1992 and 1996).

Instances of open conflicts between such aggrieved peasants/ water purchasers and water sellers are sporadic and infrequent. However, the resentment of the water buyers against the water sellers is evident in their observations on the water supply conditions. Moreover, it has been observed in some villages, that the water sellers collude with each other in fixing the price of water, which again annoys the water purchasers (Folke, Steen, 1996; Janakarajan, 1992).

CONFLICTS IN THE USE OF SURFACE WATER AND GROUNDWATER

Conflicts in the use of surface water and groundwater is yet another important dimension of the problem. Pumping of groundwater in the prohibited area (such as in the river beds or in their neighbourhood) results in the drying up of the surface water bodies or in the reduced flow in the downstream or both. A survey of the 27 tanks/villages in the Palar anicut system[5] shows that in 22 of them the spring channels have completely stopped supplying water (Janakarajan, 1993). These spring channels, which originate from the Palar river, used to supply water for 6 to 8 months each year. These were considered to be one of the important traditional sources of surface irrigation in this region. But after the restrictions against pumping from the river beds were withdrawn, well irrigation spread rapidly. This led to the progressive decline of the water flow in the spring channels. At present, most of them are either heavily silted or are encroached upon by the farmers.

More glaring is the impact of the spread of groundwater pumping in the tank commands. Large-scale rural electrification and the adoption of high-yielding bio-chemical technology have contributed in a great measure to the spread of well irrigation in the anicut. The fact that high-yielding varieties require more assured, controlled, and timely application of water and since the available tank water was inadequate to raise three short-duration HYV crops, the growth of well irrigation in the tank commands gave an added stimulus to the process. Since these wells derive most of their supplies through seepage from the tanks, conjunctive use, wherever it is significant, discourages the adoption of incentives for community efforts in the management of tank supplies (Vaidyanathan and Janakarajan, 1989).

There is thus a correlation between the rapid growth of well irrigation and the decay of traditional tank irrigation systems. The traditional irrigation institution was found to be defunct in 6 out of the 17 tanks studied in the Palar anicut system. These are also the tank commands in which well density was quite high. In one of the tanks, the tank sluices are kept closed permanently with a view to facilitating the recharge into the wells located in the tank commands.

On the contrary, in the rest of the tanks, the traditional irrigation system was reasonably unimpaired but these were also the tank commands in which the well density was very low (Vaidyanathan and Janakarajan, 1989; Janakarajan, 1993; Rajagopal and Vaidyanathan, 1997).

A more recent study of tanks in the Periyar-Vaigai system (Vaidyanathan and Sivasubramaniyan, 1998) shows that while the spread of well irrigation in the tank commands does not always lead to a total collapse of the institution of tanks, its effectiveness varies according to well density. A similar finding is reported in a large-scale study undertaken by the Tamil Nadu Agricultural University (Palanisamy, Balasubramanian and Mohamed Ali, 1996). The close association between a high well density and the weakening of the tank system is also in conformity with several other village studies carried out in Tamil Nadu (Harriss, 1982; Janakarajan, 1986; Chinnappa, 1977; Janakarajan, 1996).

The most affected group in this changing environment are the small peasants who do not have access to their own well irrigation. This category of the peasant community, though resentful, are entirely helpless, as they are reduced to the status of dependants on the big farmers/water sellers for their irrigation needs. The net result is the further polarization of the already differentiated peasant society.

Indiscriminate expansion of well irrigation has also focused attention on how individual rationality conflicts with collective rationality, eventually leading to the erosion of common pool resources to unsustainable levels. An individual rationalizes his disassociation from the collective action of tank/canal maintenance and resorts to indiscriminate pumping of groundwater, resulting in the progressive lowering of the water table. The government's policy of supplying free electricity to meet agricultural needs has aggravated this problem. The net result is the general environmental degradation in which not only has the water table been progressively declining, but the wells are also not adequately recharged due to the drying up of the surface-water bodies such as tanks.

CONFLICTS DUE TO INCREASING URBAN WATER NEEDS

Conflicting interests between rural and urban needs arise because of two main reasons: (1) rapid urbanization and ever-increasing urban water needs for industrial and domestic purposes; (2) the environmental damage precipitated by the discharge of untreated effluents by industries (such as tanneries, textile dyeing units, chemical industries, etc.) and the discharge of domestic sewage into the open surfaces or into the river, thereby polluting the groundwater as well as the surface water bodies.

Tamil Nadu ranks second in terms of the degree of urbanization in India, but stands first on the basis of a composite index which covers both the important features of urbanization, namely town density and the degree of urbanization. About a third of the state's population is in urban areas. In many districts, the degree of average for urbanization is much greater than the national average of 23 per cent in 1981, as well as the Third World countries as a whole, which was 31 per cent in 1980. The process of rapid urbanization coupled with speedy industrialization, has exerted enormous pressure on the basic needs of towns, most important of which is water.

In recent times many industrial and commercial establishments in the urban areas have been increasingly using groundwater (either through their own deep bore wells or by purchases to meet their water needs). Notable among the cities and towns in the state which depend upon the pumping of water from their rural neighbourhood are Chennai and the Coimbatore-Tiruppur-Erode corridor (constituting the erstwhile undivided Coimbatore district). The sale of groundwater has been quite extensive in this area. Perhaps, the largest consumer of purchased groundwater is Tiruppur in Coimbatore district, a major centre of the knitwear industry which has earned a place in the industrial map of the subcontinent as one of the large foreign exchange earners.

In this town, out of the total 93 million litres of water used per day (mld), private water supply (from wells) alone contributes about 60 mld or 64 per cent of the total water consumed. The water is transported by truck tankers from a radius of up to 30 km. from the town. A rough estimate puts the number of truck tankers which transport water to the town at 450, of which about 250 are not even registered with the Regional Transport Office (Appasamy, 1994). A large number of farmers in the villages around the town sell water to the Tiruppur industries. That the government is actively encouraging the process is evident from the fact that the Tamil Nadu Electricity Board has authorized 230 agricultural wells around Tiruppur, Palladam, and Avinashi for selling of water by issuing them a separate service connection.

Overall, the quantity of water consumed for urban domestic use and industrial purposes is small relative to total water availability and what is used for irrigation purposes. But in particular locations (like in Tiruppur) the competition can be very severe. Moreover, water that is pumped and transported to the urban areas is of a good quality compared to what is used for irrigation purposes. Therefore, there is a high concentration of pumping of water from a few selected villages where the quality of water is relatively better. As a result, the people living in these villages, particularly farmers, have experienced rapidly depleting groundwater conditions. Water levels in and around Tiruppur have fallen due to extensive pumping by industrialists, and shallow wells have dried up, resulting in the scarcity of water supply even for domestic purposes (Blomqvist, 1996).

In recent times, there have been widespread protests against the sale of groundwater from the villages to the urban consumers, in particular the industries. Instances of picketing, protest demonstrations, and processions have been reported in many parts of Coimbatore and Periyar districts, in particular Palladam, Avinashi, Pongalur, Sultanpet, villages around Dharapuram, and in several villages around Tiruppur town. The extreme scarcity in Palladam town has resulted in the sale of water even for drinking purposes at Rs. 2 to 3 per pot. Most of the wells in this area have either dried up, or those with water are making money by selling water to Tiruppur town. Farmers, individually and collectively, have sent many memoranda to the concerned district collectors and to the ministers, urging them to ban the sale of water from the agricultural wells to the towns. Since all their representations have yielded no positive result, the outraged farmers, in particular the women, have started blockading the water trucks in several villages around Palladam town.

Widespread demonstrations are also reported to have occurred in the Dharapuram-Tiruppur road, one of the main water transport corridors.[6] 'Depletion of the groundwater level in places outlying Tiruppur like the Mandapam, Parinthulavu, Orathapalayam areas have affected agriculture, and farmers registered their protest against groundwater extraction for selling/industrial purposes (Appasamy, 1994, p. 71).

One would expect that the widespread sale of groundwater to urban consumers might affect the village societies adversely in two ways. One, as more and more water is pumped from the deep bore wells, the wells located in their neighbourhood might get dried up, and two, since many farmers have found it lucrative to sell water to the urban consumers, there might have been a decline in the agricultural activities, thus causing an increase in rural unemployment and large-scale migration from rural to urban areas. These aspects deserve a more systematic and careful examination.

CONFLICTS ARISING DUE TO ENVIRONMENTAL DAMAGES

Conflicts arising due to the pollution of water bodies by industrial effluents have emerged as a major problem and this one issue causes much concern to policy makers as well as to the general public. The main sources of polluting effluents are tanneries, textile dyeing units, and chemical industries. They discharge their effluents onto land or into the rivers/streams which eventually affect groundwater as well. Contamination of groundwater has been reported in several parts of the state where tanneries and dyeing and bleaching units are concentrated. Notable among these are: Pallavaram near Chennai, Visharam, Walaja, Ranipet, Arcot, Vellore, Ambur, Peranampet and Vaniyambadi (a stretch of about 100 km. in Vellore district), Dindugul (in Dindugul district), Erode (in Erode district), and Tiruppur and its neighbourhood (in Coimbatore district). The most important among the river basins very badly affected due to the discharge of industrial effluents are Bhavani and Noyyal (both tributaries of the river Cauvery) and the Palar. The industrial units located close to these rivers consume much water, and the effluents discharged by them have contributed quite heavily to the pollution of both surface and sub-surface waters.

Bhavani

The Bhavani river, the mainstay of a part of Coimbatore and most of Erode districts, is getting polluted with chemicals and heavy metals. The main source of pollution is a paper pulp and fibre factory of the South India Viscose (SIV), started in the early 1960s, in the upstream of the river. It produces 180 tons of wood pulp per day. The latest available information indicates that the industry consumes on an average 40 million litres of water per day and discharges back almost an equal quantity of water, without much treatment, thereby contributing quite significantly to the pollution of the river water, both surface and sub-surface (Appasamy, 1994).

The farmers' associations and the Bhavani River Environment Committee are apprehensive that the water stored in the Bhavani Sagar Dam, located 10 miles below the SIV, is already heavily polluted, and they claim that the area irrigated by the dam water is slowly but steadily losing its fertility and that it contributes to the salinization of groundwater in this area.

While the veracity of all these claims has yet to be verified objectively, the Madras High Court found sufficient evidence to hold the SIV factory guilty of polluting the river for over three decades and ordered it to stop letting effluents into the river, whether treated or untreated. The SIV had to shut down its plant after the court order.

There are a significant number of other industrial units (especially tanneries, dyeing and bleaching units, and sugar mills) which contribute to the pollution of water in the Bhavani river. Notable among them are Tan India and United Bleachers. A large number of tanneries along the 600-year-old Kalingarayan canal (near Erode town) contribute heavily to the pollution of water in this region. The Kalingarayan Canal Farmers' Association have expressed anxiety that the effluents discharged by the tanneries into the canal have not only affected the yield but have also polluted the groundwater around this area.

Palar

The worst affected area, however, is the Palar basin, where a large number of tanneries are concentrated in a 100 km. stretch along both sides of the river, from Visharam to Vaniyambadi town on the Chennai-Bangalore national highway.

A study undertaken by the Soil Survey and Land Use Organization of the Government of Tamil Nadu in the early 1980s to examine the impact of tannery effluents indicates that a large number of these tanneries dispose of their effluents directly into the irrigation tanks and into the Palar river. Moreover, solid wastes (like lime, hair, and leather bits) heaped near the tanneries slowly get washed away by rainwater and collect in the various surface-water bodies. The affected areas were demarcated on the basis of water quality (both surface-water and groundwater) measurements. It showed that about 16,000 ha. of land was affected by the tannery effluents. As this study was conducted in the early 1980s, the area affected presumably must have gone up significantly since then.

A more recent study funded by the Asian Development Bank to assess the environmental condition in the state concludes that the Palar river is one of the most polluted rivers in the Tamil Nadu (ADB, 1994). The study identified 3,226 small, medium, and large industrial units which primarily contribute to the pollution of the eight major river basins. About a fifth of these are located in the Palar basin. They are assessed to have a major impact on groundwater quality in the basin; impact on soil, land use, surface water, and public health is rated as 'minor'.

Narayana Murthy, who works for a voluntary organization, has probed the issue of groundwater pollution in the Palar basin due to tannery effluents. While the tanning industry in this region has registered an impressive growth over the past four decades, the quality of life has been on the decline. The number of tanneries had gone up from 108 in the 1960s to 240 in 1975, in Vaniyambadi taluk alone. During the same period, land rendered unfit for use is estimated to have gone up from 240 ha. to 6,400 ha. The number of polluted wells increased from 48 to 520. By 1984, their number had increased to 10,000, spread over 200 villages. This has resulted in decreased agricultural yield, severe scarcity of drinking water, and in several villages farmers have stopped cultivation (Narayana Murthy, 1987). Most important of all is the serious health problem caused by the contaminated groundwater.

The incidence of cholera, skin ailments, and gastro-enteritis are some of the most common health problems found in this area. Narayana Murthy writes, 'The biological oxygen demand (BOD) level in the polluted water has already crossed 20,000 mg. per litre, when the safe limit for drinking water is 3 mg. per litre' (p. 97).

The detrimental effect of tannery effluents on groundwater quality is documented in several studies (Narayana Murthy, 1987; Rajarathinam and Santhanam, 1996). A more recent study carried out by the Water Resources Organisation, Government of Tamil Nadu (Rajarathinam and Santhanam, 1996) shows that the untreated effluents discharged by the tanneries into the Palar river and into the open surface for the past three decades or so have quite seriously affected the groundwater quality. The study concludes that in 10 out of 12 sample locations, heavy metals like copper, zinc, chromium, iron, and manganese were detected in the groundwater. The total dissolved solids (TDS) level was quite significant, the value ranging from 4,905 to 10,172 mg. per litre in the sample locations. The chloride values exceeded the normal prescribed limit in all the locations. The pH value ranged from 7.8 to 8.9, indicating that the groundwater is more alkaline. Most important of all is the presence of chromium.

> The ions like sodium, magnesium, chloride, sulphate, and total hardness are also on the higher side in all the wells corresponding to high TDS. Chromium is below the detectable limit in 2 out of 12 locations but is present in all other sites in moderate to high proportion. This indicates that the contamination is mainly from tanneries, the intensity of pollution depending on the location of well sites with reference to the industry. (ibid., pp. 28-9)

Noyyal

Noyyal, a tributary of the river Cauvery, originates in the south-west of Coimbatore, passes through Tiruppur town, feeds several check dams built to recharge groundwater, before reaching the Orathapalayam dam below Tiruppur town. The planned command area of this dam is 8,000 ha. The river eventually joins the Cauvery, north of the Tiruchi district border. The river, which once provided safe water for drinking and irrigation, at present looks miserable and is dreadfully full of chemicals and salts, thanks to the flourishing textile and knitwear industry of Tiruppur.

The growth of the hosiery industry in Tiruppur has been extraordinary. The number of knitting mills in the town has gone up from 22 in 1941 to 2,800 in 1991. The direct export value of hosiery products from Tiruppur has gone up tremendously from Rs. 190 million (US$ 5 million) in 1985 to Rs. 20,000 million (US $ 530 million) in 1995. Dyeing and bleaching, an important process in this industry, has also expanded *in tandem*. The town had hardly any dyeing and bleaching units in the 1940s. In 1994, as per the latest information provided by the Tiruppur Dyers' Association, there were 713 such units functioning in the town. This figure excludes several unregistered units about which no reliable information is available.

Dyeing and bleaching is both water and chemical intensive. On the average per 100 kg. of clothes the process uses wetting oil (100 per cent) 500 g., caustic soda 4 kg., hypern 750 g., sodium peroxide 4 kg., hydrochloric acid 8 kg., soda ash 15 kg., acetic acid 3 kg., common salt 10 kg. and petroleum oil 2 kg. It takes 40,000 litres of water to dye and bleach 100 kg.

An average dyeing unit can process 20 tons of clothes per month or about 700 kg. per day, consuming about 8 million litres of water per month or 280,000 litres of water per day (Palanichamy, and Palanisami, 1994). The estimated water requirements of the bleaching and dyeing units in Tiruppur are about 92 million litres per day (mld). Almost the entire quantity of this waste water is let out along with the chemicals into the Noyyal river, causing permanent damage to the river, top soil, and groundwater.

A 1989 Government Order (G.O. No. 213, dated 30 March 1989) prohibits the establishment of any polluting industry at a distance of 1 km. adjacent to the rivers. The Noyyal is one of the notified rivers in the G.O. The Tiruppur Dyers' Association wanted exemption from this G.O., since according to their claim, the Noyyal river is dry and that whatever water flows in the river is not used for irrigation. However, the government of Tamil Nadu subsequently constructed a dam across this river (called the Orathapalayam dam) in the year 1992, about 10 km. below Tiruppur town, with a view to providing irrigation for about 8,000 ha. This has turned out to be a mockery. It stores much more water than the natural water flow in the river; the difference is largely accounted for by the effluents let into the river by the Tiruppur dyeing and bleaching units. Thus, the dam stores contaminated water which is unfit for irrigation. At the same time, it is contributing significantly to the pollution of the environment, including the groundwater in the adjoining areas.

Blomqvist's (1996) field visits confirm that the groundwater on both sides of the Noyyal river has become brackish and considerably harder in the last 10-15 years. It is reported that the water quality remains bad even at a depth of 300 feet, rendering it unfit even for irrigation. Jacks et al. (1994, p. 5) write:

> Downstream of Tiruppur at Orathapalayam, there is a newly constructed dam, aimed at arresting flash floods and utilizing them for irrigation. The water in the dam was brackish (7,000 mg/l TDS) and had a SAR (Sodium Absorption Ratio) in between that found in the effluent and that in the groundwater in Tiruppur town, indicating a mixing of effluent from the textile industries discharged directly into the Noyyal and groundwater from Tiruppur town. Downstream of Tiruppur, however, the salinity was excessive, definitively, rendering the water unsuitable for almost any purpose.

Chemical testing shows that FDS, BOD, COD, and other measurements of water quality of both effluents and groundwater in this area are well above tolerable limits (Jacob, 1996). Farmers in several villages along the Noyyal complain about the hardness of the water and the drastic decline in the yield of crops. Several reports have appeared in the local newspapers about people in this region who have to walk miles to get a pot of drinking water.

What needs further investigation, however, are the claims of the farmers' associations that (a) several agricultural wells had become dry due to the indiscriminate deepening and pumping of water by those who sell water to the industries; (b) there is a general slackening in the agricultural activities; (c) due to declining yields of water from wells a large number of people, in particular landless agricultural labourers, have migrated to the towns, notwithstanding the increasing unemployment in the villages; and (d) the industrial effluents have contributed to the decline in the soil quality, and consequently to poor agricultural yields.

PUBLIC RESPONSE

The idea of entering into a detailed discussion pertaining to the environmental damage caused to the groundwater is to show that it is not occurring in one or two places; rather, damage to groundwater is quite heavily concentrated in several river basins, notable

QUALITY OF EFFLUENT AND GROUNDWATER IN THE TIRUPPUR AREA

Characteristics	For Effluents		For Groundwater	
	Tolerable Limit	Actual	Tolerable Limit	Actual
Electrical Conductivity	<2,000 Ms	4,780 to 23,300 Ms	<2,000 Ms	2,910 to 16,400 Ms
TDS	2,100 mg/l	4,500 to 19,920 mg/l	1,500 mg/l	2,541 to 14,010 mg/l
Chloride	1,000 mg/l	3,750 to 7,273 mg/l	<1,000 mg/l	1,007 to 4,490 mg/l
Sulphate	1,000 mg/l	168 to 1,413 mg/l	<1,000 mg/l	156 to 821 mg/l
Magnesium	NA	NA	50 mg/l	100 to 766 mg/l
Sodium	NA	NA	200 mg/l	352 to 2,800 mg/l
BOD	30 mg/l	37 to 365 mg/l	Zero	1 to 128 mg/l
COD	250 mg/l	230 to 1,786 mg/l	NA	NA
Iron	NA	NA	0.3 mg/l	107 mg/l

NOTE: TDS: Total Dissolved Solids; BOD: Biological Oxygen Demnad; COD: Chemical Oxygen Demand; mg/l: milligram per litre; NA: Not Available.
SOURCE: Derived from Jacob, 1996.

among them being the Palar, Noyyal, Bhavani, Kalingarayan canal, Amaravathi, and Kodaganar. The last five are not only contiguous, but they are also the constituents of the larger Cauvery river basin.

There have been mixed responses from the people in the affected regions. On the one hand, some sections feel that since these industries generate a good deal of employment, both directly and indirectly, taking an extreme view of the matter might result in their economic misery. A large majority, however, express their anguish and resentment over the damage caused to the environment, including groundwater depletion and pollution, land degradation, decline in agricultural productivity, and so on. A good deal of study is necessary to substantiate all these issues. However, responses to this effect by the affected rural population have been spontaneous, which are reflected in processions, demonstrations, picketing of government offices, hunger strikes and even the physical impounding of the water trucks. What strikes one more is the participation of women in large numbers in all these protest demonstrations. On the other hand, the political parties in the region are not willing to mobilize the masses against the polluters who are an important source of their funds. Also, since most of them have their support bases among the workers employed in these industries, they are more interested in organizing wage struggles through labour unions rather than in organizing them to fight on environmental issues.

Nevertheless, many NGOs are active in these affected regions, creating awareness among the people, representing their cases to the authorities, representing them in the courts of law, assessing the damage caused to the environment and to individuals, and so on. In fact, the cases filed by individuals and supported by the NGOs have led to the closure of polluting industries like tanneries and dyeing and bleaching units in various parts of the state. On the whole, the pressure exerted on the government by the people, the NGOs, and the courts has been quite enormous. But what effective measures the government will take to tackle the situation is yet to be seen. At the moment, whatever measures the government takes are through the Tamil Nadu Pollution Control Board. But the Board, which is supposed to be an agency to control and monitor pollution, does not go beyond issuing reminders, guidelines, etc. In other words, the Board's role as a pollution control agency is far from effective.

The ineffectiveness of the Board may be traced to the ambiguous government policies. On the one hand, the government has been quite blind to the pollution problems in the export promotion zone in which its only motivation is to earn export dollars. On the other hand, the government under the threat of closure of these polluting industries has been trying to encourage and finance, through subsidies, the installation of common effluent treatment plants. But it would not like to become unpopular by resorting to any drastic policy decision as that may lead to the massive retrenchment of workers. The government is in a dilemma and does not seem to have any long-term policy to handle the situation. It does not collect any systematic information and nor does it encourage any public debates on these critical issues. It is, however, effective in discouraging the flow of information to the public.

SUMMARY AND POLICY SUGGESTIONS

This chapter seeks to highlight the nature of conflicts over the use of groundwater and provides insights into the forces which limit the effectiveness of efforts to seek remedies. In rural areas, conflicts occur between different users essentially because of the restricted access to this important resource. We have also seen that the conflicts reflect the ambiguous property rights over groundwater.

Conflicts occur for many reasons: (a) As land gets fragmented, the related well also becomes fragmented into as many shares, posing serious managerial problems. Eventually, such disputed wells either remain in disuse, or are purchased by a sharer who has better access to resources. (b) Conflicting interests among well owners surface essentially because of excessive extraction. As a consequence, wells are deepened in a competitive manner. This eventually results in skewing the distribution of access to groundwater, favouring richer and more powerful farmers against the small and resource-poor ones. (c) Spread in groundwater trading improves access to small farmers without a well, but access remains unequal and many conflicts occur between the parties involved in the trade, namely, the water seller and the water purchaser. The divergence in the resource positions and the resulting unequal positions of the parties involved put the buyer at a disadvantage in resolving the conflicts. (d) Conflicts in the use of surface water and groundwater are noticed in several tank commands. The aftermath of unregulated pumping in the tank commands and near the river beds leads to drying of spring channels and reduction in supplies to tanks. Rapid growth of wells in the tank commands is one of the principal reasons for the weakening of the traditional tank system.

Conflicts over the use of groundwater also exist between rural and urban areas and have become acute in some parts of the state where a large quantity of groundwater is transferred from rural areas to urban areas to meet the increasing municipal and industrial water needs. And then there is the growing problem of environmental damage caused to the groundwater as a consequence of untreated industrial effluents discharged into the river or into open-surface water bodies. The farmers are quite aggrieved and angry about the depletion of groundwater in the village wells used for both irrigation and drinking purposes. The resulting damage to the top soil and groundwater aquifers has reduced their agricultural activities, and the farmers are naturally furious.

Appropriate legal measures to protect and regulate this precious but eroding resource are essential: (a) The first and foremost legal measure should be to explore the possibility of disassociating the tie-up between land and groundwater; it is high time that legal experts made an attempt to tackle this problem. The ambiguity of property rights in groundwater has raised a major dilemma for policy makers and for all those concerned with this problem. (b) Some legal coverage is necessary to regulate water sale in the rural areas for both agricultural and non-agricultural purposes. It may even be useful to consider the possibility of banning the sale of water from agricultural wells for industrial uses. (c) Water extraction beyond a critical depth, when it is no longer rechargeable, should be prohibited in the best interests of the society. (d) The official policy regarding the supply of free electricity for agricultural purposes needs a rethinking. A uniform electricity tariff law for agricultural purposes should be evolved for all the states, essentially to curb the wastage of and misappropriation of subsidies for power as well as to discourage the attitude of the political parties which resort to measures of 'competitive populism' (in subsidizing electricity and eventually supplying it completely free) with a view to capturing the farmers' vote bank at the cost of the exchequer (MIDS, 1988; Lindberg, 1996).

And finally, appropriate legal cover is necessary to extract payment or impose penalties on from those who are responsible for polluting the environment in general and the groundwater in particular. It is imperative because the damage caused by, for instance, tanneries and dyeing and bleaching industries is permanent, causing a severe negative externality on the future generations. Since the social cost involved is enormous, the guilty parties cannot simply escape by erecting a treatment plant (which very often does not function).

In Tiruppur, for instance, many dyeing and bleaching units have started erecting treatment plants quite reluctantly only after the court order[6] was issued, and that too after they themselves were personally affected by groundwater pollution.[7] It is extremely important to devise some legal measures to enforce what is called the 'polluter pays principle'. This must be ensured not only to internalize the cost of the externalities caused to others, but also to recover the cost of permanent damage already caused to the general environment and to the groundwater in particular. It is also necessary to undertake a public awareness campaign to educate the public about the signs and consequences of over-exploitation and the dangers of environmental damage.

ACKNOWLEDGEMENTS

I am grateful to Professor A. Vaidyanathan for his extensive editing and constructive suggestions for improving the quality of this paper. I thank Professor K. Nagaraj who helped to clarify some key points.

NOTES

1. As part of the resurvey programme, 11 villages were studied in Tiruvannamalai-Sambuvarayar district by the author jointly with Dr. Barbara Harriss-White, during the years 1993-5. The study was funded by the ODA, UK.
2. Wet and dry lands are basically revenue classifications, made with a view to differentiate between the area commanded by the surface sources of irrigation and the area which does not have access to surface sources. While the former type of land is called 'wet', the latter type is called 'dry'.
3. See also Folke, Steen, 1996 for evidence in the other parts of the state. This is in conformity with what Ruth Meinzen-Dick, in her study *Groundwater Markets in Pakistan*, has observed. She found that water purchasers enter into informal contracts with water sellers as a part of the water deals, involving exchanges such as sharecropping for water, having the water purchasers bring fuelwood, and so on (Meinzen-Dick, 1996).
4. The Palar anicut system is an age-old surface irrigation system in which water from the river Palar is diverted from a weir through a network of unlined channels to 317 irrigation tanks.
5. Blomqvist (1996) observes that the occasional sight of water tankers full of good-quality water in a drought-affected, water-starved region triggered off demonstrations and the blocking of roads. 'In 1993, a wave of agitations against the water lorries was organized along the road between Tiruppur and Dharapuram, one of the main water transport corridors' (pp. 70-1).
6. It was only after the intervention of the Madras High Court, in March 1997, that the dyeing and bleaching units in Tiruppur took up the issue of erecting treatment plants seriously. The report submitted by the Pollution Control Board to the Madras High Court, in June 1997, stated that many units had not taken steps to erect treatment plants and that they were running these without the consent of the Board. The report stated that only 31 units had fully completed the work. While most of the others (of about 700 units) had just started the work, 44 of them had not even begun the work. The Madras High Court ordered, on 23 June 1997, the immediate closure of these

44 units, and in respect of 708 others (which were in various stages of erecting the treatment plants), the High Court granted four months time to complete the work (*Hindu*, 24 June 1997).

7. The detailed interviews carried out by Blomqvist (1996) with the owners of the dyeing and bleaching units clearly reveal this point. Many of them expressed a deep concern about the polluted groundwater in Tiruppur and in the neighbourhood. One of the large dyers admitted, 'All dyers are well aware that we are polluting. We know that we ourselves will get affected. We have invested in the industry here, and all those investments are going to be ours for ever. If it is risky to live in Tiruppur, if it gives health effects, then what will be the use? Even now, people who can afford to send their children to school in Ooty will do so. My daughter goes to a school in the Nilgiris. And when the children come back home for a vacation, they have to go to the doctor every week, which is not necessary when they are in school. If it is like that, everything will be in vain, all the hard work. . . .' (p. 156). This statement from a polluter is quite revealing, and it clearly shows that the problem that a common man would encounter is much more serious.

REFERENCES

Appasamy, P., 'Sharing Common Water Resources: Urban Study A Review', Background Paper for Policy Advisory Committee Meeting, Chennai: Madras Institute of Development Studies (mimeo), 1994.

Asian Development Bank, 'Tamil Nadu Environmental Monitoring and Pollution Control', a study carried out by Stanley Associates Engineering Ltd., Canada, for the Tamil Nadu Pollution Control Board, Madras, 1994.

Bhatia, Bela, 'Lush Fields and Parched Throats: Political Economy of Groundwater in Gujarat', *Economic and Political Weekly*, vol. 27, nos. 51 and 52, 19-26 December 1992.

Blomqvist, Anna, 'Food and Fashion: Water Management and Collective Action Among Irrigation Farmers and Textile Industrialists in South India', Linkoping: Linkoping University, 1996.

Chinnappa, B. Nanjamma, 'Adoption of the New Technology in North Arcot District', in B.H. Farmer (ed.), *Green Revolution?: Technology and Change in Rice Growing Areas of Tamilnadu and Sri Lanka*, Macmillan, 1977.

Folke, Steen, 'Cauvery Delta: Canal and Well Irrigation and Conflicts Over Land and Water', unpublished paper (mimeo), 1996.

Harriss, J., *Capitalism and Peasant Farming: Agrarian Structure and Ideology in Northern Tamil Nadu*, Bombay: Oxford University Press, 1982.

Jacob, C. Thomson, 'Impact of Industries on the Groundwater Quality of Tiruppur and Ethical Implications', Ph.D. thesis, University of Madras, Chennai (mimeo), 1996.

Janakarajan, S., 'Aspects of Market Inter-relationships in a Changing Agrarian Economy: A Case Study from Tamilnadu', Ph.D. thesis, University of Madras, Madras (mimeo), 1986.

______, 'Interlinked Transactions and the Market for Water in the Agrarian Economy of a Tamilnadu Village', in S. Subramanian (ed.), *Themes in Development Economics: Essays in Honour of Malcolm Adiseshiah,* New Delhi: Oxford University Press, 1992.

______, 'In Search of Tanks: Some Hidden Facts', *Economic and Political Weekly*, vol. 28, no. 26, 26 June 1993.

______, 'Complexities of Agrarian Markets and Agrarian Relations: A Study of Villages in Northern Tamilnadu', paper presented at a workshop on 'Adjustment and Development: Agrarian Change, Markets and Social Welfare in South India, 1973-1993', Chennai: Madras Institute of Development Studies (mimeo), 1996.

______, 'Consequences of Aquifer Over-exploitation: Prosperity and Deprivation', *Review of Development and Change*, vol. II, no. 1, January-June 1997.

Janakarajan, S. and A. Vaidyanathan, *Conditions and Characteristics of Groundwater Irrigation: A Study of Vaigai Basin in Tamilnadu,* a report of a research project sponsored by the Planning Commission Government of India, Chennai: Madras Institute of Development Studies (mimeo), 1997.

Kolavalli, S. and D.L. Chicoine, 'Groundwater Markets in Gujarat, India', *Water Resources Development*, vol. 5, 1989.

Lindberg, Staffan, 'While the Wells Went Dry: Tragedy of Collective Action Among Farmers in South India', in *From Wild Pigs to Foreign Trees* (mimeo), 1996.

Meinzen-Dick, Ruth, 'Groundwater Markets in Pakistan: Participation and Productivity', *Research Report* 105, IFPRI, Washington, D.C., 1996.

MIDS, *Tamilnadu Economy: Performance and Issues* (Chapter 6), New Delhi: Oxford & IBH Publishing Co. Pvt. Ltd., 1988.

Moench, Marcus, 'Chasing the Water Table: Equity and Sustainability in Groundwater Management', *Economic and Political Weekly*, vol. 27, December 1992.

———, *Selling Water: Conceptual and Policy Debates Over Groundwater Markets in India*, VIKSAT—Pacific Institute for Studies in Environment, Development and Security, Natural Heritage Institute, 1994.

Narayana, Murthy, 'Murder by Poisoning—Death of the Palar', in Anil Agarwal, Darryl D'Monte and Ujwala Samarth (eds.)., *The Fight for Survival: People's Action for Environment*, New Delhi: Centre for Science and Environment, 1987.

Palanichamy, N.V. and K. Palanisami, 'Groundwater Pollution in Tiruppur Town,' Background Paper for Policy Advisory Committee Meeting, Chennai: Madras Institute of Development Studies (mimeo), 1994.

Palanisami, K., R. Balasubramanian and A. Mohamed Ali, *Management of Tank Irrigation by Government and Community: An Economic Analysis of Tank Irrigation Performance in Tamilnadu,* A Report, Coimbatore: Water Technology Centre, 1996.

Rajagopal, A. and A. Vaidyanathan, *Conditions and Characteristics of Groundwater Irrigation in the Palar Basin: Some Preliminary Results*, Chennai: Madras Institute of Development Studies (mimeo), 1997.

Rajarathinam, N. and K. Santhanam, 'Industrial Pollution Affecting Upper Palar Environ', paper presented in the seminar on 'Ecology, Environment and Water Management', Tiruchi: IMTI, 18-19 October 1996.

Rao, D.S.K., 'Groundwater Over-exploitation Through Bore-Holes Technology', *Economic and Political Weekly*, vol. 38, 25 December 1993.

Rukmani, R., *The Process of Urbanis North Arcot District*, Soil Survey and Landuse Organisation, Government of Tamil Nadu, 1982.

Vaidyanathan, A., 'Depletion of Groundwater: Some Issues', *Indian Journal of Agricultural Economics*, vol. 51, nos. 1 and 2, January-June 1996.

Vaidyanathan, A. and S. Janakarajan, *Management of Irrigation and Its Effect on Productivity Under Different Environmental and Technical Conditions: A Study of Two Surface Irrigation Systems in Tamilnadu,* a report submitted to the Planning Commission, Government of India, Chennai: Madras Institute of Development Studies (mimeo), 1989.

CHAPTER 6

Managing and Utilizing the Kosi Waters: Experience and Prospects

T. PRASAD

INTRODUCTION

The Kosi is a major, perennial tributary of the river Ganga with nearly 78 per cent of its catchment mostly in Nepal and partly in Tibet where little of its water is used. The plains of north Bihar, having 22 per cent of the basin area, are densely populated and heavily cultivated. The river had long been notorious for the human miseries caused by its frequent floods, deposition of coarse, infertile silt on the flood plains, and the continual shifting of its course in the plains. But it also inspired visions of harnessing abundant hydropower and ample water for year-round irrigated agriculture in the basin. Starting from the Indian proposal of a Kosi High Dam in Nepal formulated in the early 1950s, the existing Kosi project was implemented in the late 1950s and the early 1960s as an immediate and intermediate-term measure in view of the people's sufferings. Nearly four decades of experience of this project has brought out its basic deficiencies as well as its inappropriateness, and has given rise to distorted perceptions regarding the benefits or otherwise on both sides of the border. While Nepal feels that the present Kosi project hardly delivered any substantial benefits to it and that thus it was cheated, the Indian experience has been that no effective and reliable flood protection measures and sustainable irrigation systems were provided by this intermediate-term measure.

Thus, the management and utilization of the Kosi waters have suffered in effectiveness and sustainability not on account of any scarcity of the water resources as such in the system or the non-availability of appropriate civil engineering solutions, but mainly on account of a lack of appreciation of the historical context in which certain actions were adopted as interim measures and certain other actions were envisaged for more enduring solutions, which still remain elusive. While both the co-basin nations have been endeavouring to move forward towards securing 'enduring solutions' in terms of optimum development, effective management, and efficient utilization for irrigation and allied benefits in a sustainably productive manner, conflicts in perspectives and crises of confidence engendered during the past few years have been stalling further progress.

THE KOSI BASIN

The Kosi basin lies between 85-degree and 89-degree longitudes and 25-degree 20 minutes and 20-degree north latitudes. It is bounded on the north by the ridge that separates it from the Tsangpo (Brahmaputra), on the south by the main channel of the Ganga, on the east by the ridge separating it from the Kankai/Máhananda system, and on the west by the ridge separating it from the Bagmati and the Gandak/Burhi Gandak basins.

The catchment with an area of 92,538 sq. km. drains the Himalayas in the central and eastern parts of Nepal and Tibet. Some of the highest peaks in the world, including Mt. Everest and Mt. Kanchenjunga, lie in its catchment. The catchment has three distinct parts: one (covering 30,800 sq. km.) lying in Tibet across the great Himalayan range, the other to the south in Nepal (extending over 41,333 sq. km.), and the third (20,405 sq. km.) in India from the Nepalese border up to the Kosi's confluence with the Ganga.

The major part of the river's catchment (nearly 59,500 sq. km.) lies in the mountains. The trans-Himalayan portion is a high plateau which is not as high and not as jumbled as the mass of hills in the Himalayan mountain ranges. The hills run generally in the east-west direction. All the peaks and valleys above 5,000 m. are perennially snow covered except the steep slopes and rocky outcrops. In winter, the snowline remains at 3,000 m. while in summer it rises to 4,600 m. It is these snowfields that contribute to the river flow in spring when the thaw sets in. The area under permanent snow in the basin is 5,700 sq. km.

The River System

The Kosi, the main drainage channel of the basin, and its tributaries rise in the Himalayas. It is the third longest of the Himalayan rivers, next only to the Indus and the Brahmaputra. The Kosi is considered to be one of the holy rivers and is referred to as 'Kaushika' in ancient Sanskrit literature. In Nepal the river is called the 'Saptkosi', deriving its name from its seven tributaries, i.e. the Sun Kosi, the Bhote Kosi, the Tama Kosi, the Dudh Kosi, the Barun, the Arun, and the Tamur. Of these the Sun Kosi, the Arun, and the Tamur are the three major tributaries. The Sun Kosi comes from the west, draining the region east of the Kathmandu valley; the Arun from the north; and the Tamur from the east. The three tributaries meet at Tribeni, about 5 km. upstream of Barahkshetra and form the 'Saptkosi'. The Arun is the longest tributary, and cuts through deep gorges in the Himalayan range. The river is called Phung Chu in Tibet, and drains the region west of Mt. Everest in Nepal. The Tamur drains the region of Kanchenjunga. The catchment areas of the Sun Kosi, the Arun, and the Tamur at their confluence at Tribeni are 19,000 sq. km., 34,500 sq. km., and 5,900 sq. km., respectively.

The river flows in a narrow gorge for about 10 km. below Tribeni, past the historic temple at Barahkshetra, and emerges through the gorge of about 900 m. width into the plains at Chatra. Further down, the river flows through the flat plains of Nepal called the *tarai* which extends from the foothills to Nepal's southern boundary. The Kosi flows through Nepal for 50 km. below Chatra to Hanuman Nagar. From Chatra to Galpharia, the river flows in a south-west direction. Here the westward swing of the Kosi is restricted by the ranges which the river hugs at places right up to the Belka hill nose. Below Hanuman Nagar, the river enters Indian territory, and after flowing for 125 km. joins the Ganga at Kursela.

Hydrology

The hydrology of a basin is about the most important of its physical features, characterizing and indicating its water resource related problems, as well as its potentialities. Three main aspects—surface water flows, sediment transportation, and groundwater—constitute the hydrology, and are indicated by appropriate measurements.

Data on water flows in the basin are available from gauge and discharge sites located at several places, which were all set up after 1947. As the upper catchment, constituting about 85 per cent of the total, accounts for the bulk of the flow in the basin, observation sites are situated on the three principal tributaries—the Sun Kosi, the Arun, and the Tamur—located upstream of their confluence at Tribeni as well as on the main river below this confluence. The last one was subsequently shifted downstream of the proposed Barahkshetra dam site. Another site was established at Chatra in the Nepal *tarai* where the river enters the plains. The Central Water Commission of the Government of India has a gauge and discharge site at Baltara in the downstream reaches of the Kosi.

FLOW BEHAVIOUR

On an average, 4.7 million hectare meters (mhm) of water flows annually through the Kosi and its tributaries. While the Kosi is a snow-fed perennial river redundant, the flow varies from less than 0.1 mhm in the winter months of February and March to 12.7 mhm in August. Nearly 70 per cent of the annual flow occurs during the months of July, August, and September (Table 1).

The flood-flow feature of the Kosi is given by the fact that while

TABLE 1: AVERAGE MONTHLY FLOWS OF THE RIVER KOSI (AS MEASURED AT CHATRA)

Month	Flow		per cent of the total annual flow
	Volume in lakh ham.	Average rate in cumecs	
January	1.109	414.23	2.33
February	0.873	360.93	1.84
March	0.920	343.37	1.93
April	1.080	417.97	2.47
May	1.944	725.67	4.09
June	1.080	417.97	2.47
July	9.353	3,492.15	19.67
August	12.675	4,731.06	26.65
September	9.860	3,804.41	20.74
October	4.850	1,811.07	10.20
November	2.166	835.58	4.56
December	1.454	542.84	3.05

TABLE 2: TYPICAL FLOOD-FLOWS OF THE RIVER KOSI (MAXIMUM DAILY DISCHARGES IN CUMECS)

Time in Days	1979	1980	1981
1	3,744	1,281	4,726
2	4,859	1,683	4,430
3	4,250	1,606	5,761
4	4,369	1,580	4,820
5	5,550	1,883	5,460
6	4,923	1,924	4,605
7	9,294	8,843	6,531
8	13,463	10,145	11,021
9	9,909	8,053	5,929
10	6,864	8,559	5,199
11	6,103	7,380	5,929
12	7,707	6,697	6,975
13	7,141	6,697	7,365
14	6,841	7,188	7,086
15	6,826	6,757	6,791

its average minimum flow is around 300 cumecs, it carries a maximum daily discharge of more than 6,000 cumecs for one to two weeks every year during the monsoon months. In 1979-81, the maximum daily discharges exceeded 6,000 cumecs for more than 9 days (Table 2). Annual peak discharges are much higher, ranging across the years from 0.19 million cusecs (5,400 cumecs) to 0.9 million cusecs (26,000 cumecs) (Table 3). When such huge flows passed through the river, they overflowed its banks and flooded the vast plains of north Bihar with its rather flat topography in the pre-embankment period.

SEDIMENT TRANSPORT

The Himalayas constitute a geologically young and active mountain system. Rivers originating from its friable slopes tend to carry substantial quantities of sediment. The tributaries of the Kosi, particularly its eastern tributary, the Tamur bring in considerable sediment resulting from sheet erosion as well as landslides along the unstable banks into the river flow. The annual sediment load of the three tributaries at their confluence at Tribeni (given in Table 4) is very high.

This sediment gets transported through the channel system,

TABLE 3: ANNUAL PEAK DISCHARGE SERIES OF THE RIVER KOSI (AS MEASURED AT BARAHKSHETRA)

Year	Date	Peak Discharge (in cusecs)	R.L. of Gauge (in feet)
1948	13.07.1948	478,422	405.26
1949	19.07.1949	395,640	401.05
1950	20.08.1950	340,661	401.96
1951	24.08.1951	256,284	399.96
1952	24.09.1952	306,443	400.46
1953	30.07.1953	191,413	396.96
1954	23.08.1954	855,237	420.16
1955	07.08.1955	250,000	400.46
1956	29.08.1956	192,000	394.66
1957	12.08.1957	266,000	400.16
1958	25.08.1958	373,000	404.86
1959	10.08.1959	211,000	396.46
1960	28.09.1960	254,000	398.36
1961	20.08.1961	293,000	400.46
1962	02.08.1962	371,000	401.46
1963	16.08.1963	270,000	397.76
1964	22.09.1964	380,000	407.86
1965	11.08.1965	235,000	400.06
1966	24.08.1966	382,000	409.46
1967	09.07.1967	312,000	404.56
1968	05.10.1968	913,000	433.66
1969	28.07.1969	287,326	403.66
1970	15.07.1970	489,800	411.26
1971	12.07.1971	430,000	407.76
1972	28.07.1972	378,205	407.26
1973	13.10.1973	347,850	405.86
1974	05.08.1974	403,280	409.36
1975	28.07.1975	324,950	403.96
1976	23.08.1976	334,830	402.36
1977	26.08.1977	274,640	404.76
1978	28.07.1978	347,106	405.06

TABLE 4: ANNUAL SEDIMENT LOAD OF THE KOSI TRIBUTARIES

Tributaries	Annual Sediment (mm^3)	Annual Sediment Yield (m^3/km^2)
Sun Kosi	54.2	2,818
Arun	34.6	947
Tamur	29.6	5,016
Total	118.4	8,781

primarily through the Kosi, undergoing changes in shape and size as well as in the mechanism of transport. The transported sediments settle in, and are eroded from, the beds and banks of the channel, depending on the hydraulic conditions of the flow, and resulting in the aggradation or degradation of the river. Also, the transported sediments get deposited in the flood plains, leading to the building up of the plains and to the formation of deltaic fans in the lower reaches of the Kosi. The deposition of coarse grades of sediment on the sprawling flood plains renders them infertile; cut-off avulsions and consequent deviations or changes of the river's course have been causing damage, devastation, and misery to the people.

GROUNDWATER

Detailed hydro-geological investigations of the Kosi basin have not yet been carried out. Comprehensive and quantitative information about the aquifers in the basin is, therefore, readily available. Investigations about the spatial and temporal variations of the water table have been carried out in the plains of north Bihar and the Nepal Terai portions of the basin since 1948. These consisted of pre- and post-monsoon observations of the subsoil water levels in a selected number of wells lying on various lines of observation covering the region and of monthly observations in about 10 per cent of these selected wells. Similarly, computations of recharge from rainfall, return flow of irrigation, and seepage from the canal have been done for the existing command of the Kosi irrigation project.

These data reveal that several aquifers at various depths and of varying thicknesses occur in the thick alluvial deposits of the Nepal Terai and north Bihar plains of the basin in confined, semi-confined, and unconfined conditions. The depths at which these aquifers exist may vary from a few meters for the topmost unconfined aquifers to more than 1,000 m. for deep confined aquifers. The depths to the water table from the ground vary from 10 m. to 15m. The fluctuations of water table level from the pre-monsoon to the post-monsoon periods range from 1.5 m. to 5.0 m. The water table gradient is primarily from north to south, and secondarily from east to west. The hydraulic gradients generally follow the ground slope and vary from about 0.8 per thousand in the Nepal Terai to 0.1 per thousand in the lower reaches of the basin.

The vertical recharge of the groundwater from rainfall (estimated at 15-20 per cent of the rainfall) may be of the order of 25 cm. annually. In the canal-irrigated area, groundwater is further recharged from canal seepage as well as from the return flow from irrigation. The rate of horizontal recharge to the semi-confined and confined aquifers from the upstream recharge zone has not been quantitatively determined, but is considered to be substantial. The quality of groundwater throughout the basin is excellent to good. The Kosi basin is considered to be a high groundwater-yielding zone, with its annual safe yield estimated at around 40 cm. in the alluvial plains.

DEVELOPMENT OF THE KOSI BASIN

The Kosi Project

GENESIS

With 1,300 mm of annual rainfall, 4.7 mhm of water flowing in the river, and being rated as a high groundwater-yielding zone, the Kosi basin has an abundance of water. The problem is one of extreme and severe flooding during the monsoons when the river carries a huge amount of water, causing extensive damage to both crops and people. Problems also arise from the sediments carried by the river. On the other hand, water flow in the river during the non-monsoon months drops to very low levels.

The basin is densely populated; the population depends overwhelmingly on agriculture; the present productivity of land is low, and the incidence of poverty is among the highest in the country. Recurrent and extensive floods cause extensive damage to life, property, and agricultural production. On the other hand, the basin has fertile soil, a climate which permits year-round cropping, and has immense potential for increasing land productivity. Essential to realize this potential is the effective means to control and contain flooding during the monsoons, and to augment the water supply during the rest of the year.

This was the understanding and the assumption behind the Kosi project mooted by the Government of India in the mid-1940s. It envisaged a 239 m. high dam at Barahkshetra in Nepal to provide flood moderation from 24,088 cumecs to 5,668 cumecs; to irrigate

1.55 million ha. in Nepal and India; to generate 1,890 MW of electricity; and to provide facility for inland navigation. Although this proposed project did not reach a stage of detailed investigation, planning, and design, and though the required agreements between the two concerned governments to implement the project have remained elusive, the dream of transforming the life, living, and economy of the region by training and harnessing the Kosi continued to entice the people ever since.

While the pros and cons of the high dam proposal were being examined by the national and concerned state governments another series of catastrophic floods struck the Indian plains of the Kosi basin. Primarily motivated by the desire to mitigate the suffering of the people, and keeping in view the inevitable delay in finalizing the high dam proposal and the long gestation period in implementing it, it was decided at the highest levels of the Government of India that something must be done as an immediate and intermediate-term solution.

This led to the formulation, in 1953, of a project to construct embankments along both the banks of the Kosi, spaced 5 km. to 16 km. apart, extending upstream from its confluence with the river Ganga and tied up by a barrage across the Kosi at a location about 8 km. inside Nepal, in order to provide flood protection to 280,000 ha. of land outside the embankments in the Kosi flood plains. In addition, two canals taking off from the barrage were envisaged to irrigate 740 thousand ha. on the eastern side and 312 thousand ha. on the western side, mostly in India. An inundation canal taking off from an upstream location at Chatra was to irrigate land in the *tarai* region of Nepal. Taking advantage of a fall in the eastern main canal, a 20 MW hydropower plant was designed to provide electrical energy to both Nepal and India. The exceptionally severe flood of 1954 gave an impetus to serious negotiations on this project between India and Nepal, culminating in an agreement between the two countries. The project was thereafter implemented in the latter half of the 1950s.

PERFORMANCE

The performance of this project has been reviewed and evaluated on several occasions during the last 30 years (Sain, 1965; Government of Bihar, 1979; Sanyal, 1983). They show that the impact of the project has belied expectations and has given rise to new problems. The sector-wise performance revealed in these reports is as follows:

Flood Protection

The project, as noted earlier, was designed to protect only the lands situated outside of the two embankments. The area between the embankments—about 50,000 ha.—was not expected to be protected. However, due to the capricious behaviour of the river, breaches often occurred at vulnerable points in the embankments, and this caused several devastating floods and damage to areas outside the embankments. As the flood plains between the embankments were considerably aggraded by sediments, the threat of overtopping, and breaches had increased.

Due to the wide spacing between the embankments on both sides of the river, considerable areas of the embanked land, which was earlier under cultivation, have been condemned to more intense and more frequent flooding, to more deposition of largely coarse silt, and to more risks due to the shifting of the river's course. Although the owners of these lands, mostly poor and marginal farmers, have been paid compensation, they continue to harbour a sense of resentment at having been made to sacrifice their interests for the benefit of others. Of course, in the hills, flash floods due to cloudbursts, glacial lake outbursts, and landslides continue unabated.

Similarly, cultivable lands adjacent to and outside the Kosi embankments have been subjected to both seepage from the aggraded river and to obstructed drainage. Farmers owning these lands, now rendered permanently uncultivable due to inundation or waterlogging, perceive themselves to be the victims of the Kosi project. Constant aggradation of the embanked flood plains has made the periodic strengthening and raising of the embankments necessary to contain the flood flows. Since this was not done, they became more susceptible to breaches even at less than the design discharges, making the protected villages vulnerable to more serious devastation from floods in the event of breaches than what they had suffered in the past without embankments. There have been several such instances of devastations in recent years.

Irrigation

Even in the case of irrigation, a major secondary benefit of the project primarily designed for flood protection, the project is far from being an unequivocal success. According to a 1979 report of the Government of Bihar, the average area annually irrigated during 1973-8 was only 25 per cent of the planned 1 million ha. Data for subsequent years show a distinctly declining trend in the irrigated area. This is attributed to the non-completion of the micro irrigation infrastructure, the reduction of the carrying capacity of canals, water courses, and field channels due to silting, the undulating topography of the command, and poor maintenance.

The areas being satisfactorily served both by the Kosi flood protection and irrigation systems (about 280,000 ha.) have definitely enhanced their agricultural performance. Fortunately, a large number of farmers in such areas, having been exposed to the benefits of irrigated agriculture, have taken to the use of groundwater available at low depths through cheap improvised shallow tube wells for irrigation. In fact, it is this unplanned development of groundwater that has saved the Kosi irrigation system, based on only canal water use, from being an unmitigated failure.

Power

Generation of hydropower to the tune of 20 MW through a canal fall, against an estimated 10,000 MW feasible in the Kosi basin, was the most incidental and marginal benefit planned under the Kosi project. Hardly 60 per cent of even this planned generation is ever achieved due to siltation and other operational problems.

Ecology

More important are the adverse ecological effects that have emerged over time as a consequence of the Kosi project. The aggradation of the river-bed between the embankments has enhanced seepage, and caused waterlogging and submergence of the areas outlying the embankments. This has either reduced the productivity of, or has rendered uncultivable, a large part of the Kosi command. The systematic rise of the water table has been observed, and waterlogging conditions have been created or exacerbated in considerable areas of the Kosi command, reducing its agricultural productivity and yield. The embankments have also interfered with the drainage of the Kosi basin, necessitating a massive drainage scheme for the basin at a huge cost.

PERCEPTION

Thus, in the perception of the people of the lower Kosi basin lying in India for whose primary problems of floods and secondary benefits of irrigation the project was planned, designed, and implemented with the agreement of the Government of Nepal, the Kosi project has been at best a mixed blessing. Though a large number of people have benefited from the project, a sizeable population has also suffered from its adverse consequences. Among the people who have become better off in the post-project era, there are those who have benefited more from other collateral developments than from the project itself.

In Nepal, the Kosi project has been perceived in different ways in different contexts in the post-project period. The intended benefits to Nepal from the project consisted of irrigation from the Chatra inundation canal and the western Kosi canal, 50 per cent of power from a fall on the eastern main canal, some flood protection and river-training benefits, as well as an all-weather communication link with the barrage serving as a bridge across the Kosi in Nepal. There has been hardly any experience of adverse consequences of the project. However, except for the last item of benefit, the benefits on other counts have fallen short of expectations for a variety of reasons. This experience has led to some dissatisfaction with the project on the part of the intended beneficiaries. However, this does not fully explain the negative perception of the Kosi project in Nepal as a whole.

As water constitutes the most important and vital natural resource of Nepal, all matters concerning it are invested with considerable sensitivity. Because of this sensitivity, water resource issues tend to be political and emotionally volatile which is a veritable spawning ground for creating and nurturing myths. In the case of the Kosi, a perception has been generated, largely in pursuance of political purposes, that in the Kosi project very little benefit has been given to the people of Nepal compared to the people in India, that Nepal got a raw deal, or the Nepal has been 'cheated'. This perception, examined in the light of the context in which the Kosi project was conceived, agreed upon, and implemented, is not well founded. It is more a reflection of the history of a complex and difficult relationship between Nepal and India as two neighbouring countries, one small and the other large. Yet this perception, or myth is assiduously kept alive, out of the correct context, to gain political mileage.

PROBLEMS AND PROSPECTS

The Kosi High Dam Proposal

FEATURES

Partly as a response to this dissatisfaction and disillusionment and partly in a delayed action on the agreement which considered the Kosi project as an intermediate measure till more comprehensive measures were investigated and implemented, the original 1950 Kosi High Dam proposal was revived in an updated and modified form in 1980 by the Central Water Commission, Government of India. This revised version of the project envisages a 269 m. high dam 1.6 km. upstream of Barahkshetra, another barrage and canal system at Chatra to provide for additional irrigation of 143 thousand ha. in Nepal and 74 thousand ha. in India with enhanced intensity of irrigation in the existing Kosi command, a flood moderation from a peak of 42,500 cumecs to 14,000 cumecs, and a hydropower generation of 3,300 MW (Central Water Commission, 1980).

This project is expected to go a long way in realizing the high development potential of the Kosi basin for the benefit of both Nepal and India, and in achieving a better balance in the sharing of benefits between the two principal co-basin nations from their shared Kosi water resources. It was also expected to provide some (as yet unidentified and un-assessed) benefits to the lower riparian nation of Bangladesh in terms of both flood-flow moderation and lean-flow augmentation. Yet the progress towards its implementation can at best be described as halting. At any rate, the enthusiasm and urgency that this apparently win-win project should generate are missing. There are problems and misgivings on grounds of technical feasibility, environmental impact, and of course on political grounds which may be indicated as follows:

Extreme Hydrology: The planning and design for the harnessing of any river for the development, management, and utilization of its water resources have to take into account and contend with its extreme hydrology. The seasonally skewed precipitation occurrences in the Kosi basin lead to equally skewed discharges into the river Kosi. While maximum, minimum, and average discharges into the river at Chatra are 27,142 cumecs, 231 cumecs, and 1,507 cumecs respectively, about 76 per cent of the annual flow of 47,364 ham passes in the four monsoon months of July-October. Such high variability poses severe problems in the harnessing, management, and utilization of the river flows for various purposes. First, run-of-the-river projects cannot adequately and reliably meet the various demands which hardly ever conform to the hydrology of the river. This is compounded by topographical features marked by the abrupt change of gradients from very steep to very mild over relatively short distances and the hydraulic factor of the limited conveyance capacity of the river channels, resulting in recurrent floods in the sprawling flood plains of the river. Although the spread of these flood waters has been confined to the area between the widely spaced embankments since the late 1950s, the flood flows in the river have, as noted above, been causing other problems. Further developments in the basin have to take into account the constraining hydrological consequences resulting from the past actions.

The Kosi being a snow-fed and perennial river, the minimum flows obtaining in it do not pose any threats or problems to the riverine communities or to the hydraulic structures constructed or still to be constructed to deal with the flows. Of course, such minimum flows may affect, to varying extents, the different uses to which the Kosi waters are to be put. From the historical record of its flows, there is no appreciable persistence in minimum-flow occurrences and hence any adverse effect on the use of the Kosi waters on account of its minimum flows will be transient.

The other extreme hydrology, i.e. maximum or peak discharges into the river constitutes veritable threats and hazards in various forms, both to the riverine community as well as to the safety of hydraulic structures. They have to be conservatively estimated, carefully planned for, and adequately provided for in the design. From an examination and analysis of the historical record of the Kosi peak flows, given in Table 3, the variance of peak-flow occurrences is not very high, the coefficient of variation being 0.44, and hence flood flows of various return periods, as given in Table 5, are not markedly different. This is fortunate in the sense that more confidence at reasonable incremental costs may be ensured in selecting design values of peak-flows for various structures and measures.

High Seismicity. Earthquakes are among the most devastating natural disasters. Vibrations in the earth's crust are likely to endanger the safety of structures. Where, as in the case of the Kosi, water resources development requires massive structures to be built, considerations of earthquakes are also quite critical. This is particularly so in the case of high dams. The Kosi basin, and especially the sites for the high dams most of which are in the sub-montane Himalayas, are marked by high seismic activity. The probability of earthquakes of high intensity in this region is high. This is one of the major reasons for concern about the feasibility of constructing safe dams in the region.

This danger is widely recognized. Some experts are of the view that the belief that dams should not be built in highly seismic zones is not tenable. This is also unacceptable from the point of view of national development and is not justified by the trends of seismic activity *vis-a-vis* the experience of high dams already built in similar regions elsewhere.

Over the years, a great deal of work for the scientific monitoring and measuring of seismic activity has been done in the region. A seismological observatory was established at Chatra in Nepal in 1950. It has been modernized gradually with the addition of sophisticated instruments. The Earthquake Research Centre at Roorkee in India and other such centres in countries like the USA, Japan, and Russia have done considerable research work on structures in highly seismic areas. It is noteworthy that scientific

TABLE 5: KOSI DESIGN FLOOD-FLOWS OF VARIOUS RETURN PERIODS (BASED ON PEAK DISCHARGES AT BARAHKSHETRA USING GUMBEL'S METHOD)

Return Period	10 Years	50 Years	100 Years	500 Years	1,000 Years
Design Flood-Flows (cusecs)	604,100	839,600	939,500	1,170,000	1,269,100

developments and the advancement of knowledge have enabled high dams to be built with confidence at Nurek and Rogun (325 m. high) in the Soviet Union, Sussodha in Alaska (280 m. high), and Mica in Canada (245 m.)—all of them located in highly seismic zones. There is thus a much better basis now, in terms of knowledge and proven experience, to design structures that can withstand severe probable earthquakes in the region.

Despite the fact that many dams have stood for years and that others have been built in the most complex terrain and seismically active regions, the possibility of dam failures remains. Inadequate investigations invite trouble. The factor of risk, whether from seismicity or extreme flood-flows or any other damaging probable events, is always to be contended with in the design of dams, or for that matter, of any other structure. This, by itself, is no reason to avoid construction of dams provided the technology exists to take care of it at an acceptable level of risk. It implies a higher order of design safety and preparedness to bear the attendant cost.

Erosion and Silting. Soil is eroded from the slopes of the young, fragile, and friable mountains of the Himalayan and sub-Himalayan ranges. It is carried by water in the form of sheet and channel flows, and is ultimately transported to the river Ganga which carries it to the Bay of Bengal. In the process, the silt spreads over the flood plains, taking part in the geo-morphological process of the formation as well as the building up of the alluvial plains and causing the shifting of the river's course in these plains. The fact that the mountains are young and still growing and that their steep slopes experience heavy precipitation and extreme variations in temperature causes a great deal of loss of soil. In the high relief areas of the Himalayas, weathering is mostly due to block disintegration by alternate freezing and thawing and the production of scree due to frost action. The transportation and erosion are by the glaciers and glacier rock flows, respectively.

A considerable movement of the scree also takes place simply through the force of gravity. In the outer Himalayas and the Siwaliks, sub-aerial denudation is the major natural process contributing to the instability of the hill slopes, the silting of the river valleys, and flooding because of the obstructions in the river courses. These ranges mainly comprise soft tertiary sediment, sand stones, silt stones, shells, and clays, and are readily subject to mechanical disintegration. These ranges also face the full force of the torrential rains during the monsoons. Hence, loss of soil mass and its transportation in the form of silt through the rivers is a natural phenomenon (Carson, 1984).

This phenomenon is significant in the context of the development of Indo-Nepal water resources in two important ways. One is the sedimentation of reservoirs which may be built as part of this development. The rivers and streams which bring the waters to these reservoirs will also bring in silt, partly as bed load and mostly as suspended load, which will get deposited in the space assiduously created for the storage of water. This reduces the life of the reservoir, and consequently adversely affects the economic viability of the project.

The second aspect of the silting problem *vis-a-vis* water resources development is the ecological impact of the latter on the former. This impact may be better considered in two parts, the upstream impact and the downstream impact. The upstream impact mostly relates to the concern whether water resources development measures will aggravate soil erosion. Compared to the magnitude of the natural forces which are inevitably at work leading to the erosion of soil, the contribution by water resources development measures must be too localized and insignificant. Moreover, if soil conservation and other watershed management measures are also taken up as part of the water resources development programme, the impact may be positive, rather than negative, in this regard.

The downstream impact emanates from the change in the silt-water regime of the river below the impoundment, affecting the processes of aggradation and degradation of the river beds, of the building up of the flood plains, and of the migration of the river courses. Again, these effects can be both damaging as well as beneficial, in the short-term as well as in the long-term, and have to be carefully studied and accounted for in each case. It may be understood, however, that through suitable river-training measures, the damaging effects can be contained or minimized and the beneficial effects enhanced. It may be mentioned here that as part of the natural geo-morphological process, the shifting of the river courses, such as the westward swing of the Kosi through 112 km. over the past 130 years, has brought misery and devastation to the affected population.

Environment and Ecology. Water resources development in the basin entails large-scale intervention in the natural regime. A dam, for example, may alter the entire regime of the lower reaches of the river by regulating flows and changing its natural pattern. It may stem the flow of the river to create a new lacustrine condition above the dam. It will, therefore, be necessary to do an ex-ante evaluation of the various impacts resulting from the developmental activities. The impacts may adversely affect the environment and the natural eco-system, but there may also be some beneficial impacts leading to the improved quality of the environment. What is required is to identify and evaluate various impacts, both positive and negative, as a result of the envisaged developments and to formulate measures to enhance the former and diminish the latter.

In the case of the proposed multi-purpose project in the Kosi basin, the crucial issues are displacement of population, reservoir-induced seismicity, submergence of historical monuments, reservoir sedimentation, erosion and morphological changes downstream of the dam, damage to landscape, submergence of forests, disturbance to flora and fauna, impediments to fish migration and downstream aquatic life, as well as waterlogging and salinization in the proposed irrigation commands. As against this, the beneficial environmental impacts include moderation of hazards like recurrent floods, reduced waterlogging and drainage congestion in the plains, arrest of silts in the reservoir, and the consequential beneficial impacts.

Forests constitute the basic component of the ecosystem. Their contribution to the quality and efficiency of nature's life-support systems is critical. A considerable part of the upper catchments of the Indo-Nepal rivers was at one time densely forested. The situation is, however, now very different from what it was in the past. Great concern these days is being expressed over the environmental degradation in the Himalayas under the increasing pressure of population and livestock, and reclamation of marginal slopes for cultivation, setting off a vicious spiral of deforestation, erosion and increased sedimentation, and flooding in the plains downstream. It has been assessed that the area under forests in Nepal was reduced by 5.70 lakh ha. between 1964 and 1985 alone. The reservoir projects must, therefore, be planned to submerge the minimum forest area. There must be adequate provision of compensatory afforestation in the catchments, besides large-scale soil conservation measures in the upper catchments to prevent further degradation of the environmental and ecological condition.

It is worth mentioning here that in the developed countries, a good deal of concern is voiced at present regarding the preservation of natural systems and the improvement of the environmental quality which may have been degraded in the past due to rapid and large-scale economic development. More or less on the same pattern, there is an increasing emphasis on these aspects in the developing countries also. While developing countries must take into account the emerging environmental problems and issues in planning, designing, executing, and operating their development projects suitably and effectively, they have to strike a balance, in the ultimate analysis, between the concerns of the environment and the imperatives of development. This is particularly significant for the region under discussion in the prevailing socio-economic context. Steering a course between the two extremes of 'environment without development' and 'development without environment', this region has no choice but to tread the path of 'development with environment'.

Perceptional. The above-mentioned factors are undoubtedly important and must be carefully analysed. They are significant not only to the economics of the project but are also inextricably related to both the risk and the quality of life and living. However, these factors are not unique to the Kosi. There is worldwide experience and appropriate technology to help deal with them adequately. Technological and environmental issues, important as they are, have not been the factors stalling the implementation of the proposal. An analysis of the problems and prospects in implementing the proposed project *vis-a-vis* the progress of negotiations over it shows that the physical problems—either of hydrology, or of seismicity, or of silting, or of environment and ecology—have not been the principal impediments. In fact, these problems are neither prohibitive nor unique to the Kosi; they are tractable with the available technology. Rather, it is the people's perceptions generated, engineered, and/or nurtured over the years, particularly in the wake of the implementation of the existing project, that have been at work in slowing down the process.

The people in the plains of the lower Kosi basin in India have viewed the implemented Kosi project as at best a mixed blessing, having provided some benefits, caused certain adverse consequences, and generated a feeling of having been victimized by it among some sections of the people. Similarly, the people in the upper part of the basin in Nepal also feel that they got a raw deal out of it. In this perceptional gap the initiative of the Government of India regarding the high dam proposal is viewed with suspicion and wariness by the Government of Nepal.

At the government-level negotiations, the proposed project is viewed by Nepal as an 'Indian project'. The perception in Nepal about having got a raw deal under the existing Kosi project makes the Nepali people wary of the larger project. While India seems to be on the defensive, it also feels that it is being confronted with a certain degree of intransigence as well as unreasonable demands on the part of its upper riparian neighbour. In spite of the political will expressed at the highest levels to move forward with the project, progress has been frustratingly slow. It is thus the perception and the psychological milieu in which the proposal is being considered by the two governments that lie at the root of the problem of halting progress.

Role of Science and of People

It is apparent that the present approach towards realizing the high potentials of the Kosi—based exclusively on the governmental negotiations which are subject to political vicissitudes and constrained by the bureaucratic mindset—is inadequate to break the impasse. The immensity and the mutuality of benefits to the peoples of the two co-basin nations and the opportunity cost of delay in terms of realizable benefits is extremely high to both nations. Therefore, it is imperative to explore other ways to bring the process of water resource development in the Indo-Nepal region in general, and in the Kosi basin in particular, onto the right track, build-up momentum, and move forward. To this end, it is important to bring two elements—namely scientific knowledge and people's involvement—into, the process. This would facilitate a more informed debate on the merits of the project, the problems involved, and the ways in which they can be overcome, and thereby create an atmosphere to move the project forward.

Science provides the base for all human activities and developments, including water resources development. Thus, science must not only inform all endeavours for water resources development but must also guide them. This is particularly relevant in the case of water resources development of international basins, where science plays the role of a unifying force among the co-basin nations by bringing out and emphasizing the hydrological integrity of a basin and the cardinal principle of a basin being the rational unit of development and management of water resources. In the case of the Kosi, science will not only be a unifying force but will also be a great facilitating factor in view of the immensity of benefits to, and the mutuality of interests of, India and Nepal in the multipurpose development of the Kosi basin. It makes cooperation between the co-basin nations in water resources development logical, imperative, and, hopefully, inevitable.

The second critical element missing from the present process is the involvement of the people concerned. The professed aim of water resources development is to effectively tackle the water resource related problems of the region and to help realize the immense potentialities for the benefit of its people. It is not only right but also desirable that these people be associated with this process. Such an association will provide an added incentive and impetus to the governments concerned to overcome any obstacles to the progress of the process, as the delay will then be directly perceived as a denial of feasible benefits to the people.

Thus, what may be called the track-two approach—in which science provides the driving and guiding force and the people's involvement imparts the required momentum—will considerably facilitate and expedite the process of water resources development by strengthening the track-one effort being made by the governments through negotiations. The almost total absence of any track-two effort in the process of water resources development in the Kosi basin has been about the main inhibiting and stalling factor in the process.

Role of Institutions

The moot point is how to bring the track-two approach into play in the matter. For this, it is essential that suitable institutions are in place, at work and participating in the process in order to ensure that it is on the right track and moving in the right direction. If the milieu in which this process has been taking place so far is appraised, one would be struck by the utter institutional deficiency.

The governments have been almost the sole institutions participating in the process. Political compulsions, the bureaucratic mindset, and technocratic self-righteousness characterizing governmental institutions have over the years conditioned the process, not allowing it to acquire a momentum of its own and resulting in the frustratingly slow pace of forward movement.

Several types of institutions, apart from those of the governments, are needed to play their respective roles in the process. First and foremost, suitable institutions where water resource development can be pursued in scientific and multidisciplinary perspectives, in an atmosphere of objectivity and freedom of enquiry, must be brought up and nurtured in the region. Scientists who are more committed to their disciplines than to any narrow and myopic concept of nationality have to come to the fore and play their role through these institutions, shedding their monumental indifference to their societal responsibility. This will help the process acquire a desperately needed scientific temper. On the one hand, these institutions have to play an intermediary role between the governments and the peoples, and have to be sensitive to the limitations of the former and mindful of the concerns of the latter. At the same time, these institutions have to influence the decision-making process in the governments as well as educate and motivate the people as per the dictates of science.

To create public awareness aimed at informing, involving, and motivating people, and thereby generating cooperation with good science as one of the vehicles, intermediary institutions at the horizontal contact level should be given more opportunity to play their part and to facilitate changes (Dixit, 1993). NGOs and the media have distinct and positive roles to play in this process.

When the perceptional problems have been overcome, new and appropriate institutions will have to be brought into being at the stages of inception, implementation, and operation. The existing institutions and organizations of the two governments may not be conditioned and geared for undertaking and accomplishing new tasks of water resources development in an international basin. Guidelines formulated on the basis of worldwide experience in comparable situations are available in the matter. In the Indian subcontinent itself, there are a few useful precedents. The Permanent Indus Commission, set up under the Indus Waters Treaty concluded between India and Pakistan in 1960 under the good offices of the World Bank, has eminently facilitated the development of the water resources of the Indus river and its major tributaries shared by the two countries for their mutual benefits. In spite of the fact that political relations between the two countries are fraught with tension, the functioning of the Commission was hardly affected, and the development and utilization of their shared water resources went on unhindered. Provision of a similar institution, the Mahakali River Commission under the treaty recently concluded between India and Nepal for the integrated development of the Mahakali river, which is a boundary and shared river between the two countries, is expected to facilitate the multipurpose development of the water resources of the basin for mutual benefits. The essential idea behind setting up such institutions is to distance and insulate the process of development of the water resources of river basins shared by more than one nation from the vicissitudes of political relations between or among the co-basin nations and providing an ongoing machinery for consultation and conflict resolution. In the case of the Kosi, such an institution must come up at an appropriate stage and under a mutually acceptable framework.

CONCLUSIONS AND FUTURE OUTLOOK

The Kosi is not a water-scarcity basin as such. In fact, if all the surface water and groundwater potentials are taken into account *vis-a-vis* the various uses, including agricultural use, to which the Kosi basin water can feasibly be put, it would usher in prosperity for the region. Unfortunately, the reality is far from this happy prospect, in spite of a major water resources project implemented in the late 1950s and early 1960s. This project was planned primarily to secure flood protection through embankments and secondarily to provide irrigation through a network of canals, with a barrage located 8 km. inside Nepal serving as a tying-up end for embankments and diversion structures for canals. Hydropower generation using a canal fall was planned for as a very minor and incidental benefit. More than four decades of experience has shown that the region is still not free from the actual occurrence and definitely not from the perpetual threat of devastating floods through an increasing likelihood of embankment breaches. As far as irrigation is concerned, not more than 25 per cent of what was originally planned to be achieved has been actually achieved. Along with such poor accomplishment of the flood and irrigation objectives of the Kosi project, serious adverse consequences of waterlogging due to seepage from aggraded beds of the embanked river, impeded drainage, and percolation of irrigation water have emerged. A redeeming feature of the post-project development, not planned for in the project, has been the increasing use of groundwater for irrigation through shallow tube wells, greatly facilitated by high-yielding top unconfined aquifers at relatively small depths and cheap improvization of what are locally called 'bamboo borings'. But for this development, the irrigation performance of the Kosi project would have been an even more dismal failure (International Workshop on the Kosi, 1975).

For an enduring solution to both pre-project and post-project Kosi problems as well as to realize very high multi-purpose potentials of the river, a high dam at a well-identified location at Barahkshetra in Nepal has long been considered to be imperative. In fact, the high dam proposal was mooted and a feasibility report prepared in 1950 for solving the problem of recurrent and devastating floods and the shifting of the river's course, as well as for providing the benefits of hydropower of 1,800 MW and inland navigational transport. These were colossal benefits by any standards and represented a dream for the people of the concerned Indo-Nepal region plagued by misery and poverty. For various reasons of expediency, lack of foresight, and lack of response this proposal was shelved in favour of what was considered at that time an immediate and intermediate-term measure, that is, to provide primarily flood protection and secondarily irrigation benefits.

The high dam proposal was revised and updated in 1980 by the Government of India, and has been under consideration by the Government of Nepal since then. In spite of political wills expressed at the highest levels, particularly after the advent of democracy in Nepal in 1990, the progress in negotiations for the eventual implementation of this proposal has been frustratingly slow. The extant situation is confounding in view of the immensity of benefits to, and the mutuality of interests of, the principal co-basin nations of India and Nepal and even the positive advantages to the lower riparian country Bangladesh, in terms of the attenuation of flood flows and the augmentation of lean flows.

In order to retrieve the process of desired water resources

development in the Kosi basin from the sole domain of governmental negotiations, it has to be brought on to a new track by introducing the two missing elements into the process: the infusion of science and the involvement of the people. While the problems, concerns, and opportunities for action are typically different for the peoples in the upper and lower parts of any international basin, it is science which emphasizes the hydrological integrity of the basin as well as the immensity of benefits with the mutuality of interests in the basin-wise development of the water resources, hence making cooperation among the co-basin nations logical and rational. The involvement of the people suffering the tremendous opportunity cost of delay will make cooperative action imperative and urgent. Thus, while science will provide the guiding force, the people's involvement will impart the required momentum for the process to move ahead on the new track.

For the infusion of science and the involvement of people in the process, appropriate types of institutions have to be set up. One, institutions where water resources development can be pursued as a science in an atmosphere of objectivity and freedom of enquiry must be set up in the region, both to influence the decision-making process of the governments as well as to educate the people. Second, intermediary institutions like NGOs and the media have to play their roles in raising public awareness and motivation using good science as one of the vehicles of communication and persuasion. Third, beyond the perceptional stage and at the operational levels, institutions that are insulated from the pressures of political vicissitudes have to be created to carry on the task of water resources development of the international Kosi basin under a mutually agreed upon framework.

In order to facilitate further, firmer, and faster movement towards the solution of the chronic Kosi problems and the realization of the high Kosi potentials, systematic and scientific analyses and studies of the performance of the existing Kosi project have to be done. Similarly, the options and alternatives for flood control, irrigation, energy production, and inland transport have to be studied in an integrated and scientific manner in order to arrive at optimum measures or a combination of complementary measures for effective solutions and sustainable benefits. It will be helpful if the results of such studies are properly documented and are made accessible to all concerned parties.

The optimum development, effective management, and multipurpose utilization of the water resources of the Kosi basin are no doubt tasks of challenging proportions, but they are immensely prospective and technologically tractable. Intermediate or midway solutions have proved to be self-deceptive and counterproductive. Once the perceptional problems are suitably taken care of, the process will acquire a dynamism of its own and the task will be well on its way to being accomplished, leading to the eradication of poverty, ushering in prosperity, and the eventual realization of the 'Kosi dream' for the people of the concerned Indo-Nepal region.

ACKNOWLEDGEMENTS

Studies, investigations, analyses, and other academic activities related to water resources development in the Indo-Nepal region, in general, and in the Kosi basin, in particular, have been on the research agenda of the Centre for Water Resources Studies (Patna University) for more than the last two decades. These activities include an International Workshop on Post-Fact Evaluation of the Kosi Project organized in 1975; an Indo-Nepal Workshop on Cooperative Water Resources Development in 1992; and a study based on a series of interactions, field visits, and investigations undertaken in collaboration with the Royal Nepal Academy of Science & Technology, Kathmandu during 1992-4. This paper is heavily based on the proceedings and outcomes of these activities. Technical assistance provided by R.K. Prasad, Research Associate of the Centre in the preparation of this paper, the help of Vinod Kumar, Computer Programmer/Operator in word processing and related D.T.P. work, and the involvement of other concerned personnel in related work are hereby appreciated and gratefully acknowledged.

REFERENCES

Carson, B., *Erosion and Sedimentation Processes in the Himalayas*, Kathmandu: ICIMOD, 1984.

Central Water Commission, Government of India, 'Feasibility Report on the Kosi High Dam Project', New Delhi, 1980.

Central Water Commission, Government of India, '1953 Project Report (Kosi Project)', New Delhi, 1953.

Centre for Water Resources Studies, 'Theme Paper for the Workshop on the Co-operative Development of Indo-Nepal Water Resources: Prospects, Opportunities and Challenges', 1992.

Dixit, A. and D. Gyawali, 'Building Regional Co-operation in Water Resources Development', *Water Nepal*, vol. 3, nos. 2/3, October 1993, Kathmandu.

Ghose, J.P., 'An Assessment of the Problems and Potentials in the Water Resources Development of North Bihar', M.Sc. (Engg.) thesis in Civil Engg. (Water Resources Engg.), Patna University, 1982, pp. 42-3.

Government of Bihar, 'Report on the Kosi Project and Its Irrigation Potential', Patna, 1979.

International Workshop on the 'Post Fact Evaluation of a Water Resources Project: The Case of the Kosi Project', held at the Bihar College of Engineering, Patna, 1975.

Prasad, T. et al., 'Co-operation for International River Basin Development: The Case of the Kosi Basin', *Integrated River Basin Development*, edited by Celia Kirby and W.R. White, John Wiley & Sons, 1994.

Sain, Kanwar, 'Report of the Kosi Technical Committee', Government of Bihar, Patna, 1965.

Sanyal, N., 'Report of the High Level Technical Experts' Committee for the Kosi Barrage', Government of Bihar, Patna, 1983.

CHAPTER 7

The Legal Framework for Water Management and its Working

K.V. RAJU

Water, being the common pool resource par excellence, raises particularly complex and difficult issues concerning the basis on which the claims of competing claimants will be determined and the means by which they will be articulated and implemented in specific situations. The principle is widely accepted that there can be no absolute private ownership rights over water and that the state is entitled to regulate its exploitation and use in the public interest. The state has everywhere assumed a substantial responsibility for the development and management of water resources directly or through public agencies. Even where exploitation by the private sector is allowed, it is subject to a number of regulations legislated by the government. This chapter provides a brief review of the legal framework in which Indian irrigation systems operate, the characteristics of the rules and regulations concerning water allocation in a few selected systems, and the manner in which they actually operate. In the process we hope to highlight the deficiencies in the law and its implementation.

STATE IRRIGATION ACTS

In India the right to harness and distribute surface water vests entirely with the state. No private individual or agency can impound or divert water from streams and rivers without prior

*The author is extremely greatful to Prof. A.Vaidyanathan for his very useful comments and suggestions on the earlier version of this paper.

permission from the government. The Constitution and several enactments define the general legal framework for this purpose. Though irrigation is a state subject under the Constitution, clearance by the central government is essential for undertaking projects on inter-state rivers. The central government is also empowered to intervene to settle inter-state water disputes through mediation or adjudication. Under the River Boards Act, it can also set up institutions for the integrated development of water resources in inter-state river basins. States have their own Irrigation Acts, mostly based on legislation enacted during the period of British rule and amended subsequently at various times. These acts are not all state specific; groups of states belonging to different regions have chosen to follow antecedent legislation pertaining to their region.

Thus, the northern region (comprising the states of Uttar Pradesh, Punjab, Haryana, Delhi, and Jammu and Kashmir) has adopted the Northern India Canal and Drainage Act, 1873 as amended by each state from time to time. The Bombay Irrigation Act, 1879 has been adopted by Gujarat, Maharashtra, and Madhya Pradesh in the western region. The Bengal Irrigation Act, 1876 regulates irrigation practices in the eastern region. The system of irrigation management in the southern states largely follows the pattern envisaged in the Mysore Irrigation Act, 1950. These laws enunciate the general principles guiding the performance of various functions related to irrigation such as land acquisition, water supply, water rates, and dispute settlement.

In all the four regions, the government has the power to acquire land for irrigation works and the landowners the right to receive compensation from the government. States in the north and the east generally assign the duty to construct and maintain watercourses and field channels to village panchayats. However, in Rajasthan, Jammu & Kashmir, and Orissa, the responsibility lies with the individuals or shareholders. In the western region the task may be assigned to individuals, groups, or village panchayats. In the south the responsibility for the construction and maintenance of watercourses lies with the government and that for field channels with the beneficiaries. The government has the right to acquire village labour in emergencies for irrigation-related works. In the north this responsibility for mobilization is given to the panchayats. Beneficiaries have the right to seek the help of civil courts in matters relating to irrigation disputes.

Irrigation water is supposed to be supplied to individuals or shareholders in the northern and southern regions and to groups or associations in the eastern and western regions. In all states, it is envisaged that the government will distribute water according to contractual arrangements made with the beneficiaries. Maharashtra and Kerala are exceptions in that water distribution is expected to be managed by associations and village panchayats respectively. The *lambardar* or *sarpanch* of the village in the north does the collection of water charges. In Maharashtra and Madhya Pradesh dues may be collected through associations or village panchayats. In other states the revenue department collects irrigation charges.

Water Distribution Rules

The Irrigation Acts are of a very general nature. More detailed formal rules for system operations are usually formulated for each specific project. In some cases, 'Rules of Regulation' have been decided and are issued by the state cabinet of ministers while in others as executive orders issued by the designated officials. In order to illustrate the nature and content of these rules and the different ways in which allocation is managed, we briefly describe the practices in respect of three systems: the Tambraparani system in south India; the Eastern Gandak system in the eastern region; and the Mahi Right Bank Canal in the western region.

The Tambraparani system is one of the 18 major systems of Tamil Nadu, for which Rules of Regulation have been issued: The rules governing the distribution of water to the 11 major channels of the system were first prepared during the nineteenth century and amended periodically. The district collector made the most recent change in 1935 after a major expansion and restructuring of the system. After Independence, the Rules of Regulation were formulated for newly constructed reservoirs at Papanasam, Servalar, and Manimuthar and amended from time to time. Several agreements between the system and industries/municipalities are now part of the formal distribution rules. In 1969 rules regarding the authorization of irrigated rice cultivation in the Advance Kar season were notified by the PWD, and in 1987 the government issued an order permitting special release during April-May to cultivators in the tail reach.

As in most systems, the formal rules cover allocation at major divisions: anicuts, head of main and branch canals, and distributaries. Allocation at the tertiary level is supposed to be determined by the extent of area by season and crop for which different segments are authorized to get irrigation. In the Tambraparani project, irrigation is meant only for rice. Operating rules distinguish between three crop seasons—Kar (June-October), Pishanam (November-February), and Advance Kar (March-June). The entitlements of different channels are determined on the basis of authorized area for use in each season and certain norms regarding irrigation water needed per acre per season. There are a large number of tanks—most of which are several centuries old—and their water entitlements are defined by convention.

In the Tirhut main canal of the Eastern Gandak system operates in a different manner: The likely demand for water in different parts of each sub-division is estimated by the officer-in-charge and passed up through several hierarchical layers to the divisional engineer and then to the chief engineer. This is supposed to be the basis for the release of water in the main canal. Indents are made every two weeks, 10-12 days in advance of the time when water is needed. The command is divided into upper and lower portions. The standing orders of the system require the main canal to be opened according to a prescribed schedule, namely, *kharif* from 20 May to 20 October; *rabi* from 1 December to 10 March; and hot weather from 10 April to 19 May. Headwork canal gates are operated on the basis of the demand placed by the executive engineer of the Tirhut Canal Division.

Water to the lower portion is made available on the request based on the estimated demand with a safety margin of 10 per cent made by the officer responsible for this segment. But he has no control over the actual supply available at the cross regulator from which he has to draw the water. This depends on the decision of the officer-in-charge of the main canal and the appropriations in the upper portion, which lies above the regulator.

The distribution of water in the Mahi Right Bank Canal System in Gujarat is based on what is known as the *shejpali* system. Before each season, water resources are estimated, and the area that can be irrigated for the ensuing season is determined on the basis of the experience of the previous years. Farmers, having been informed of the proposed cropping pattern, make individual applications to receive irrigation water and are given an application number indicating their sanctioned surface irrigated area and cropping pattern. This then entitles them to receive water on request during each of the rotation periods specified by the Irrigation Department. Normally, the main and branch systems run continuously, as do most distributaries, thereby leaving the task of matching supply and demand to the field-level functionaries with the assistance of four to five *chowkidars* (watchmen) who are responsible for passing up requests for water and recording which farmers receive water for which crop during each two-week period. This information is subsequently used for billing purposes.

The seasonal water allocation procedure starts from top to bottom. Based on past records as well as on the collective advice of the Command Area Development (CAD), the Irrigation and the Agriculture Departments, routed through the 'Project Canal Advisory Committee', the CAD Commissioner sets the area target for the three seasons and informs the superintendent engineer of the system. This official estimates the volume of water to be allocated to each season on the basis of certain parameters (ostensibly based on experience) relating to cropping period, area irrigated per cusec supplied at the head regulator, supplies from other sources (in this case, the Kadana reservoir) and 'irrigation efficiency'.

During the *kharif* season the rainfall and cropping pattern are such that not every acre under a crop needs to be irrigated in each watering. Based on their previous experience, the Mahi-Kadana project authorities have developed area coefficients for different crops and the number of waterings. The total water requirement of each rotation of 16 days for different crops is then worked out for the command as a whole. This is then allocated between different canals based on the proportion of Culturable Command Area (CCA) under each canal in that particular division. It is noteworthy that while the computation of total water requirement is based on the projected area and cropping pattern, the actual distribution of that water among the different canals is in proportion to the Command Area under each canal.

Operation of the Rules in Practice

In the operation of these rules, managers have to reckon with, and adjust, allocations in the light of variations in rainfall and water availability in the system between and within seasons, and from year to year. Most systems, therefore, give considerable discretion to their managers to decide allocations and scheduling on the basis of actual rainfall, crop area, and system supply in each season; to adjust daily deliveries to particular channels to match demands/needs of different segments; and to change channel allocations under certain circumstances (for instance, when the crop in a section is in danger of withering). This flexibility does not, however, always work in a manner consistent with the authorized entitlements and their underlying rationale.

There are several reasons for this. In the first place, the assumptions and projections regarding the likely quantum of water available for distribution, effective command area, crop patterns, and irrigation efficiency may not be—and typically are not—realized. They are in any case variable. Second, in the absence of reliable information on rainfall, area sown to different crops, and their condition in different segments of the command, adjustments are based on the officials' judgement rather than on reasoned analysis. Persistent discrepancies between planned and actual values of various aspects ought to induce a systematic review of the allocation rules to bring them in line with the ground reality. But this seldom happens. When decisions are left to judgement without adequate facts, arbitrariness is unavoidable and it also becomes difficult to differentiate between genuine errors of judgement and decisions biased in favour of one or the other segment. Third, there is invariably a hiatus between the prescribed localizations and crop patterns (on which allocations are supposed to be based) and the patterns adopted by the irrigators.

In this situation farmers have a strong inducement to evade or subvert the allocation rules and to pressure or persuade the managers and/or local functionaries to effectively change the rules in their favour. That all these elements are present in the Indian context can be seen from the experience of several projects.

System Deficiencies

Thus, in the case of the Tirhut canal, though water is abundant at the head works during the *kharif* season, tail-end farmers, nevertheless, suffer due to the defects in the water distribution network. The carrying capacity of the upper portion of the main canal is inadequate to carry simultaneously the indents of the upper reach and the lower reach. The result is that the upper reach receives most of the canal water during the day and the lower reach during the night. Farmers not being used to night irrigation, most of the water either goes waste or floods the fields.

During the *kharif* season of 1993, allocation was well under the amount requested. In the early part of the season (June), demand for irrigation being small, there was no short supply. In the rest of the season users upstream got the water when they wanted it while downstream users got it when they did not want it. Thus, in August when demands of flowering rice are the highest, the Muzaffarpur division got less than 60 per cent of the requirement, while in September when irrigation demand drops sharply as the rice crop ripens, the lower reach got over 150 per cent of the requirement!

The problem is aggravated because the lower portion was planned to irrigate a cultivable command area of 154,000 ha. But distribution networks have been provided only for 40,412 ha. Because ongoing construction was stopped in 1985, canals and outlets have not been constructed in many places. Where there are no distributaries or minors, farmers cut the canal bunds to obtain water, especially during *kharif*, and this weakens the canal bunds.

Also, canal alignment in many places was altered under public pressure towards natural depressions (*chaurs*) and this has resulted in heavy fillings. Soil in these areas, being mostly sandy loam, gets saturated when there is water in the canal. Consequently, it becomes unstable, and bunds are liable to be breached if not managed properly. To overcome this problem it has been proposed that canals should be lined but this has not been done for want of funds. Floods from breaches occasionally damage crops because neither the Water Resources Department nor the farmers maintain their canals in good repair. Fear of breaches causes the system managers to limit the flow in many canals, including the Tirhut main canal.

Another problem is that along certain reaches, earth was removed from the canal bed to form bunds. It was presumed that during the initial stages of canal run these depressions would be filled with silt. Since the Tirhut main canal has not been operated at full design discharge, cross regulators had to be operated in the head reach and only silt-free water has been carried out in the lower reaches. This has left the excavated pockets unfilled.

In the Mahi-Kadana project, water deliveries are continuous but are controlled to match the anticipated demand during each season.

This requires continuous adjustment at various cross regulators and at head regulators of canals served directly by the main system. At secondary levels and below, the *chowkidar* (watchman) is supposed to inform the *karkoon* about how many farmers are demanding water during the next one or two distributions who then matches the demand with the available supply.

This matching, generally, is on an ad-hoc basis, particularly when demand is high in the *kharif* season. Furthermore, unauthorized irrigation, currently estimated at 60 per cent of the total irrigated area, makes it impossible to assess the actual irrigation demand. Nor can the officials ensure that deliveries are really at or close to the designed level of one cusec.

The sandy loam soils in the upper portion of the system are susceptible to erosion, so that many secondary canals are much wider than designed. It is also hard to maintain adequate head in the upstream of the off-take. Many outlet gates are broken or missing and this further undermines the capacity of the *karkoons* to effectively match demand and supply. Under these conditions, the rules and the reality are far apart. Thus, the defects of the original design, unanticipated problems, lack of coordination in completing different components, wide divergence between designed and actual crop patterns, the spread of conjunctive use (or some combination of these) may often make it impossible to strictly observe the original rules. Unpredictable supply and unequitable distribution are the inevitable result.

Defective Rules

In some cases, there are also inconsistencies in the operational rules themselves. In the case of Tambraparani, for instance, the rules require that flows in the main system be adjusted daily to meet demands from the channels. However, the rules for operating the channels rigidly prescribe the delivery schedules and the levels to be maintained. Rigid schedules with varying flows mean that those lucky enough to get their turns when the flow is high get extra water while those whose turns come when the flow is low get less. This inconsistency forces the *laskars* responsible for the operation of the channels to choose between ignoring the channel-operating rules or framing their demands with a view to facilitating the channel-operating rules. In the latter case, their ability to implement the rigid channel-operating rules depends critically upon the ability of the main system manager to deliver water according to demand.

There are also inconsistencies in the operational rules for different reservoirs in the system. Thus, the rules for the Papanasam reservoir require that discharge from the reservoir for use by the Papanasam powerhouse be 'generally' limited to 1,400 cusec even though this is far below maximum irrigation demand. The rules also place some water in the reservoir under the control of the PWD executive engineer to meet irrigation demands above the 1,400 cusec level but operation is by the Tamil Nadu Electricity Board (TNEB). If the TNEB insists on interpreting the term 'generally' so as to restrict the releases to 1,400 cusec, the ability of the PWD executive engineer to respond to the daily demands of the *laskars* is severely limited.

The Rules of Regulation for the third reservoir (Manimuthar) require releases to be so regulated as to assure sufficient flow in the Tambraparani river. The rules also set an absolute annual limit of 51 million cubic meters of water that may be taken from the Manimuthar Reservoir for the Tambraparani system. If flows in the Tambraparani river are low, it is not possible to satisfy both of these rules.

Adjustment of Rules

Apart from inconsistencies and gaps in operational rules, managers also have to cope with highly variable supply and demand. Formal rules often do not provide clear guidelines regarding scheduling at the tertiary level or the principles of allocation when the water available falls short of demand in a particular season or during a part of it. The problems are compounded by the divergence of actual crop patterns from those assumed in the formal rules.

In the Mahi-Kadana project, droughts, and inadequate water supply during the mid-1980s led the then superintending engineer to introduce a rotation system in 1985-6 and this became the normal operational plan from 1986-7. From then on till 1992-3 detailed plans for rotational irrigation for both *rabi* and hot weather seasons were prepared and incorporated as part of the operational schedule.

The proposed rotation was 10 days on and 5 days off during

both seasons. The command area was divided into two zones, each comprising the main canal and a few branch canals, with supplies being rotated on a '10 days on and 5 days off' basis. This is reported to have worked well, resulting in smoother operation, greater reliability, and more equitable distribution.

The basis of rotation was changed in 1992-3 from the rotation of a group of canals at a time to the rotation of the entire system based on 7 days off and 7 days on. Rotation of the entire system meant a minimum gap of 15 days between two irrigations, which farmers found inconvenient. The change to 10 days on and 10 days off, implemented in 1994-5, made matters worse, leading to the return to the 10 days on and 5 days off rotation.

While in this case the changes did respond to the farmers' reactions, it is doubtful whether they were based on a systematic and objective analysis of the data and whether they were taken after discussion of allocations with the affected *ayacutdars*. By not specifying the criteria and the process by which rational changes in rules should be made, they leave far too much to the discretion and judgement of the officials, with scope for much arbitrary decisions and abuse of power on their part.

There are, of course, instances where major changes in the configuration of a system (following, for example, the construction of new reservoirs and the extension of commands) have triggered modification of formal rules after due deliberation. Such changes were made in the Tambraparani system during the 1930s; when a major expansion of storage was completed, the system's distribution network was rationalized and extended. In some cases, persistent droughts and chronic water shortages have induced managers to change the rules. But in point of fact these rules have been honoured much more in the breach than in the observance.

The rules say that during the Kar season, the PWD executive engineer makes an initial allocation before the season with the possibility of increasing it later. Adding to the Kar season allocation is common; PWD records show that it was done eight times over the 14-year period from 1979 to 1992. Over this period, the average initially authorized area was 19,974 ha. but the average final authorized area was 25,121 ha. Government records indicate that the areas authorized are very close to the areas harvested.

During the Advance Kar season, according to the rules, the PWD executive engineer is to allocate water for rice to specific areas in the tail reach of the system in a defined priority order. PWD records show that the authorized area for the Advance Kar has varied greatly. From 1977 to 1989, the Advance Kar authorization varied from 0 to 18,660 ha. The average was 5,167 ha. However, the actual area cultivated and irrigated was generally much greater than that which had been authorized. Thus, during 1994-5, a year with above normal water availability, Advance Kar rice was authorized for 4,500 ha. the maximum area in the South Main Channel permitted by the rules. But official figures indicate that rice was planted in 5,166 ha. of the South Main Channel along with over 2,800 ha. of bananas. Other sources suggest that rice was planted in much less than this area.

The total area officially reported under Advance Kar rice in the last four channels entitled to Advance Kar water was 11,127 ha. less than the 13,617 ha. officially authorized. On the other hand, adding the banana areas under the last four channels to the reported Advance Kar rice area gives a figure of 19,777 ha., well above the authorized area. It is common knowledge that the tail reaches irrigate a much larger area than was officially authorized. *Assuming that the latter figures are correct, a great deal more area was irrigated during Advance Kar than was authorized.*

This is done partly by authorizing a larger area for rice in the Advance Kar season than is likely to be cultivated and partly by manipulating irrigation schedules by tail-end irrigators to get additional water to irrigate unauthorized annual crops like banana and betel. These crops are all concentrated in the tail of the system and give rise to a large demand for year-round water supply. Except during Pishanam, when the north-east monsoon brings heavy rains and the system has ample supplies, water supply is inadequate relative to demand. This scarcity is acute from April to June when rainfall is scanty. Tail-reach farmers who grow Advance Kar paddy and soya also have bananas and betel and are most affected by this scarcity. They use a variety of strategies to overcome the shortage.

Groundwater is an obvious possibility. But actual use is limited because few farmers have access to good quality groundwater. The more common practice is to pump canal flows meant for other uses such as industry or from the special releases discussed below. Such pumping is against the rules, and can be done only with the connivance of the Irrigation Department functionaries. In this context, it is significant that in order to meet its obligation to supply 40 cusec daily to the Tuticorin industrial park, the PWD releases everyday 150-60 cusec in the North Main Channel head. Most of this huge difference is accountable *not* by real transmission losses but by the (illegal) pumping of water from the North Main Channel by some 300 farmers.

Command area farmers in the tail reach have also been able to exert effective pressure for special releases of water. Getting Advance Kar allocations is a major way of getting water for bananas in the tail of the system. To influence seasonal allocations, command area farmers in the tail reach send delegations to persuade government officials to authorize Advance Kar issues, particularly for the second and third priority lands. The water so obtained, although ostensibly released for rice, is then used for betel and banana crops. Betel growers also request special releases for their crops under a 1987 government order; farmers under the Avudayar Tank request special releases for the Thiruchandur Temple; others request special releases for 'domestic purposes', i.e. drinking water. In all cases, the special release is used for irrigation, particularly for betel and bananas.

When special releases are required, the command area farmers under each tank meet to mobilize resources needed to secure a favourable decision from the bureaucracy. Rates are determined by the intensity of the problem and the likely expenditure. The usual rate is Re. 1 per banana plant or Rs. 150 per acre of betel or banana. These funds, collected from irrigators, are used to send their leaders to Tirunelveli and Chennai to persuade legislators and senior bureaucrats to grant special releases and to some extent for gifts to these persons. For example, in the South Main Channel in 1993 the command area farmers under 6 tanks on the third priority land sent a person related to a Member of Parliament to visit officials at the district and state levels. Through his connections, he convinced the State PWD Minister to grant Advance Kar authorization for these tanks.

Tail enders spend a considerable amount of money to get Advance Kar allocations or special releases to get water to their tanks during the shortage period. According to an estimate from a leading farmer, who is also a senior government official, resources collected each year by one WUA for this purpose range from Rs. 65,000 to Rs. 135,000. Of this, it is estimated, the WUA leaders share 20 per cent, 65 per cent is distributed to officials at different levels, and the remaining 15 per cent is spent on travel and other expenses. Local water users associations generally provide water obtained through such interventions only to farmers who contribute to the fund.

Occasionally the tail enders also threaten to conduct, or actually carry out, public demonstrations to influence government officials. A popular means is to block highways, particularly when an important official is scheduled to pass. Blocking a minister's car in 1992 resulted in a special release demanded by the farmers.

The success of the tail-reach farmers in getting water has adversely affected the farmers in the head and middle segments. Excess allocations for the Advance Kar limit water availability during the Kar season and 'special releases' deliver more water to the tail than is allocated. By getting Advance Kar allocation, the tail-reach farmers have been growing crops on their field by the time the farmers in the head reaches begin Kar sowing in June. If there is a water shortage in June or July, when the two seasons overlap, the tail areas successfully argue for, and obtain, priority for water on the plea that their standing crops are at risk. These strategies have cut into supplies available to the head and middle segments to levels well below their entitlements under the rules.

This led to the formation, in 1987, of the 'three channel association', representing the farmers of the first three channels of the Tambraparani system. The association was formed with the specific purpose of protecting the allocation priority of the head area channels. The association brought suit against the district collector, alleging that the district collector and the PWD had issued more water to the tail areas for Advance Kar than is allowed by the rules. They claimed that this practice resulted in the delay of issues for Kar cultivation. A 1992 court judgement dismissed the farmers' allegations but conceded their key claims by ordering that water must be reserved for Kar water releases as specified in the 1969 rules and that Kar releases for the first three channels must begin on 1 June.

DEFICIENCIES IN THE LEGAL FRAMEWORK

Many of the difficulties in formulating and implementing allocation rules, which meet the criteria of optimal use of water and equitable distribution of benefits, reflect the inherently complex nature of the problem. The difficulties are compounded by lack of reliable information on actual water deliveries, areas and crops for which they are used by different segments and classes of farmers, and their ultimate impact on the incomes of the command area farmers. There is hardly any attempt at collecting such information, not to speak of analysing it as the basis for a rational and consistent restructuring of rules and procedures in the light of the changing ground reality. Even if this were done, lacunae in the legal framework defining the basis for and content of 'property rights' in water, the nature of the relations between the government, system managers, and users, and the criteria and procedures for the settlement of disputes are difficult aspects of the problem that must be reconciled in order to reach a satisfactory solution in this area.

DISPUTE RESOLUTION

Many state irrigation acts have few or no provisions to settle disputes between the system management and irrigators and among irrigators. Those that do have provisions generally refer the disputes to an irrigation agency officer. If the irrigation agency officer fails to settle the dispute, the irrigators can go to a civil court.

All states have court systems as general conflict resolution mechanisms. However, in the case of irrigation disputes, it is essential that the dispute be settled quickly so those crops are not lost. State court systems cannot provide this kind of quick resolution. Under the Northern India Canal and Drainage Act, 1873, currently used as the basic irrigation law in Haryana and Punjab, it is expressly forbidden to take irrigation disputes to the courts simply to ensure a rapid resolution (Aryan 1992).

Persistent conflicts and disputes do get to the court but litigation is both time consuming and expensive. The alternative recognized in most states is to submit irrigation disputes to the officials of the state irrigation agency. To allow for the resolution of complaints against the irrigation agency, the laws also allow for review of a decision by some other bureaucrat, in some cases a higher level irrigation official, in others a district collector or other civil administrator.

A key weakness in the present dispute-handling mechanisms is that the legally preferred ones include only bureaucrats, mostly those with the state irrigation agencies (Aryan 1992). This has unfortunate consequences:

- Bureaucrats generally see disputes from their own point of view which tends to be dominated by a concern for efficient water use or equity of water distribution to the exclusion of concerns with customary rights. This can make their decisions seem unfair and arbitrary to farmers, even when decisions are made as fairly as possible.
- Given the widespread feeling that government bureaucrats can be influenced by bribes and political power, there is real concern that it is not possible to get fair decisions.
- Finally, in disputes between WUAs and the irrigation agency, it is expecting too much to ask bureaucrats to find fault against themselves.

No state has provided the means of dispute resolution other than reference to bureaucrats and courts, such as referring disputes to local bodies like panchayats, except in cases of locally managed irrigation systems. The existing laws show that the lawmakers had too much faith in the bureaucrats and too little in local institutions. If rapid and fair resolution of disputes is needed, it may be necessary to explore other means of dispute resolution such as special water courts or refer disputes to particular local bodies that have user representation and which are acceptable to complainants.

IRRIGATOR'S WATER RIGHTS

A major legal impediment to the resolution of disputes is that water rights of the users are not clearly defined. Most states in India have enacted laws giving the state government the right to allocate surface water to uses and users as it judges best adapted to the public interest. For surface water, then, the legal situation is that the state can decide what the water rights of an irrigator are. Every state attaches surface irrigation water rights to specified pieces of land, although the rights specified are often quite different. But these can be unilaterally changed by the state. Nor is the relationship between the user and the government based on an explicit and well-defined contract. In any case, the courts do not seem to view this as a contractual relationship. At best, they recognize rights of customary or prior usage; but even this has been greatly weakened over time.

Present laws give the government almost complete power to

decide who should have access to surface water for irrigation. Although a few court cases have attempted to limit this power (Singh, 1991), the government's right to regulate irrigation is paramount and sovereign in character; however, it should not be exercised arbitrarily.

First, the government's right should be subject to the irrigator getting the same quantity which he or she is accustomed to and which must be treated as part of a contractual relationship. Second, any changes made should be based on clearly enunciated and publicly accepted principles, such as equity of distribution, or priority for drinking water. Security of water rights is essential to give confidence to irrigators so that they will have trust in investing in irrigated agriculture. Also, when changes in water rights occur, they should be seen as principled and as following well-established rules rather than as arbitrary and subject to private or personal influence.

No such laws exist for groundwater usage. Implicitly, then, groundwater is legally controlled by the holder of the land overlying the groundwater. Since water rights are tied to land, water rights can be transferred from one individual to another only through the transfer of land. This is an unsatisfactory situation since landholders have an unrestricted right to pump and no one has the assurance that groundwater will be available in the future. Given the rapidly falling water tables in some areas and the documented inequities in the availability of groundwater, there is a need in the public interest to provide some legal limits to the power of landholders to exploit groundwater (Devi, 1991; Moench, 1996).

The arrangement of tradable water rights to individual users or association of users, which has been tried in the US and in some other countries, is increasingly advocated as another solution (Bruns and Ruth, 2000). This is expected to reduce the burden on the judicial or administrative settlement of disputes by allowing parties to explore mutually profitable arrangements and at the same time ensuring that water is used effectively. Tradeable water rights are rights to use water which can be transferred, entirely or in part, separately from the transfer of land. While tradable water rights themselves should be permanent or very long-term, to ensure their security, the transfer of water rights need not be permanent; water rights can be leased for a season, for a year, or for many years.

ACCOUNTABILITY OF IRRIGATION AGENCIES

The state Irrigation Acts contain:

(a) General provisions for the supply of water,
(b) Procedures for water supply and conditions under which they may stop the water supply, and
(c) The rights of the government.

There are no provisions to deal with situations in which irrigation agency officers specifically fail to ensure the proper delivery of water.

- There are generally no rules requiring the government to inform users about the availability of water and its supply dates in different parts of the command area.
- There is no requirement that the government abide by its scheduled dates of supply and quantity.
- There are no provisions to compensate farmers for the failure of the irrigation agency to live up to its commitments to deliver water on schedule and in adequate quantity.

Any such legal provisions or rules must, of course, allow for the failure to deliver water for reasons beyond the control of the irrigation agency such as lack of water or failure of irrigation infrastructure.

One key point is that the existing irrigation laws contain provisions to punish farmers if they waste water or fail to maintain field channels, but there are no provisions to punish irrigation agency personnel. The laws are very one-sided (Devi, 1992).

In addition, in most government-managed irrigation systems, farmers have no role in deciding the water delivery schedules. There are no consultation processes with farmers. In those states that have implemented canal advisory committees (e.g. Kerala, Karnataka), the advisory committees have generally failed to incorporate true farmer representatives (cf. Bhaskaran, 1980; Sakthivadivel et al., 1995). The problem is aggravated by mismatch between the crops grown by farmers and the timeliness and quantity of the water supplied; agricultural conditions have changed over a period of time, but the system operation rules have remained unchanged (Brewer et al., 1997).

Farmers have a reasonable concern regarding water delivery to their WUAs or to the outlets. However, neither the National Water Policy (MOWR, 1987) nor the recently issued state water policies (e.g. Kerala Water Policy, 1992; Water Policy of Tamil Nadu, 1994) nor most of the other policy documents recognize this legitimate concern. The guidelines issued by Maharashtra for the promotion of water users' associations are an exception. These guidelines say: 'There will be a guarantee to get the sanctioned quota of water at predetermined intervals as per the legal agreement entered with the government' (Government of Maharashtra, 1994: 3). However, the guidelines do not say what recourse the farmers have if the government agency fails to deliver water on schedule. The Gujarat Irrigation and Drainage Bill, 1996 also subscribes to this view (Section 46.3), and includes provisions for compensating farmers whose crops are damaged by failure to deliver water properly.

Accountability of the irrigation agencies towards the farmers and more balanced rights and powers with regard to delivery of water to WUAs in government-managed irrigation systems are essential for successful irrigation management transfer. These features should be reflected in law.

REFERENCES

Aryan, Rita, 'Dispute-Handling Process in Water Laws', in C. Singh (ed.), *Water Law in India*, New Delhi: Indian Law Institute, 1992, pp. 209-80.

Bhaskaran, P., 'Emergence of Farmer Organisations: A Survey', in A. Sundar and P.S. Rao (eds.), *Farmers' Organisations for Efficient Water Use in Irrigated Agriculture*, Bangalore: Indian Institute of Management, 1980.

Brewer, J.D., R. Sakthivadivel and K.V. Raju, *Water Distribution Rules and Water Distribution Performance: A Case Study of the Tambraparani Irrigation System*, Research Report No.12, Colombo: International Irrigation Management Institute, 1997.

Bruns, Bryan Randolph and Ruth S. Meinzen-Dick (eds.), *Negotiating Water Rights*, London: Intermediate Technology Press, 2000.

Bottrall, Anthony F., *Comparative Study of the Management and Organization of Irrigation Projects*, World Bank Staff Working Paper No. 458, Washington, D.C.: World Bank, 1981.

Coward, E. Walter, Jr., 'Irrigation Development: Institutional and Organization Issues', in E. Walter Coward, Jr. (ed.), *Irrigation and Agricultural Development in Asia: Perspectives from the Social Sciences*, Ithaca, NY: Cornell University Press, 1980, pp. 15-27.

Devi, P. Rema, 'Groundwater Development and Legal Regulation', *Journal of the Indian Law Institute*, vol. 33, no. 4, 1991, pp. 614-34.

______, 'Accountability in Water Resources Management', in C. Singh, (ed.), *Water Law in India*, New Delhi: Indian Law Institute, 1992, pp. 90-113.

Government of India, *Constitutional Provisions Relating to Village Panchayats and Municipality in India*, Lucknow: Eastern Book Company, 1993.

Government of Maharashtra, *Co-operative Water User's Association Guidelines*, Pune: Irrigation Department, 1992.

Jacob, Alice, 'Farmer's Participation in Irrigation Management and the Law', in *Proceedings of the National Conference on Participatory Irrigation Management*, Vigyan Bhawan, New Delhi: Ministry of Water Resources, 19-23 June 1995, pp. 81-3.

______, 'New Demand Policy', in K.K. Singh (ed.), *Farmers in the Management of Irrigation Systems*, New Delhi: Sterling Publishers, pp. 217-34.

Jacob, Alice and S.N. Singh, *Law Relating to Irrigation*, New Delhi: Indian Law Institute, 1972.

Maloney, Clarence, and K.V. Raju, *Managing Irrigation Together: Practice and Policy in India,* New Delhi: Sage Publications, 1994.

Moench, Marcus, *Groundwater Policy: Issues and Alternatives in India*, Colombo: International Irrigation Management Institute, 1996.

MOWR, *National Water Policy*, New Delhi: Ministry of Water Resources, 1987.

Pathak, R.S., 'Participatory Irrigation Management and Legal Support', in L.K. Joshi (ed.), *Management of Irrigation: A New Paradigm Participatory Irrigation Management*, New Delhi: Ministry of Water Resources, 1997, pp. 231-7.

Patil, R.K., 'Participatory Irrigation Management: Legal Dimensions', in L.K. Joshi (ed.), *Management of Irrigation: A New Paradigm Participatory Irrigation Management,* New Delhi: Ministry of Water Resources, 1997, pp. 239-42.

Pundarikanthan, N.V., R. Mohandoss, L. Jayasekar, P. Gomathinayagam, K. Jayachandran, R. Kulandaivelu and S.R. Sundarajan, 'Main System Performance and Management Control Processes: Tamiravaruni System', in *Proceedings of the Workshop on IIMI-India Collaborative Research in Irrigation Management*, Colombo: International Irrigation Management Institute, 1992, pp. 9-35.

Rosegrant, Mark and R.G. Schleyer, *Tradable Water Rights: Experience in Reforming Water Allocation Policy, Irrigation Support Project for Asia and the Near East*, Washington, DC.: United States Agency for International Development, 1994.

Repetto, Robert, *Skimming the Water: Rent-Seeking and the Performance of Public Irrigation Systems*, Research Report No. 44, Washington, D.C.: World Resources Institute, 1986.

Sakthivadivel, R., J.D. Brewer, R. Bhatia, S. Narayanmurthy, and P.S. Rao, *Evaluation of Schemes under NWMP-I: Bhadra Scheme in Karnataka and Sathanur Scheme in Tamil Nadu*, 2 vols., Colombo: International Irrigation Management Institute, 1995.

Singh, Chhatrapati, *Water Rights and Principles of Water Resources Management*, New Delhi: Indian Law Institute, 1991.

Sampath, R.K. and Robert A. Young (eds.), *Social, Economic and Institutional Issues in Third Irrigation Management, Studies in Water Policy and Management,* No. 15, Boulder, CO: Westview Press, 1990.

CHAPTER 8

The Krishna Valley Development: An Approach to Equitable and Efficient Use of Water

K.R. DATYE, SUHAS PARANJAPE,
SEEMA KULKARNI AND K.J. JOY

In many parts of India the development of water resources has not kept pace with the increasing needs of a rapidly growing population. Even where unutilized supplies are available, the inadequacy of investment (especially public spending) constrains the expansion of supplies. In some areas, the scope for harnessing more water is limited. Everywhere harnessed water is not utilized to the best advantage.

The combination of these circumstances has created numerous problems, including acute and growing water scarcity in some areas even for basic needs. Access to water has become a sensitive issue between communities, regions, states, and even countries. The Krishna valley in Maharashtra is one among such regions. This chapter addresses issues related to creating wider and more dispersed access to water for the rural poor and their livelihood, with an emphasis on the rapid enhancement of the productivity of degraded lands and an entitlement for the poor to the produce of common and private lands developed at public cost.

EMERGENCE OF WATER SCARCITY IN THE KRISHNA VALLEY

The Krishna valley, especially those parts (the districts of Pune, Satara, Sangli, Kolhapur, and Solapur), which were under British rule, attracted a large share of investment for famine relief after

First World War. A large number of tanks deriving water from low-rainfall catchments along with major irrigation projects with catchments lying in heavy rainfall areas on the eastern slopes of the Western Ghats were also built. However, the expected positive impact of these irrigation works on food production and famine relief was not realized.

Farmers responded indifferently to the availability of irrigation facilities for two main reasons. First, highly water-retentive black cotton soils were prevalent in the command areas and the main crops were less water-demanding ones like coarse cereals, pulses, and oilseeds. In good years, therefore, the irrigation requirement was very small. Second, farmers had to apply and pay for water in advance in cash. Due to the low price of coarse cereals very few could afford to pay in advance and most farmers preferred to wait until an actual failure of the rains before applying for water. When that happened, however, the system was incapable of delivering water on demand and that too at an affordable rate. For these reasons, the system suffered from gross under-utilization of the irrigation facilities; but there was no conflict amongst farmers with regard to water.

The post-Independence period witnessed the rapid spread of the cultivation of sugar cane—a water-intensive cash crop—on irrigated lands. This was made possible by massive investments in large irrigation projects and the availability of low-cost hydropower from the Koyna hydropower project for lift irrigation. Both contributed to expanding the irrigated area. Nonetheless, until about 1970, sugar cane cultivation was confined to the narrow strips supplied by gravity-irrigation canals and to low head lifts from the Krishna river and its tributaries. Therefore, in this period, despite the expansion of irrigation demand, the expansion of supply matched the growth in demand and there was essentially no severe conflict over water.

After the early 1970s, aided by the rapid proliferation of cooperative sugar factories, sugar cane cultivation spread even to the areas served by smaller streams and minor irrigation projects. The consequences became apparent in the aftermath of the severe drought year of 1972 and the succession of below-normal years which followed. It began to be recognized that the sugar cane farmers were appropriating a major portion of the groundwater as well as the surface water flows. The production cycle of sugar cane cultivation—which starts in July with a 12-16 month first crop followed by one or two rattoon crops—results in the commitment of water for anything from 24 to 40 months. Once a planting decision is made, the available water supply is pre-empted for up to two or two and a half years.

Sugar cooperatives—which have become such a 'successful' and powerful force in Maharashtra—were unconcerned about issues of equitable access, in spite of the fact that the initiative for forming the first cooperatives was rooted in a vision that viewed them as part of the social movement for the emancipation of the relatively poorer segments of society from the domination of moneylenders, middlemen, and Brahmin landlordism. The sugar cooperatives of Satara, Sangli, and Pune districts, which have been a source of much of Maharashtra's political leadership, have hardly ever placed equity and sustainability issues at the heart of their water policy. They were insensitive to the pressures for equitable sharing of water and have, in fact, aggressively opposed any attempt at reducing their members' access to water.

The state government policies further sharpened the conflict by pushing massive irrigation investments but encouraging high-water-demanding systems such as sugar cane cultivation initially, market-oriented horticulture subsequently and floriculture of late. These systems have proved manifestly unviable because of poor cost recovery and have already created an unbearable burden on state finances. More importantly, by continuing the process of concentration of benefits at greater and greater cost, this strategy has accentuated the conflicts. The government has failed to address the major issues of cost recovery and resource mobilization and also failed to prevent wasteful use by those with privileged access to the scarce water resources. Nor has it been concerned to ensure productive use of 'doles' to create assets with entitlements for the poor.

Not only is state policy blind to the dimensions of equity and sustainability, but its policies have also become the active instruments of private accumulation at public cost. Much of the agrarian surplus, especially that appropriated by the politically influential segment in the rural areas, has been siphoned out from the rural areas and invested in urban areas in real estate and, to some extent, in industry. Government assistance for drip irrigation and processing of high-value crops and for cold storage and infrastructure development was largely cornered by grape growers and horticulturists from non-farming groups.

As a result, the poor availability of water for subsistence crops and even for domestic use has been further adversely affected not only during drought but even in normal years. This has sharpened the conflict over water between cane growers and others, as well as between larger commercially-oriented, farmers and the small/ regional cultivators.

Meanwhile, the pressure of non-sugar cane farmers to restrict water allocation to sugar cane was building up, and an internal conflict developed within the ruling party between the so-called sugar lobby and the farmers outside the command of the water irrigation systems. In response, the state government formulated a policy of limiting canal irrigation to 8 months and curtailing water delivery during the summer months. The sugar lobby responded with threats of bloodshed if water allocation to the sugar cane growers was reduced. The government capitulated in the face of this opposition and the policy was never implemented.

The appropriation of water for sugar cane continued unabated. Indeed, the trend was aggravated by the active support of the state for investments in wells and a policy of flat annual electricity rates based on crop and the installed horsepower of pump sets. Sugar cane cultivation began to cut into groundwater resources and post-monsoon surface flows in a big way. Water shortage in years of drought was aggravated by the depletion of the groundwater reservoir, and acute scarcity was experienced even of drinking water and fodder. Small and marginal farmers on the verge of destitution were haunted by the spectre of famine, a total loss of crop production, and the destruction of cattle due to shortage of fodder. The resource-poor farmers started demanding an equitable share of water. There were even stray incidents of threats to the lives of sugar cane growers who were monopolizing the water and depriving the others of water for their domestic and cattle needs.

RESPONSES

Maharashtra has a long tradition of community management of irrigation systems. The ancient *phad* system, for example, effectively separated the right to summer irrigation from landownership, encouraged grouping of farmers according to different farming systems depending on irrigation, and helped equitable distribution by enabling water shares of a farmer to be used for raising crops on the fields of other share owners by mutual consent. More recently, the growing consciousness of water scarcity has stimulated several significant local initiatives to achieve more efficient and equitable use of water.

There is thus considerable experience accumulated from a number of successful community efforts to increase groundwater recharge, persuading beneficiary farmers to share the costs, and inducing them to shift to water-saving techniques of irrigation (especially drip irrigation) and cropping. One of the earliest and best-known examples is the Pani Panchayat (PP) movement. Ralegan Siddhi, Ankal Kop, Ozar, Takari, and Khudawadi village are cases in point.

The Pani Panchayat Movement

The PP movement was aimed at creating solidarity among the resource-poor by restructuring water rights based on a recognition of the right of the landless to water. It allocated water on a per capita basis with minimum water assurance for half an acre per capita for each family, to be utilized to stabilize production of two crops of low-water-using varieties (mainly subsistence crops like cereals in the *kharif* season followed by pulses and oilseeds in the *rabi* season).

The PP movement was essentially confined to lift-irrigation schemes in upland areas on existing reservoirs. It promoted a consensus on equitable water allocation as described above and an obligation on the part of the users to fully meet the operation and maintenance (O&M) costs of the lift scheme (including electricity charges) and a part of the capital cost. Initially, it evoked a favourable response from the farmers, and the movement spread rapidly to more areas. However, the pace of expansion soon stopped because the State Electricity Board was indifferent to the needs of the PP groups and the increasing shortage of electricity supply, on which the groups crucially depended, restricted their spread. Nonetheless, 55 groups are still functioning in the Pune district.

The Khudawadi Experience

The Khudawadi minor, which is at the tail-end of the Bori tank, has received very little water over the past 20 years. A women's group in this village took the initiative in working out a scheme for using limited quantities of variable water to re-establish vegetation on privately-owned wastelands. The group has negotiated a long-term produce-sharing agreement with the owners. The water users' society in the village has also agreed to allocate 15 per cent of the water they receive for use by women's groups from landless families. These lands are being developed by channelling funds under the employment assistance programme through the women's groups. Tenancy issues have been resolved by locally negotiated, informal agreements between the women's groups and the landowners. The women have also been able to negotiate produce-sharing agreements on a 40-40-20 basis between themselves, the landowners, and the society. However, the legal framework for such agreements is not well established. The state government has so far not addressed this issue at all, though they have a formal agreement for leasing government wasteland to corporate enterprises.

For the time being, the main benefit to the women is the availability of land for intensive cultivation, and the landowners are demanding only a small share from high-value crops such as vegetables and fruits. The women's group working on about 16 ha. of wasteland has been able to verify some of the norms for farm forestry and for diversified crop production, which have been described by Lundquist and others cited below. For example, they have shown that an allocation as small as 200 to 250 m^3 of water per family allows the establishment of more than 2,000 trees annually per family.

At a later stage, there are good prospects of fuel self-sufficiency being reached which can become the basis for a more ambitious solar/biofuel-based energy production. This would benefit the landowners and simultaneously create livelihood assurance for the landless artisans. There is a prospect of improving cost recovery by the production of oils or resins on wasteland and value addition by decentralized processing with the use of renewable thermal energy, mostly recovered heat from electricity-generation plants using biofuel.

The Khudawadi society has been able to bring about solidarity among different sections of the village community, including the resource-rich landowners and the resource-poor women's groups. This has enabled the society to get allotment of development funds for building local storage of capacity equivalent to about 10 per cent of the annual water allocation. Another remarkable achievement has been that the society has, after a long wait of 20 years, been able to get its allocation of water by reduction in sugar cane irrigation.

The institutional structure of the water users' society and its MoU with the irrigation department, achieved after considerable effort, has contributed to its success. Another contributory factor perhaps is the weakening of the political dominance of those controlling the sugar cooperative factories in state politics. The legislators from the region, who used to be largely from this class, have lost in recent elections.

The Mukti Sangharsh Response

These experiences acquired added importance and relevance from the early 1980s onwards. The droughts between 1980 and 1984 coincided with the failure of the mill strike in Bombay, an event of some importance for the Krishna valley since this area supplied a substantial chunk of the textile industry's labour force. The failure of the 1982 textile strike effected a drastic reduction in the existing employment in the industry as mill owners took the opportunity to either rationalize the workforce or simply to shut down plants. The prospects for expansion were also dampened. This resulted in a reverse migration from Bombay to the drought-prone areas of the Krishna valley.

The Mukti Sangharsh (MS) movement emerged in the specific circumstances shaped by these developments. The movement was spearheaded by activists and textile workers returning from Bombay who initially came together to organize the drought-affected people in the area in search of relief work and to demand fair wages. Slowly they were drawn into the larger issues of drought eradication and people's mobilization. Some of their earliest demands focused on getting productive rather than unproductive drought relief work, creating fodder banks, and minimum water assurance. As early as 1985, they had carried out an investigation into the problems in the area and had organized two *vigyan jatras* (science fairs) to study the problems in the area in collaboration with the People's Science movement in Maharashtra.

This subsequently spread and become part of a much larger and broader movement towards an exploration into and a search for an alternative development strategy. It has taken over the PP concepts and developed them into a more comprehensive framework. It has also been instrumental in initiating a valley-wide mobilization that demands an equitable distribution of water with a priority right to the drought-prone areas in respect of all new resources in the valley; and restructuring of all existing projects so as to provide minimum water assurance for the drought-prone regions in the valley. The movement is gearing up to organize an extensive mass mobilization with a clearly defined agenda for change in the water policy and development strategy.

WHAT CAN BE LEARNED FROM THESE EXPERIENCES

A comparative study of Ralegan Siddhi and Adgaon, the two most famous and pioneering watershed development projects, brings out clearly the importance of social regulation of water use and the need for sharing the 'new' water that becomes available in the wells as a result of watershed development. In Adgaon, there has been remarkable improvement in productivity of horticulture resulting from improved ground and surface water availability. But benefits have been concentrated within a section of the community. Conflicts were temporarily contained due to initial employment generation in the development programme.

But they are now getting sharpened because there is no specific allocation of water for the resource-poor and they do not have assured access to biomass in adequate quantities for their basic needs or for access to livelihoods in animal husbandry or other non-farm employment. With an initial increase in groundwater availability, a pumping race has developed, leading to the depletion of this source to such a degree that recently Adgaon had to be supplied with drinking water tankers by the state in the summer.

In contrast, special efforts have been made in Ralegan Siddhi to avoid a pumping race as groups of farmers draw their allocations from strategically placed common wells, and a fair degree of equity is ensured in the matter of access to water. Another factor in the better performance in Ralegan has been the use of local labour to rectify previously constructed percolation tanks and other structures, and this labour has had a high marginal productivity. When project implementation is controlled by the bureaucracy, such efforts of productively used labour are not possible; taking up rectification works amounts to an admission of poor supervision of previous work and consequently there is tremendous resistance from the bureaucracy to such work.

Ralegan demonstrates that small improvements in technique through local initiative can make significant contributions to better water use. It also shows how local water resource development helps to build capabilities to obtain water from large systems (Ralegan now gets water from the Kukadi canal) and use it efficiently and productively.

That equity is a prerequisite of participation is also underscored by the Pani Panchayat experience. Users have taken full responsibility for the recovery of operation and maintenance costs, including electricity charges for relatively high lifts, and have also accepted liability for 20 per cent of the capital cost. Although the water allocation norms were based on water requirement to stabilize the two-crop system to meet subsistence needs, once the equity approach was internalized, additional water available for summer crops has been used more widely for water-efficient crops instead of being appropriated by a few. The prospects of getting access to water through lift irrigation and the piped delivery system combined with management and technology support would facilitate the acceptance of equitable sharing principles, as well as the responsibility for sharing a substantial part of the costs. The replicability of the Pani Panchayat on a wider scale has been restricted mainly because of the indifference of the state and the inadequate attention paid to many of the technological and policy issues.

Clearly, new opportunities have emerged as local people and experts have together grappled with the problem of developing techniques and institutions suitable to the area and offering various alternatives. The local groups have created new institutional arrangements of community management to ensure that minimum water needs are met, and informal, but stable, produce-sharing arrangements facilitate access for the land-poor to at least a part of the fruits of irrigation. They show how even macro policy changes and macro choices of technology can be foreshadowed by local initiative. They have simultaneously thrown into sharp relief the dead end reached by state policy even as it continues to subsidize inherently unviable and unsustainable policies with no prospect of a shift towards equitable and sustainable water use and energy self-reliance.

The knowledge and experience gained from these local non-governmental initiatives suggest that a more equitable, efficient, and sustainable use of water resources in the Krishna valley is feasible. The principal elements of such an approach should include first, an integration of large and small sources as well as technological innovations in the use of water to help increase and diversify the biomass production and value addition to this biomass. This is essential to increase the value generated per unit of water utilized, especially for the resource-poor whose combined share of water and land is relatively much smaller. A second element is the acceptance of decentralized community management of water in a manner which combines considerations of equity with those of efficient use. The third is the need for major changes in the institutional framework to facilitate the implementation of the total plan.

Turning Conflicts into a Positive Sum Game

The main basis for the resolution of the conflict lies in turning the situation into a positive sum game. Detailed calculations show that if sugar cane growers and growers of water-intensive crops in command areas are induced and assisted to use water-saving techniques, substantial saving in water is possible. If such assistance is made on condition that the water thus saved from existing sources and the potential additional supplies from new local and external sources will go to hitherto water-starved areas, it would be possible to ensure livelihood for all, including the resource-poor. Water must, however, be used in an efficient and sustainable manner, to create biomass pools for value-adding/income-generating activities. Considerations of viability and affordability make it important to avoid undue emphasis on horticulture. It is, therefore, important to diversify.

A superior alternative would be to give a definite allocation of land and water to farm forestry with limited but assured water for establishment and variable water for yield augmentation. An integrated development of renewable energy and materials technology utilizing the biomass surplus produced from the farm forestry area also opens up avenues for generating sizeable non-farm incomes and stabilizing them. There is an increasingly positive grass-roots response to the biomass strategy, and with the further advancement in renewable materials and energy technology the outline of a viable and affordable option is emerging.

Technology Base

The techno-economic viability of the alternative strategy has been established by studies and also by field tests. For over 15 years, NGOs in active interaction with local communities have been exploring various aspects of the conservation, development, and productive use of water resources. A number of useful contributions have been made by individuals, scientists, and technologists from the People's Science movement of Maharashtra as well as associated organizations, and concerned scientists in institutions such as the Indian Institute of Science, Bangalore and the IITs in Madras and Delhi. Professional technologists and artisans have made it possible to carry out applied research work and to develop prototypes with a very low budget research programme and small private contributions.

It is not possible to narrate the history of this development and furnish details of the outcome. (The interested reader is referred to the short bibliography at the end which lists publications and chapters which contain some of these details.) Briefly stated, technology upgradation is envisaged in integrated watershed development with supplementary water from larger systems to reduce conveyance losses of water by using pipes made of local material as well as increasing application efficiency by adopting modern techniques like drip irrigation; encouragement of farm forestry and growth of high-value oil- and resin-bearing crops; use of biofuels, along with solar energy in combination with external energy (oil, coal, and electricity) to sustain processing of biomass; and use of local biomass-based material for construction of roads, buildings, and other infrastructure.

These technologies are distinguished, first and foremost, by a sizeable saving in energy costs, of the order of 80 per cent in direct and indirect energy consumption. They also provide the basis for local income generation, with relatively small quantities of external and energy-intensive inputs, using local and renewable materials (including biomass with short regeneration periods) and easily learned techniques not far removed from artisanal skills.

Crop diversification and inclusion of low-value, bulk biomass as part of the production system affords the opportunity for extending the scope and coverage of the water service and to make the most productive use of various components of water such as local and exogenous water, variable as well as assured water, surface water as well as groundwater. It also makes possible cost recovery in kind in the form of biomass materials. This concept and that of earmarking a part of the biomass output to constitute a community pool have been found to be *prima facie* acceptable in field-level discussions. This opens up avenues of combining the funds available in the social sector for employment assistance, training, education, and other related programmes and integrating them with loan and capital schemes.

An important immediate concern is to persuade the sugar cane growers to shift either to water-saving techniques of sugar cane cultivation or preferably to more sustainable and diversified cropping patterns. Introduction of soybean or gram inter-culture in the sugar cane field between the rows or pits and macro basin irrigation with improvement of the soil by ploughing in the biomass-enriched compost in the furrows have been field tested. The acceptability of this system is restricted by the fact that presently subsidies are available only for drip-irrigation systems and also by the fact that any improvement in the sugar cane area will call for common group management of groundwater storage, as in the system in Ankalkop.

More and more evidence is now accumulating from studies as well as field experience that there is a good prospect for value addition to low-value bulk biomass, such as small-dimension non-forest timber produce, bamboo, fibres, as well as oils and starchy products, by use of renewable energy and materials technology. It should be noted here that it is possible to match the composition of this biomass to regional characteristics, and agricultural production is not ruled out for well-endowed areas. Formation of regional pools of biomass materials, taking full advantage of low-cost electricity from the grid and combining it with renewable sources for process energy as well as electricity generation, the targeting of low-interest finance for renewable energy generation or process equipment used for energy-efficient and high value-adding processes, together create the basis for a wide range of products and processes which have an inherent capacity to cope with the diversity of local situations.

Energy from locally generated biomass, producing processed materials for road constructions, buildings, and other infrastructure from such bio materials, and production of high-value liquid fuels and chemical intermediates from specialized crops will generate a substantial amount of additional income and employment in the locality in an energy-efficient way. The range of alternatives is extremely wide. The appropriate combinations have to be chosen and fashioned to suit the specific conditions of each region. The technology choice will be finally based on optimization studies with multiple criteria, and our studies indicate that an appropriate choice can be made in each of the chosen options which would not compromise functional adequacy, technical feasibility, economic viability, and adaptability to the local situation.

It is important to note that the suitability and viability of a wide spectrum of technologies have been field tested in the spheres of biomass production with optimized inputs: technology for the water system itself and technologies for rising products of farm forestry for the roads, buildings, and other infrastructure needs. This work has given the basis for defining performance parameters with regard to input-output of labour, material, and investment cost of equipment at different sites. Evaluation criteria for these techniques are sought to be established on the basis of system studies rather than individually.

The policy shift with regard to the financing of the long-term investment improvement, employment assistance, and cost recovery in kind in the form of biomass can go hand in hand with a withdrawal of fertilizer subsidies. Avenues for increasing incomes can be created by providing assistance for processing equipment with process energy for decentralized chapter pulp manufacture, solar thermal hybrid energy generation, and processing of soybean or other oilseeds and pulses. This will call for a concerted effort of participative technology development and operational research so that the transition can be facilitated and the adverse impact on incomes could be avoided by diversification combined with better services for processing, marketing, and technology support.

Equity

Equity implies ensuring a minimum amount of water to everyone to meet their basic needs of water, for domestic use, for their livestock, for productive use, and to meet minimum fodder and food requirements from commons or common pools. It has been established that on an average for most areas this minimum water allocation needs to be only about 2,000 m^3 of water per family of five. Studies in the Krishna valley have shown that integrated watershed development combined with limited but assured water allocation from the large projects through canals or lifts can bring water availability to this level. By combining less water-intensive higher value crops on public and private wastelands and decentralized processing of biomass and use of renewable energy, incomes of the poor and the marginal groups can be increased with given amount of water. The second aspect of equity is cost recovery from users. The huge losses currently incurred by the government in effect pre-empt resources which could be used for fresh investments to benefit less developed regions and population groups.

A recent committee set-up by the Planning Commission suggested a two-part tariff where water is made available for basic service at a low flat rate, and a higher variable tariff is charged for extra service at economical rates. The committee recommended that while full cost recovery should be the ultimate goal, the immediate aim should be to cover O&M costs. This may be possible if this recommendation is restricted to basic service since cumulated old investments had cost considerably less than the current cost of borrowing, and it is matched by a concurrent steep increase in water charges for the economic service. This too would involve a significant rise in water rates but the increase would be acceptable to the users if concurrently credible steps are taken to raise the quality of service and productivity of water.

Measures to induce farmers to resort to water-saving techniques are also essential. Water saved from existing sources and the potential water development from new local and large sources can then be used for water-efficient horticulture, farm forestry, and agricultural crops with high value-addition and material and energy-replacing potential, making it easier for local communities to contribute a part of the capital required to improve water management and to pay for the cost of continuing service.

Participative experience in the management of water systems by water users' groups shows that if the water users' group is empowered to determine its internal tariff structure and is assured of assistance for upgrading the system to enable joint use of exogenous and endogenous water as well as the joint use of groundwater and surface storage to create a system capable of on-demand delivery, most of the water users would respond favourably to the introduction of water-efficient techniques and low chemical input farming systems which can also be more profitable. Priority, therefore, needs to be given to creating decentralized social instruments for better management of water and for sustainable enhancement of the productivity of ecosystem resources.

The challenge is how to extend the coverage of the service to the extent needed for assurance of livelihood to the poor while simultaneously keeping avenues open for use of water for commercial crop production so that flow of funds on commercial terms can augment the resources available in the public domain. Funds available outside the irrigation system, for example those for watershed and wasteland development or for employment guarantee programmes and training, can be combined with irrigation sector funds to meet a substantial part of the cost, since new technologies not only bring down costs, but also increase the labour component and skill levels.

Success, however, depends on large-scale use of renewable energy and materials-based technology with an organized effort on the part of the village community and on their initiative to acquire management capability. Success also depends on a supportive state policy aimed at targeting and prioritizing funds for communities who take the initiative to work out an appropriate financing package and accept conditions relating to equitable and sustainable water use and energy self-reliance.

Privatization is obviously not consistent with these aims. It is not a feasible or viable solution because the water sector cannot afford the massive investments needed for constructing and even improving the surface water system. There is little likelihood that the cost of previous investment can be recovered through privatization. The supposed commercial viability of present private investments is illusory since it largely depends on privileged access to water facilities and electricity networks and large subsidies, i.e.

cheap credit and power for groundwater and this cannot hold for a generalized condition. Unrestricted privatization of groundwater will reinforce the existing bias in favour of large farmers. There is also a high risk that privatization will result in exploitation of monopoly positions and encourage rent seeking based on privileged access rather than promote more efficient and dispersed availability of water supply. Social regulation of water use is imperative in the interests of equity and is consistent with the search for efficiency.

Coping with Diversity

An important issue is whether the input-output parameters and the functional adequacy of the technological alternatives could be assured when the techniques are applied to the wide range of socio-economic and physical conditions such as prevail in the Krishna valley. The important part here is to work with the right combination of biomass productivity and renewable energy. A judicious combination can cope with the diversity involved. For example, it is not difficult to achieve an availability of 6 tonnes of biomass in the form of fuel and bulk biomass material in the well-endowed regions in the valley. In poorly-endowed areas, a greater input of solar-thermal process energy and technological upgradation with more sophisticated equipment would be needed to provide for greater value-addition possibilities. This may require providing marginal lands with small quantities of assured water, priority in variable water allocation, and producing high-energy value-added liquid oils, fuels or petrochemical substitutes after upgrading through reclamation, and labour-intensive soil improvement and amelioration.

It is possible to match the composition of this biomass to regional characteristics, and agricultural production is not ruled out for well-endowed areas. Formation of regional pools of biomass materials, taking full advantage of low-cost electricity from the grid to combine it with renewable sources for process energy as well as electricity generation, the targeting of low-interest finance for renewable energy generation or process equipment used for energy-efficient high-value-adding processes—together create the basis for a wide range of products and processes which have an inherent capacity to cope with the diversity of local conditions.

The main issue that needs to be addressed is that of building up the management capabilities of the village organizations and upgrading the technological capability of the resource-poor families, including the women and the unskilled landless labourers. The institutional forms as well as structures of the village organizations would vary according to local conditions. In fact, large organizations which are subject to rigidity of functioning and are susceptible to concentration of power and political manipulation or to the domination of bureaucrats need to be replaced by small functional groups and organizations which can have a much greater solidarity and cohesion due to an understanding regarding the implications of equity, efficiency, and sustainability in resource management. It has been found that use of *in situ* conserved water and locally harvested rainwater and external water are manageable in groups with about 50 members. They may be federated into larger entities whenever the need arises.

NEED FOR A PARADIGM SHIFT

To sum up, the experience shows that the success of efforts based on charismatic individuals placed in favourable local conditions cannot be replicated in a linear manner and that there are several instances of failure of such linear generalizations of successful strategies. Important components from instances of success can only be assimilated meaningfully within a paradigm shift which not only includes present needs of secure livelihood but also has a vision of the ecosystem beyond subsistence. This vision must be capable of creating a sustainable dispersed industrial production and have the potential to accelerate the growth of the energy and infrastructure sectors.

More seriously, it needs to be emphasized that without a paradigm which is capable of carrying an ecosystem beyond subsistence, alternative arrangements too are likely to remain restricted or are likely to breakdown when subjected to the market push resulting from indiscriminate privatization of water resources coupled with the opening up to the global market. What is needed, whether in the Krishna valley or elsewhere, is a clearly-defined policy of minimum water assurance combined with technology choices and a development strategy which can take the ecosystem further along a path of sustainable development of renewable energy and materials resources relying on stronger grass-roots initiatives for sustainable and equitable resource use.

In post-Independence India, it has been clear that community organizations could not withstand the onslaught of the formal structures of representative democracy and large irrigation systems managed by bureaucrats. Centralized decision-making and the lack of participation of various rural interest groups in deciding pricing and allocation policies has been a major cause of poor cost recovery and mounting fiscal deficits. In the absence of a consensus on the terms of financing and the mode of cost recovery, it has not been possible to create enduring institutions for participative management of water, energy, and infrastructure services. A dead end has been reached in the development process due to mounting deficits and decline in the quality of service.

The development departments of the state need to recognize the importance of decentralizing the decision-making process regarding choice of technology and priority of programmes. If a performance orientation is given to allocation of public funds it should be possible for village organizations to work out appropriate financial packages and to improve the cost recovery. One important avenue is to recover at least a part of the materials component of investment in the form of contribution in kind of non-perishable biomass with a high value-addition potential. This can go a long way towards improving service and making it affordable without compromising economic viability.

The only way out of the failure of the state and the market, is to create a community of users-producers and empower them to use the funds to develop their local and regional resources. Optimum solutions would have to be location specific and have to be worked out at the grass-roots level even while avoiding localism. Experience shows that it is possible and that the hazards of localism can be avoided if there is a consensus on important issues and policies:

1. Water allocation and pricing to create secure and sustainable livelihoods for all while keeping avenues open for use of water for profitable economic use.
2. Establishing norms of water use and allocation for various needs grouped according to water uses as well as various social segments of the users and establishing priorities.

3. Establishing a pathway of development where common resource pools can be built and progressively enlarged by creating new resources by saving water and targeting the use of public funds for building and progressively enlarging the common resource pool.

Recent experience shows that public funds and concessional institutional finance can be channelled through self-help groups of women. The experience of the Gramin Bank in Bangladesh strengthens the case for relying on local organizations. Various sources of finance can be combined to build an optimum package to fulfil basic needs according to agreed priorities, and cost recovery can be made to sustain the flow of funds according to the terms of concessional and commercial finance. A performance orientation will have to be given to the financing operations and disbursement control.

A major shift is needed to establish a link between sources of revenues and targeting the collection for priority uses. Examples of these are the use of professional tax to mobilize funds for employment guarantee schemes, fuel tax collections to provide capital subsidy for energy conservation and renewable energy generation. There are many such options that which can be identified which, when implemented, will progressively enlarge resource availability and enable building and enlarging the common resource pools.

The main issue is whether accountability will be compromised in the use of public funds as a result of decentralization. A system of performance-oriented financing can work only when there is a consensus on policy for financing and performance norms regarding sustainability and equity. The productivity consideration can best be fulfilled by the judicious use of commercial finance in combination with concessional finance for specific components where the concessions are justified on the basis of clearly defined social and ecological criteria. If this is done, there is a large scope for decentralization in the actual operations. The limitations of the legal framework and the formal structure of representation also need to be noted in this context. In view of the experience of southern California, there is a justification to decentralize the process of making and changing the rules so long as the legal framework facilitates the implementation of the policy consensus on allocation and cost recovery and legal instruments are used only to regulate deviant behaviour and it is left to the local organizations to work out a strategy for the management of conflicts and disputes. Experience of traditional systems substantiates the above-mentioned approach, and there is also enough evidence that local initiative is forthcoming to resolve the problem where the state as well as the market have failed.

Creating a Facilitating Legal Framework

The suggested approach relies on the social energy that can be generated by a process of resource reallocation integrated with the process of development itself whereby basic needs would, by and large, be met from 'new water', i.e. water becoming available from the integrated development of small watersheds, irrigation systems (major, medium, and small), developing and raising the efficiency of use of water to create local surpluses and transferring it to areas of water scarcity by use of cost-effective conveyance and energy technology. This makes improving the distribution of access a positive sum game, while usually distribution is often a zero sum game, since the total productivity does not increase. In the alternative, adequate attention to technological aspects and to conditionalities to be imposed while providing for public finances ensure a positive sum game.

Redistributive processes need to be combined with a policy tilt in favour of public funding in the social sector, such as employment assurance, training, and education and its use to build up the bioresource and the human resource base and to entitle the poor to the new productive assets. This calls for greatly improved financial management to create new assets by use of optimized financing packages where concessional and institutional finance is combined with the productive use of public assistance in the social sector.

There is also a need to create a legal framework that would facilitate the devolution of power to make rules and change them. Wherever a conducive social environment exists, an expenditure push should be avoided and the outside imposition of physical targets should not be allowed to hamper the institution-building processes. In this context, decision-making by the state legislature has not proved to be conducive to building upon local experience.

The state legislatures have, by and large, failed to establish a water policy that addresses the issue of equity and sustainable resource management with user participation. This is a major issue that needs to be addressed in the context of present plans of the state government to hastily implement a highly capital-intensive plan of construction to meet the deadline imposed by the inter-state awards.

Review Mechanisms for Inter-state Dispute Settlement

The most visible, and most sensitive, forms of the legal provisions for conflict resolution over problems of water allocation are the Inter-State Tribunals and Awards to settle inter-state disputes. The nature of these awards, their assumptions and the subsequent rigidity in their interpretation, and their questioning in various scientific fora by experts and by mass movements in public life necessitates a rethinking of their role.

For example, losses and gains have not been correctly assessed. Can we really assess all the damage and put a value to it? It is not just the forest land which gets submerged. Crop land also gets submerged, as do cropping systems. In the upland areas a diversified cropping and biomass production system based on the low external input mode already exists. When these areas gets submerged, people from these areas are uprooted and put down in command areas where an unfamiliar system of monoculture and high-input agriculture is practised. What will these people do? How do we assess the loss of farming systems?

Then there is the issue of how to tackle the question of reliability, which is generally taken to be 90 per cent for hydropower and for industrial and urban water and 75 per cent for irrigation. In the awards, reliability of available water is normally taken at and planned for a dependability of around 75 per cent or more of average irrigation need, including water-intensive crops like sugar cane, etc. This invariably tends to concentrate irrigation and its benefits to a very small section in a very small area.

In this the awards show no awareness of the importance of making a distinction between basic service and economic service. Only the basic service needs to be provided at a high reliability. In the case of economic service for commercial crops, or for higher returns, reliability need not be a limiting factor. There would be variations from year to year, but first, basic needs would not be affected, and second, higher income generation and stability can be provided through value-added processing. Water could then be provided extensively, maximizing production through limited water use over extensive areas, and aggregate production and incomes would be larger. (See Annexure 3 for an illustrative example.)

NEW POSSIBILITIES HAVE ARISEN

While assuming and often recommending farming systems, awards have not taken into account the emerging, alternative scenarios and the dead ends reached by the high-input system based on subsidies, and nor have they recognized the fact that sustainability of projected yields is in question. No mention has been made in these awards of the need for water allocation for regeneration of the ecosystem and groundwater recharge. Experience has shown that specific allocations should be made for forestry and groundwater recharge and green manure production.

The awards have not said anything about resource mobilization. In the case of the Narmada system, it has been shown that with farming systems based on minimum external inputs of water and other energy-intensive materials and services, the same area may be sustainably and equitably served with a much smaller quantum of water or the same water can profitably serve a much larger area.

It is time to recognize that the awards underestimated the rehabilitation burden, and that considerably longer periods are needed for socially acceptable resettlement from large projects than what the awards had foreseen. Similarly, much more social experience has accumulated, and there have been very significant changes in design and technology areas since the awards were handed down.

The need to shift from the high-input systems for maximizing output per acre to low external input systems that optimize input use and scarce resources is now well acknowledged. Such integrated farming systems were not on the horizon when the tribunals handed down their awards. Further, alternative approaches are also emerging which emphasize the construction of local storages and integrated groundwater and surface water use, along with recharge to reduce the requirement of storage from the main reservoir. In the new paradigm, the livelihoods of the poor would be taken care of through common property resources. In the case of water, users collectively can use water on the basis of a MoU with an assurance of basic service to all at an affordable tariff and economic service on the basis of full cost recovery over a period of time.

Most of the construction for these systems, suitably modified with optimum cropping patterns and with dispersed storages, will consist of local works in which employment assistance can be used. Great savings in costs and increase in service area and benefits are possible. The role of hydropower can be limited to the monsoon periods with a regulatory storage which would be depleted by mid-December, thus releasing the submergence area for crops. This was the limitation of permanent submergence of crop land. None of these alternatives are even being considered. Today we have a situation where a rigid adherence to the time schedule and technical specification of the award has become a means of imposing sub-optimal design and denial of opportunity for better local participation.

The alternative development strategy relies on self-management, local resources, and energy self-reliance and has the inherent ability to strengthen the local decentralized institutions. The creation of suitable decentralized instruments to implement water allocation and for groundwater management take considerable time. It is only recently that funds are being made available for watershed and wasteland development. Awards make no mention of water allocation for household food security or to the need to implement a biomass production strategy to provide input for energy and infrastructure development. Reallocation of water within the state and the modification and improvement in land use system would take 10-20 years, while the tribunal has imposed a deadline for utilizing their respective shares of the Krishna water. The artificial imposition of deadlines is, in fact, one of the major reasons for inter-state conflict in the case of the Krishna valley.

What we need is a full review of all awards and tribunals in respect of water allocation. The restructuring process cannot even start if technical designs and parameters are rigidly enforced, and it follows that the validity of awards would have to be limited to the water allocations alone. The new tribunal should take into account all these new scenarios that have emerged. Till a proper review is completed, the present allocation should hold; even if the water share is not utilized by the respective states their water should not be taken away. This guarantee should be there so that the states can plan and use water as per the alternative scenarios.

The present manner of the implementation of the awards is creating an expenditure push, and has created a general impression that the decisions are determined by the preferences of bureaucrats and contractors rather than by a concern for the local people with regard to resettlement, more effective use of external input, greater user participation, and the creation of livelihood opportunities for the poor in the service areas.

It is quite evident that the present systems in which politicians and bureaucrats take major decisions without any direct discussion with, nor participation of, the affected people in the decision-making process will not work. Instead, there should be a process of grass-roots education or water literacy programmes. This responsibility can be entrusted to organizations like the National Water Development Agency, to be carried out with the help of organizations like the BGVS which have done a commendable job in the area of total literacy in a campaign mode. Different scenarios can be worked out and put before the people, and discussions and negotiations can be carried out between upstream and downstream, between the submergence-affected and the command areas, between the rural/agricultural and the industrial/urban users. Let this process go on for about 10 years and only then would we be in a position to evolve decisions capable of resolving conflicts successfully, equitably, and in a sustainable manner.

ACKNOWLEDGEMENTS

This paper is based on the experience of working in the Krishna valley over more than fifteen years. It is difficult to acknowledge the contributions of individual scientists, technologists, researchers, and social activists. Many of them have been associated with the People's Science movement in addressing the problems of water scarcity and conflict resolution. Special reference, however, needs to be made to our colleagues in the Society for People's Participation in Eco-System Management (SOPPECOM): Shri V.B. Salunke, Prof. R.K. Patil, Shri S.N. Lele, Shri Bapusaheb Upadhyay, and Dr. C.M. Pandit. The most important contribution has come from the active members of the Baliraja Water Users' Society, and the Mula, Vaghad, Ozar, and Khudawadi Water Users' Cooperatives. The

Mukti Sangharsh Movement and notably Dr. Bharat Patankar played a vital role in helping village communities work out an approach to a sustainable development alternative. Several officials from the Irrigation and Rural Development Departments also made useful contributions.

REFERENCES

Awasthi, Ramesh and Dashrath K. Panmand, *Ralegan Siddhi: A Model for Village Development*, Pune: Foundation for Research in Community Health (FRCH), 1994.

Dabholkar, S., *Plenty for All*, Pune: Mehta Publishing House, 1997.

Datye, K.R. and R.K. Patil, *Farmer Managed Irrigation Systems: Indian Experiences*, Pune: Centre for Applied Systems Analysis in Development (CASAD), 1987.

Datye, K.R. and Suhas Paranjape, 'Opportunities for Small and Marginal Farmers', in *Wasteland News,* New Delhi: Society for Promotion of Wasteland Development (SPWD), Aug.-Oct. 1990.

Datye, K.R., *Banking on Biomass: A New Strategy for Sustainable Prosperity Based on Renewable Energy and Dispersed Industrialisation*, Ahmedabad: Centre for Environment Education (CEE), 1997.

Joy, K.J., 'People's Management: The Case Study of the Bali Raja Dam', in M.V.K. Sivamohan and Christopher A. Scott (eds.), *India: Irrigation Management Partnerships*, Hyderabad: Booklinks Corporation, 1994.

______, 'Maharashtra Sugarcane Farmers Taught Green Skills', in *Down to Earth*, New Delhi, 31 December 1992.

Lele, S.N. and R.K. Patil, *Farmer Participation in Irrigation Management: A Case Study of Maharashtra*, New Delhi: Horizon India Books, 1994.

Lundquist, J., K.R. Datye and M. Falkenmark, 'A Biomass Strategy for Watershed Development: Best Possible Use of Local Rains for Sound Socio-Economic and Natural Resources Development', paper presented at the World Water Resources Conference, Morocco, May 1991.

Omvedt, Gail and Bharat Patankar, 'Movement for Water—Takari Peasants' Struggle in Maharashtra', *Economic and Political Weekly*, 13 April 1989.

Paranjape, Suhas and K.J. Joy, *Sustainable Technology: Making the Sardar Sarovar Project Viable*, Ahmedabad: Centre for Environment Education (CEE), 1995.

Phadke, Anant R.S., 'Anti-Drought Movement in Sangli District', *Economic and Political Weekly*, 22 April 1989.

Salunke, Vilasrao et al., 'Pani Panchayat Model of Water Management', in Nitya Rao and Rurup Luise (eds.), *A Just Right—Women's Ownership of Natural Resources and Livelihood Security*, New Delhi: Friedrich Ebert Stiftung (India Office), n.d.

Upadhyay, Bapu, 'Ozarche Pani' (Marathi), Ozar Samaj Parivartan Kendra, 1987. (An English translation will soon be available.)

Vaswani, Ramesh M., *Micro-Watershed Development—Three Success Stories from Maharashtra*, Pune: Yeshwantrao Chavan Academy of Development Administration, 1995.

ANNEXURE 1

A 2,000 M^3 NORM CAN TAKE CARE OF DIFFERENT ENDOWMENTS

1. *A family with 1.2 ha. of land*

   The allocation would cover, first, irrigation to stabilize food production with two crops over an area of 0.8 ha. which would yield 2.4 tonnes of coarse cereals and an associated quantity of oilseeds and pulses. Additional allocation would cover the establishment for green manure production and alley cropping and yield improvement with variable water over 0.4 ha. This will be apart from commercial horticulture over a limited area with appropriate terms of finance. The assured water supply would be limited to 2,000 m^3 per year.

2. *A marginal farmer with 0.4 ha. of crop land*

   Allocation for food crop area could very well be 1,000 m^3. Another 1,000 m^3 but with variable water would cover a forestry plot, access to which could be negotiated on suitable terms of produce sharing with the private landholders if common land is not available.

3. *A landless family with 0.1 ha. allocation*

   The third option would be a landless family allocated 0.1 ha. for intensive cultivation. Here again, the allocation would be 1,000 m^3 for intensive cultivation on the plot and another 1,000 m^3 for a forestry plot which could be obtained by negotiating suitable terms of produce sharing with the land-owners.

The average water requirement, therefore, remains at 2,000 m^3 allocation and the landless family with 0.1 ha. allocated would produce supplementary food for the local market. The higher income for the first family results from commercial horticulture. The non-farm income of the second family from the forestry plot would make it possible to create sufficient income to supplement the production from its agriculture plots. The high-value food produced on 0.1 ha. land allocated to the landless family would again enable it to attain food security while access to the forestry plot would assure the fuel need fulfilment.

Activities such as fisheries, cattlesheds, and domestic water would be in the nature of common services and would be included in the minimum water assurance. The issue that remains is how the project planning and implementation process can be steered under decentralized decision-making to ensure that the other basic needs described above can be met.

ANNEXURE 2

ENERGY FOR PUMPING AND BIOMASS PROCESSING: CALCULATING THE ENERGY BURDEN

2,000 m^3 delivered at the field would imply 3,000 m^3 at the source from the main system. If we assume an average lift of 30 m. the energy input required would be 500 KWh with pump efficiency of 50 per cent. Off-peak electricity and supply from existing sources can provide this quantum for all the families in the valley.

The assumption is that the range of lifts is from 10 to 50 m. with an average of 30 m. There will be a flat energy plus water service charge with an average energy price for basic service. Recovery in kind in the form of bulk biomass of 500 kg. priced at Re. 1/kg. will provide sufficient biomass fuel for the equivalent generation by the solar thermal hybrid system.

MEETING PROCESSING AND PUMPING ENERGY REQUIREMENTS

It has been demonstrated that the solar thermal technology is already in an advanced stage of development so that local capacity building, training, and establishment of institutional arrangements can be completed in about 5 years. If capital is provided at 8 per cent rate of interest, the cost of thermal energy works out to Re. 0.65 for 1,000 Kcal, delivered in the form of steam at 200-300 ^{0}C. The relevant parameters are investment cost Rs. 5500/m^2 and Rs. 3,000 Kcal/day delivered for 280 days/year on the average.

The viability of liquid fuel and chemicals production is already been established. Energy self-reliance thus becomes a realizable goal. Fabrication of the equipment and its servicing could become an income-generating dispersed industry.

This approach can also be extended to process heat energy to produce chemical intermediates and liquid fuels. The landownership constraint can be overcome by socially regulation. Allocation of plots of land with water supply and the necessary pumping energy for the irrigation delivery system has the potential to produce 2,000-3,000 litres of high value products per hectare per year. Allocations of funds for the process equipment and the solar thermal energy equipment at concessional rates of finance are justified on the grounds of energy conservation and social considerations.

ANNEXURE 3

Let us take for example the situation in village Khudawadi. The reliability of water availability is very low. The tank does not fill for 5 to 6 years during a 10-year period. In such a situation, instead of planning for a crop like sugar cane which requires a high reliability, it is better to go for crops like cassava (tapioca) which might require only a couple of irrigations.

Even if the 300 households bring about 50 ha. under cassava, the energy potential is enormous. At a production level of 4,000 litres of liquid chemical per ha. it comes to 50 ha. x 4,000 litres = 200 kilolitres/ 300 households, and at 2,000 litres/ha. it comes to 100 kilolitres/ 300 households. This is the bottom line and yet it gives enormous opportunity for income generation and energy self-reliance in the long run.

For 1 litre ethanol production we require:

Rs. 2.00 (electricity from the grid at Re. 1 per unit)
Rs. 4.00 (solar + biomass processed heat)

Rs. 6.00

As against this, the product would fetch about Rs. 20 to Rs. 30, thus ensuring an income of Rs. 7,000 to 20,000/household for an area of 0.2 ha.

Thus, for the basic service, reliability would be 75 to 80 per cent and beyond that the bulk of the production would be on variable water and reliability need not be a factor. This is in contrast to the earlier system of water-intensive crops, where even a crop like sugar cane had to be taken at 75 per cent dependability to ensure any degree of security.

CHAPTER 9

The Deepening Divide: Diverse Responses to the Challenge of Groundwater Depletion in Gujarat

TUSHAAR SHAH

Groundwater depletion has emerged as one of the most formidable consequences of agricultural development in Gujarat in the last five decades. Compared to many eastern states, Gujarat has limited water resources though, by comparison, western Indian states like Rajasthan are even worse off. Gujarat is excessively dependent upon groundwater; three-fourths of its irrigation occurs through wells and tube wells; and the last two decades have witnessed widespread over-exploitation of groundwater with powerful economic and socio-ecological fallout. A critical issue of research as well as policy relevance in Gujarat now is: will the Gujarati people and institutions be as innovative and dynamic in dealing with this challenge of restoring the region's aquatic equilibrium as they have been in devising economic institutions and technologies to exploit this resource to their advantage? It is too early to answer this question conclusively. However, enough has been happening to suggest a growing concern about the issue. In this chapter we review the reactions of three groups of actors in the arena of groundwater management, namely the government, NGOs, and popular movements. To begin with, we present an overview of the groundwater situation in the state and appreciate the fact that a resource *management* perspective is as important and critical to formulating a comprehensive response to the challenge as the resource *conservation* perspective.

GROUNDWATER RESOURCES AND THEIR UTILIZATION IN GUJARAT

Gujarat has a wide range of geo-climatic conditions. The average rainfall varies by 1:10 between Kutch and Dangs. Saurashtra, Kutch, and north Gujarat receive scanty rainfall and face acute problems of water shortage and water quality. The south-eastern hilly areas of Dangs and Dharampur receive a stable 1,600 mm. annual rainfall and yet have acute drinking water shortages in the summer. In contrast, the bulk of Gujarat, north of the Vatrak-Sabarmati rivers, has limited groundwater resources and has been suffering groundwater overdraft of severe proportions. In the absence of astute management at the local, meso, and macro levels, it is just a matter of years before many parts of north Gujarat end up as basket cases.

Topography varies from extensive low-level plains with 80 m. elevation above msl, covering about 50 per cent of the area to hilly tracts rising from 80 to 800 m. above msl. (Patel, n.d.). The south-eastern parts with heavy rainfall have topographical and geological conditions unsuited for groundwater recharge and storage. North Gujarat, Kutch, and Saurashtra have permeable geologic formations and geomorphic conditions which are excellent for groundwater recharge and storage; but these receive very little rainfall. As a result, even though rainfall variation between south Gujarat and the rest of the state is 1:10, the water resources position in the two regions varies only by about 1:6. Gujarat is thus a classic example of a region in need of better management and husbandry of water resources at both the macro and micro levels.

In aggregate terms, Gujarat has 30,000 m. cum of water resources (excluding the Narmada), and 60 per cent of this has already been developed (Hirway and Patel, 1994). The state has 186 large and small rivers; of these, south Gujarat has just 11 but these offer a utilizable potential of 1.39 m. ham.; Kutch has 97 but these offer only 0.4 m. ham. The Narmada alone offers 1.111 m. ham., two-third of the 1.780 m. ham. offered by all the remaining 185 rivers. The total utilizable water resources offered by the rivers are thus 2.891 m. ham. (Table 1). A total of 1.8 m. ha. of ultimate irrigation potential has been planned from surface schemes, of which 1.2 m. ha. was already developed by 1990.

TABLE 1: UTILIZABLE POTENTIAL OF WATER RESOURCES OF GUJARAT

Region	No. of Districts	Surface Water m. cum/year	Groundwater m. cum/year	Total Resources m. cum/year
South Gujarat	7	13,900	4,600	18,500
North Gujarat	5	2,000	2,900	4,900
Saurashtra	6	1,500	4,290	5,700
Kutch	1	400	440	840
Total	19	17,800	12,280	30,080
Narmada		11,100		11,100
Grand Total	19	28,910	12,280	41,190

SOURCE: Hirway and Patel (1994: 19).

Even if 30 per cent of the utilizable water resources (excluding the Narmada) is earmarked for domestic, municipal, and industrial needs, about 2.1 m. ham. of water will be left for agricultural use; all of Gujarat's 11.1 m. ha. of net cropped area can, in principle, be supported with a delta of 1.9 m. with this quantum of resource. But barely 30 per cent of this area is under irrigation today; and even at such a low level of achievement, 70 per cent of Gujarat's villages suffer from inadequate supply of drinking water or from bad quality drinking water, and the bulk of Gujarat has emerged as an aquatically fragile region. The quantum of water resources available to the state is not exactly plentiful but they do offer its

TABLE 2: SOURCES AND USES OF WATER: GUJARAT *c.* 1990

Region	(1) Available Water Potential m. cum/year	(2) Human and Industrial Needs m. cum/year	(3) Surface Water Use for (2) m. cum/year	(4) Ground-water Use for (2) m. cum/year	(5) Surface and Ground-water Potential for Irri. m. cum/year	(6) Irrigation Potential with Delta 0.5 m. (m. ha.)	0.7 m.
South Gujarat	18,500	474.4	91.6	382.8	18,025.6	3.6	2.57
North Gujarat	4,900	416.9	84.0	332.9	4,483.1	0.89	0.64
Saurashtra	5,700	375.3	115.9	259.4	5,440.6	1.09	0.78
Kutch	840	36.9	5.5	30.4	833.1	1.67	1.19
Total	30,800	1,303.4	297.0	1,005.4	29,496.6	5.90	4.20

people a fighting chance with tighter, more appropriate, and imaginative management at the local, sub-regional, and regional levels. Gujarat needs, more than any other region in the country, to move from the water 'development' syndrome to water 'management' regimes.

In groundwater, too, the south is much better endowed. Of the estimated utilizable potential of 1.228 m. ham., 0.460 m. ham. is located in south Gujarat (7 districts), 0.290 m. ham. in north Gujarat (5 districts), 0.429 in coastal Saurashtra (6 districts), and 0.044 in Kutch (1 district).

These offer an ultimate irrigation potential respectively of 1.065 m. ha., 0.7 m. ha., 1 .085 m. ha. and 0.088 m. ha. in the four sub-regions. Altogether south Gujarat thus has 55 per cent of the available surface water resources and 34 per cent of the groundwater; north Gujarat, Saurashtra and Kutch have 70 per cent of the state's farm lands and less than 25 per cent of its total surface and 65 per cent of the groundwater resources.

Gujarat has a net cropped area of 11.1 m. ha. (from the total surface area of 19.5 m. ha. of the state); of the net cropped area, 1.762 m. ha. (76.6 per cent) was irrigated by groundwater and 0.54 m. ha. by surface water (23.4 per cent) in 1990. Groundwater—which is less than one-fourth of the state's water resources—contributes to three-fourths of its irrigated land. The consequences of such heavy dependence on groundwater irrigation outside south Gujarat are severe and apparent: water tables have been dropping throughout north Gujarat; the Saurashtra coast has an acute and growing problem of salinity ingress; water pollution is emerging as a major issue in many parts of north Gujarat and Saurashtra.

When it is fully operational, the Narmada system irrigating 1.8 m. ha. will make the surface irrigation resources of Gujarat nearly equal to its groundwater potential. It may thus reduce the strain on the groundwater resources and also help augment them in Banaskantha and Mehsana, two of the Gujarat districts worst hit by groundwater over-exploitation. However, that would take anywhere between 10-15 years. By then, enormous damage could have been done (Patel, n.d.).

In contrast, waterlogging has emerged as an important socio-ecological issue in the command areas of the Mahi Kadana and Ukai Kakrapar systems in central-south Gujarat. The Vatrak-Sabarmati rivers divide Gujarat into two parts: the water-surplus, canal-irrigated south Gujarat using less groundwater than is added to its aquifers and the water-short north Gujarat, Saurashtra, and Kutch which have come to overly depend on groundwater for all their needs. Naturally, growing conflicts have emerged between municipal users, industries, and agriculture as the three principal consumers of water. In many areas, the quality of rural and urban lives as well as overall economic progress are endangered because of actual or potential water shortages.

Groundwater availability estimates prepared by the state groundwater department reveal that although many districts show 30-40 per cent development of the groundwater resource, these figures conceal pockets of overdevelopment in many districts. We note that 5 of the 19 districts have already developed over half of their available potential; and just 5 or 6 districts—Surat, Valsad, Dang, Kheda, Bharuch and Vadodara—offer significant scope for further groundwater irrigation.

If and when the Narmada waters begin flowing, groundwater resources in many of these districts will increase substantially. If all the assumptions about the Narmada hold, then the groundwater situation in many parts of north Gujarat is likely to improve. Mehsana district in north Gujarat with 124 per cent development, however, has already emerged as a 'basket' case. Junagadh, Banaskantha, Amreli, and Ahmedabad follow Mehsana closely. We note that 12 of the state's 184 *talukas* have already over-exploited their groundwater resources. NABARD-CGWB have declared 14 *talukas* 'dark', with over 85 per cent of the available resources already developed (CGWB, 1995).

The bulk of this development has occurred under private initiative. Gujarat has a 3,000-strong public tube well programme. However, it has been a resounding flop, and has contributed in a very limited way to groundwater development as well as over-development. The only significant aspect of its performance is that the public tube wells have performed well only in areas with scarce and deep groundwater reserves where private groundwater irrigation is very expensive. Elsewhere, private water sellers cut deep into their commands and out-compete the public tube wells despite their subsidized water rates. The growth of private water-extraction structures or mechanisms (WEMs)—implying tube wells mounted with diesel or electrically-driven pumping plants for lifting groundwater—has been rapid. Gujarat farmers were quick to take to modern water-extraction technology. When in Uttar Pradesh, the state government began its public tube well programme in the 1930s to demonstrate to farmers the technical feasibility and economic viability of pump irrigation, vibrant water markets in central and north Gujarat had already completed the first phase of development and were trying to overcome the resistance to switch from oil engines to electric pumps.

Gujarat farmers resisted the use of electricity for pumping water from wells until the 1950s, but after that they took to electric pumps with great enthusiasm. The craze has continued right up to now. The population of electric WEMs (or eWEMs) has steadily increased since 1982 from 1.5 to 2.5 lakh. The rate of growth has been steep, too, except in the last 4-5 years when there were mild signs of a let up in the preference for eWEMs. The population of diesel-operated WEMs (or dWEMs) declined in the early 1980s but to picked up later, and has in recent years been growing at a healthy rate. The total number of WEMs in Gujarat at around 7 lakh is estimated to draw some 0.717 m. ham. by the CGWB, which gives an average of 10,243 cu. m./WEM per year; at an average discharge of 20 cu. m./hour, the average WEM would be operating for around 500 hours, which tallies with the estimates we have from field studies. If the bulk of Gujarat is moving into [c] category or from [c] to [d], the private activity level in well construction or installation of new WEMs does not reflect much worry or concern on the part of the farmers.

Many micro-level studies have begun to reveal what the CGWB estimates fail to reflect: 7 per cent of Gujarat's blocks being dark does not adequately convey the pervasiveness of the over-development problem. These studies, usually by NGOs (such as VIKSAT and AKRSP) and by individual researchers (such as Bhatia 1992), often tend to overplay localized crises but have helped produce public and policy awareness about the impending problems of groundwater over-exploitation. Private pumping and water markets have been pre-eminent, and the leading players in two of Gujarat's worst groundwater catastrophes in recent years—Mehsana in north Gujarat, which had overdeveloped its ground-water resources to the tune of 125 per cent and where the water tables are down by 10-25 feet almost every year, and Junagadh in coastal Saurashtra where collective over-pumping over the 1950-70 period—snapped the intricate equilibrium between coastal freshwater aquifers and the sea water they overlie, and has caused a massive problem of salinity ingress with devastating ecological as well as socio-economic impacts.

The Gujarat Electricity Board, like all other SEBs, has the standard clause in its contract with agricultural consumers to the effect that they cannot use the said connection to sell water to others. Had the autarky desired by the SEB been enforced, both Mehsana and Junagadh,[1] Mehsana more than Junagadh, would have been in an environmentally less precarious situation. Clearly, water markets have to be disciplined to restore such regions to environmental health. But an even more interesting question is: can water markets which have been a part of the environmental problem in Mehsana and Junagadh be a part of the solution?

MACRO-MANAGEMENT OPTIONS: GRANDIOSE SCHEMES

One of the early responses of Gujarati thinkers and activists to the growing water scarcity has taken the shape of grandiose ideas of big structures, linking rivers, and creating mega-reservoirs. Most of these have remained at the idea level, but some have fired the Gujarati imagination and keep returning to the forum of public discussions from time to time. The basis of all these is the uneven distribution of water resources between south Gujarat and the rest of the state. The proposals aim at capturing some of the huge quantity of water from the numerous south Gujarat rivers that get emptied into the Bay of Khambhat and spreading this over a large area of the state which is deficit in water.

Vithubhai Patel (1995), a noted Gujarati thinker and experimenter, likens the problem of water management at the state level with that of electricity generation and distribution or with the management of food grain procurement and distribution. He advocates integrated water management in Gujarat as centrally a problem of distribution that needs to be resolved just as modern societies distribute food and electricity between producing areas and consuming areas. South Gujarat is water surplus; north Gujarat, Saurashtra, and Kutch are water deficit. Long-run solutions to Gujarat's water problem have to explore ways of storing, transporting, and distributing the surplus water of south Gujarat to the rest of the state. He views the Narmada project as the first step in doing that.

To complete this vision, Patel supports two projects in addition to the Narmada. The first is to link 7 small rivers and unload their surplus water into the Ukai dam, and then to transport that water into the Narmada through a Tapti-Narmada high level canal. The second is the Kalpasar project, conceived to store the surplus waters of the rivers north of the Narmada in a gigantic freshwater reservoir by building an earthen dam in the Bay of Cambay, and to use the water so stored for the irrigation, domestic, and industrial needs of Saurashtra.

In principle, these solutions appear feasible. Many developed countries use technologies for long-distance transport of large volumes of water. India, indeed Gujarat too, has experience in this. Although this experience has not been entirely satisfactory, it is clear that at some high level of techno-economic development, a region like Gujarat will be able to develop some of these alternatives. However, in the present conditions, these options seem impractical, what with the difficulties the state is facing in getting even the Narmada project completed.

However, some sub-regional solutions have been proposed, too. A good example of the later are the proposals of Khemabhai Patel, a farmer leader from the district of Sabarkantha in north Gujarat, about integrating Dharoi with the Narmada system to resolve the water problems of Ahmedabad city and north Gujarat. The Dharoi dam has 5 storage of 27,500 m.cu. feet, of this 7,800 m.cu. feet was planned for irrigation in Mehsana and Sabarkantha; and 11,800 m.cu. feet was planned for meeting the needs of Ahmedabad and Gandhinagar cities and for supplying to the Ahmedabad Electricity Company. The balance 5,000 m.cu. feet was to account for loss by evaporation.

The Dharoi dam has, however, unutilized storage potential. The rainfall in the catchment areas of Dharoi ranges from 300 to 1,200 mm. As a result, the inflow into the dam has varied from 28,000 to 65,000 m.cu. feet. Thus, even after providing for planned uses, there is still potential to capture and use 3,000 to 40,000 m.cu. feet of inappropriate water which runs off into the sea. In 60 out of 100 years, around 14,000 m.cu. feet of water runs off into the sea. This resource can be used to augment north Gujarat's water resources.

Then, Ahmedabad receives water from the Dharoi dam through releases in the Sabarmati river. When the dam was built, the assumption was that 15 per cent of the gross supplies would be lost through seepage. Thus, of the 11,800 m.cu. feet stored in Dharoi for dispatch to Ahmedabad, 10,030 m.cu. feet was to reach the city; but as things turned out, the seepage and other losses have risen to an average of 70 per cent. In part, this is because seepage losses were under-estimated, but more importantly, a string of lift-irrigation schemes on the 160 km. river length between Dharoi and Ahmedabad pre-empt water for irrigation. So of the 11,800 dispatched from Dharoi, just about 3,500 m.cu. feet reaches Ahmedabad. To supply the city with the mandated quantity, nearly 34,000 m.cu. feet would have to be released at Dharoi, a quantity which is clearly not available. Patel suggests that transporting Ahmedabad's water by a pipeline can save 8,000 m.cu. feet for use in north Gujarat. Alternatively, if Ahmedabad and Gandhinagar are served by the Narmada canal which is designed to pass between the two cities, the 11,800 m.cu. feet retained in Dharoi can be used in north Gujarat.

GOVERNMENT POLICY FRAMEWORK

Important departments and agencies of the government directly involved in groundwater management are:

- State Groundwater Department and the Central Ground Water Board, both of which are primarily in the business of groundwater estimation, in quality surveillance, and information gathering, are also responsible for overseeing the groundwater situation in the state;
- State Irrigation Department, including the Command Area Development Agencies, which constructs, manages, and oversees major, minor, and medium irrigation schemes;
- Gujarat Water Supply and Sewerage Board (GWSSB), which has been primarily involved in constructing and operating drinking water supply schemes;
- Gujarat State Ground Water Development Corporation (GSGWDC) which was established to operate a public tube well programme to provide groundwater irrigation service to farmers.

A number of other agencies have indirect but often substantial roles to play in the arena of groundwater management. The Gujarat

Pollution Control Board is charged with monitoring the pollution of groundwater along with other forms of environmental pollution. Public sector and cooperative banks with support from the National Bank for Agriculture and Rural Development (NABARD) are major providers of credit support to private investment in groundwater development. The State Electricity Board, too, is an indirect but key operator; it can augment or restrain overall groundwater withdrawal in the entire state or in specified regions through the regulation of the number of power connections released for agricultural uses and through the regulation of power supplied to agriculture. Finally, most of the government programmes designed to conserve or augment groundwater resources—such as the centrally-sponsored watershed development project, the state government's farm pond scheme, etc.—are implemented by rural development or agriculture departments or by government corporations like the Gujarat Land Development Corporation and the Gujarat Agro-Industries Corporation.

Government intervention in any sphere of social life occurs in a combination of three forms: direct governmental action, through use of taxes and subsidies to discourage or encourage specific behaviour amongst the people, and through a regulatory or promotional framework, which can use laws or rules to disable harmful tendencies and enable beneficial ones. In Table 3, we construct an overview of the key government policy responses to the water sector challenges facing the state. In doing so, we summarize the current priorities in operating each instrument, and offer an assessment of the actual impact of the operation of the policy on groundwater withdrawal as well as on groundwater supply. Finally, we also comment on whether the policy, as it operates at present, is on target.

Direct Action

If we confine our discussions to the state level—and ignore municipal governments and gram panchayats—then three areas in the water sector in which the Gujarat government has undertaken direct action is in operating the state tube well programme (STWP), large rural water supply schemes (RWS), and the Narmada Project. The STWP's original objective—of supplying groundwater irrigation efficiently and equitably—has now become irrelevant with rapid

TABLE 3: AN OVERVIEW OF GOVERNMENT POLICY RESPONSES TO WATER SCARCITY

Policy Instrument	Status and Current Priority (CP)	Promotes(+) Discourages (-) Groundwater Use	Augments (+) Reduces (-) Groundwater Supply	Is the instrument well targeted?
Direct Governmental Action				
State tube well programme	2,500 STWs; CP is to wind up the corporation and sell the STWs	++		No; on the contrary
Drinking water schemes	10,000 villages; CP is to improve operating performance and costs recovery	+		Yes
Narmada Project	Has the potential to ease the groundwater situation in much of north Gujarat and Saurashtra when fully commissioned	–	++++	Yes, substantially, at least as planned
Taxes and Subsidies				
Flat power tariff	GEB is cutting its losses by reducing power supplied to agriculture	++		No; flat plus metered tariff differentiated for different areas needed
Progressive power tariff	North Gujarat is up in arms against the GEB	–		Yes; makes water costlier where it is scarcest
Subsidy on public tube well irrigation	One of the causes of its current crisis	–	+++	No; it is inversely targeted
Subsidized canal irrigation	CP has improved cost recovery, farmer management	–		Inversely targeted
Groundwater law	Lip service being paid; questions about enforcement	–	+++	?

(contd.)

TABLE 3 (contd.)

Siting regulations	Effective in dark zones	–	+++	Partially well targeted
Zoning restrictions	Chief Minister hellbent on compromising these in north Gujarat	–		Well targeted if govt. enforces them strictly
Selective credit controls	Limited effectiveness; coop banks lend heavily for tube wells	–		?
Control on power connections	Can be very effective; but CM has forced GEB to issue new connections in north Gujarat	–		Well targeted if the govt. enforces zones
Regulating power supply	Effective but politically problematic. Even so, GEB has mastered the art of curtailing power supply for pumping	–		Inversely targeted; enormous scope for better targeting
Watershed development	Gujarat has given away nearly 200 watersheds to NGOs; it is too early to assess the experience		++	?
Groundwater recharge programmes	Very limited and ineffective programmes like farm ponds		+	? Insignificant presence
Support to people's well recharge movement	Very important in scale and impact; but govt. is peripheral; the movement is run by people, NGOs, and religious formations		+++	Very well targeted

development of private tube wells and of vibrant and pervasive water markets. Instead, the Corporation has been encouraging intensive groundwater use in water-scarce north Gujarat and Saurashtra owing to its abiding policy of selling STW irrigation at a uniform rate fixed on a cost-plus basis throughout the state.[2] RWSs—most of which transport groundwater through piped networks to thousands of Gujarat's so-called 'no source' villages—are the best response yet to the miseries of these villages. But they invariably fall short of their target 'reach', have massive problems of cost recovery, and are bureaucratically managed. Improving the technical and managerial performance of the RWSs can be an important item on the agenda of the state government. The third major direct action initiative is the Narmada project; if and when it gets fully completed, the Narmada system is likely to ease the overall water scarcity problems in most of north Gujarat and Saurashtra; and even where the Narmada water will not be available for irrigation, it will ease the drinking water crisis in most fluoride- and salinity-affected areas. However, the final completion of the Narmada project is a far cry, especially now that the World Bank support has been withdrawn.

Taxes and Subsidies

Intervening through taxes and subsidies essentially involves shaping and directing economic incentives through price mechanisms. In areas of water scarcity, it requires that government policy makes it costlier for users to get water; and in waterlogged areas, government policies create economic incentives for people to use more groundwater and less surface water for all their needs. The government's track record in using these instruments has been poor. Its distortionary impact is compounded by mis-targetted subsidies. Thus, canal irrigation is supplied at one-twentieth the price at which farmers in the same villages buy irrigation from private tube well owners. The STWs charge the same price for groundwater irrigation in the waterlogged upper reaches of the canal commands as they do under the progressive tariff regime. By charging a farmer a higher *rate* per horsepower as the rating of his pump increases, the latter makes it exceedingly expensive to own and operate an electric pump in north Gujarat, where it indeed should be.

The state government has a clutch of soil and water management programmes, each with a different focus. Most of these are implemented by local NGOs with funds provided by the government. These would eventually have a beneficial impact on groundwater recharge if they are properly implemented and are carried out on a large enough scale. Neither of these seems to be the case. As a result, it is unlikely that these will constitute a *significant* response to the challenge of water scarcity.

Regulatory and Promotional Framework

The worst has been the government's performance in using regulatory and promotional mechanisms in managing the water sector. Gujarat was amongst the first to formulate a groundwater bill. This was passed by the assembly in the early 1970s but never became a law because of intractable questions of enforcement and political expediency. There are other, more recent model groundwater bills which are doing the rounds in the state. However, it is unlikely that any resolute action will follow in this direction. Indeed, according to recent announcements of the Chief Minister, the government seems hell-bent on compromising the zoning system and forcing the GEB to issue new electricity connections and private tube wells in north Gujarat and Saurashtra.

In conclusion, then, the government policy response to the water sector management problems in Gujarat is diffuse, lackadaisical, and entirely unequal to the scale and seriousness of the challenge. Lack of political stability, unwillingness of the political leadership to assume tough and forward-looking stances on major problems, and the cyclicality attached to the sense of urgency assigned to water-related problems. Added to that is the general bureaucratic incompetence, lack of drive, and absence of professionalism under an uncertain political leadership.

By far the largest water conservation impact any government policy has had is of progressively declining electricity supplied to electric pumps. In aquatically fragile north Gujarat, where groundwater has to be lifted over 800-1,000 feet, diesel engines are of no use for pumping; so farmers have to depend upon the electricity supply for lifting groundwater. A sure fire method of reducing groundwater withdrawal in such a situation is restricting the supply of electricity to agriculture.

The Electricity Board has done this successfully. Although official figures about electricity supply to agriculture are far from reliable, an opinion survey of a group of farmers from 6 villages selected from different regions of Gujarat about the quantity of the power supply to agriculture is unmistakable in showing a steeply declining trend. But this restriction in power supply has occurred by default, and not as part of a groundwater management policy.

THE RESPONSES OF NGOs: DIVERSE AND EXPERIMENTAL

Gujarat has a large and vibrant NGO community. The majority of the NGOs especially in Saurashtra and Kutch, is grounded in Gandhian thought and ethic, and places a high value on the organic links between life and nature. Many of these NGOs still preach the Gandhian values of simplicity and non-violence, and actively promote a life lived in harmony with nature. Superimposed on this bedrock of small, frugal, resource-strapped Gandhian voluntary agencies is a small group of medium to large, modern, elite NGOs which believe in professionalism, scale, efficiency, and modern technology and management. Despite the differences in their work culture, the criticality of Gujarat's water sector challenge has promoted a constructive collaboration within the state's NGO community.

NGOs in Gujarat have responded to the water sector challenge in three ways:

- *Water issues in the public platform.* NGOs have tried, often in collaboration with the media, to bring water issues to the centre stage of public discussion in a manner that has not happened in many other states. Notable in this respect is the work of VIKSAT, of Mahiti-Utthan, of PRAVAH. In a somewhat different but even more significant way, this can be said about the work of the Saurashtra Lok Manch which played a pioneering role in taking the message of groundwater recharge and information about various methods of doing it to thousands of villages in Saurashtra and Kutch.
- *Institutional and technological experimentation.* Gujarat NGOs have promoted a culture of institutional and technological experimentation in water production, use, and conservation. Notable work in community management of irrigation include: the work of the Sadguru Land and Water Foundation with tank-

based lift irrigation in Panchmahal, Banswara, and Jabhua; the Aga Khan Rural Support Programme's lift irrigation schemes in Bharuch and Junagadh and their turnover of medium-scale irrigation systems to farmers' organizations for management in Bharuch; the Development Support Centre's recent experiments with and collaboration between the irrigation department and the farmers' community under joint irrigation management; and SEWA's work with pani panchayats in Banaskantha.

Many of these experiments had no direct links with managing water scarcity. These were aimed at inculcating in the water user communities a more responsible attitude towards water use and to create self-financing irrigation organizations. However, the experimental work of many other NGOs was a direct response to the problem of water scarcity. These either aimed at devising new ways of augmenting *in situ* recharge of water or of economizing on water use. Notable amongst these are the work on the *gupat bandh* (hidden dam); Mahiti-Utthan's work on tank *Chunibhan* lining; Vivekanand Research and Technology Institute's significant work on water harvesting in arid areas; Mehsana Agro-Foundation's work on *holiyas* and tank-recharge techniques in Mehsana district; and the contribution of the numerous NGOs to a range of technologies for groundwater recharge after 1991.

- *Programmes for impact:* By far, the majority of NGOs and grassroots groups have been operating in extremely small domains. As a result, their impact, even in their local context, is extremely limited, to say the least. It is seldom that small NGOs find an initiative to produce a scale of impact disproportionate to their size and resources, as the Saurashtra Lok Manch did. However, a few of the modern NGOs have zeroed in on a compact clutch of programmatic ideas and tried to produce an impact on some scale. Sadguru's lift irrigation schemes encompass several hundred villages; AKRSP's work on irrigation also affects dozens of village communities. Banaskantha DWACRA Mahila Sewa Association works on the drinking water problem through its 4,000 strong women membership. There are numerous watershed programmes. If these are implemented properly, they will have a notable impact on sizeable areas 5 to 7 years from now.

People's Movement: The Role of the Swadhyaya Parivar and Other Mass-Based Social Organizations

By far the most energetic and inspired response to the intensification of water scarcity has come in the form of a mass movement for well recharge and water conservation catalysed first by the Swadhyaya Parivar and subsequently joined by other sects of Hinduism as also by scores of NGOs and grass-roots organizations in the aftermath of the three-year draoght during 1995-7. Before we review the nature and dynamics of the well recharge movement, it is important to briefly note two of the organizations that catalysed and sustained it.

Western India, in particular Gujarat and Maharashtra, have always been keen recipients of newer, insightful reinterpretations of Hinduism. Even an erudite Kathakar (narrator of scriptures) like Morari Bapu or Dongre Maharaj can command a huge mass following and remarkable mobilizing power. It is little surprise then that some religio-spiritual schools have emerged as powerful players on the social scene, acting as opinion leaders, social animators, and conscience-keepers of large and often elite and influential sections of the population. In Gujarat, two of the most prominent of these have been the Swaminarayan Sampradaya and the Swadhyaya Parivar.

The Swadhyaya Parivar is a relatively young formation. *Swadhyaya* in Sanskrit, means the study of self, introspection; *Parivar* is family. The Swadhyaya Parivar is the growing family of individuals devoted to the ideal of a reflective and virtuous life and who deeply respect the dignity of all human beings, be it their own or others. It is essentially a movement catalysed and led by Pandurang Shastri Athavale, a Vedic scholar and teacher, whose appeal derives from his insightful reinterpretation of the *Gita* and the Vedic scriptures that makes contemporary sense to people in their day-to-day lives. While his early teachings were aimed at creating a more complete individual being, at peace with herself and her surroundings, as the movement grew, the Swadhyaya discourse adopted a broader focus on ways to create a wholesome community. A concept that held wide scale appeal amongst the Swadhyayees was of instrumental devotion as a social force. Traditional Hinduism preached passive, ritualistic devotion to God; Swadhyaya charged devotion with an entirely new meaning. Athavale taught that devoting to God—and to His children—a portion of whatever one's effort yields is devotion to God provided there is no arrogance in this giving. These apparently simple ideas caught on like wildfire amongst the Swadhyayee communities, and an important outcome was the creation of numerous 'common properties' through acts of instrumental devotion by the Swadhyayees.

All these are created spontaneously by Swadhyayee communities with voluntary labour, donated materials and funds. Similarly, these assets are operated also through community effort and management with only occasional acts of encouragement and moral support from the movement itself. What drives and sustains these selfless acts by communities is the deeply-held belief that the Swadhyayees have in their practice of instrumental devotion.

The Swaminarayan is a Sampradaya or religious sect which was founded over 250 years ago by a saint from near Ayodhya who wandered around in Gujarat and finally settled down to establish this new branch of Hinduism. Over the 250 years of its evolution, the Sampradaya has grown tremendously in its following and the influence it has wielded particularly amongst the Patidars, the affluent and enterprising farming community. The Swaminarayan Sampradaya has been a tightly-run organization with enormous resources and patronage of the most powerful and influential in the land. Its present presiding saint (or *Pramukh Swami*) is aware that the Sampradaya can enhance its following and influence by seeking out causes that interest vast multitudes and applying its formidable power and following to these causes to alleviate the suffering of the masses.

Way back in 1978, speaking at the inauguration of a common property forest (*Vriksha Mandir*), Dada had told his followers: 'If you quench the thirst of Mother Earth, she will quench yours.' Swadhyayees found this teaching prophetic, but 10 years later. The three successive drought years that Gujarat—in particular, Saurashtra and Kutch—faced during 1985-7 brought water issues to their cyclical peak in the public mind. Taking a clue from Israel, Pandurang Athavale began asking his followers why farmers in north Gujarat and Saurashtra could not adapt and improvise the techniques used the world over for harvesting and conserving rain water *in situ*. 'The rain on your field stays in your field; the rain in your village, stays in your village' was the mantra he gave to the people of Saurashtra. Keen to act on their Dada's word, many

Swadhyayee farmers began trying out alternative methods of capturing rain water and using it for recharging wells.

In the 1989 monsoon, there were isolated experiments throughout Saurashtra. But in some Swadhyayee villages, the entire community tried out such recharge experiments on all or a majority of the fields. Here they found the results stupendously beneficial. The Swadhyaya Parivar is a close-knit community. The more devoted and committed among the Swadhyayees spend a good deal of time and effort in communicating new ideas across the Swadhyayee communities. Audio and video recordings of Athavale's talks are widely in demand and are played before voluntary gatherings of Swadhyayees; these offer a powerful medium for spreading new ideas and messages.

The most convenient forum is the time when Pandurang Athavale presides over a public meeting. 3-4 lakh Swadhyayees from far and wide throng such meetings. Athavale does some of the planting of new ideas in the course of his talk. The rest is done by other Swadhyayees during person-to-person exchanges or in small group meetings. Junagadh's Swadhyayees organized a technical training workshop on well-recharge methodologies for some 525 volunteers who came forward to further extend the techniques. A loose organization was formed around the well-recharge work; a group of 11 volunteers from each village, of 15 at the *taluka* level, and of 25 at the district level were constituted to undertake the necessary coordination and mobilization (Bhatt, 1994).

Numerous villages came to know about the recharge techniques through the *bhakti feries* (devotional goodwill marches). The beneficial results of the early well-recharge experiments by Swadhyayee communities began getting communicated and shared widely during 1990. In 1991, the well-recharge experiments began multiplying in scale. The good monsoon in 1991 helped these experiments to succeed.

It was in the 1992 monsoon that these recharge experiments began taking the shape of a movement. Farmers of all kinds—Swadhyayees and others—began collecting as much rainfall as they could on their fields and in the village and channellize it to a recharge source. This was exactly the opposite of what they had done for ages so far: during the monsoon, the standard operating procedure was to divert rain channels to a neighbour's field or a common land or a pathway. But not now. Now everyone wanted to link all the natural water-carrying channels in private, public, or no-man's land to his well or farm pond for recharge.

Stories began doing the rounds within and outside the Swadhyaya Parivar about groups of Swadhyayees building check dams or deepening tanks or building anicuts or working together to recharge all the private wells of the village. By now, many small and big NGOs had joined the movement, each trying to help in its own way. The Saurashtra Lok Manch compiled information about the technologies used by the different groups of farmers for well recharge, printed it along with illustrative pictures, and made these leaflets available in every nook and corner of Saurashtra. The well-recharge movement had caught on like wild fire. Now it was not just the Swadhyayees but farmers of all persuasions also joined in.

There are no formal studies of the actual scale of the well-recharge work. However, many different sources suggest that between 1992-6, between 92 and 98 thousand wells were recharged in Saurashtra, and some 300 *nirmalneer* (farm ponds for recharge) were constructed. The Swadhyaya Parivar workers were so enthused that they set themselves a target of over 125,000 wells and over 1,000 farm ponds during 1997. It is widely believed that if 5 lakh wells in Saurashtra are recharged, we can solve the irrigation as well as the drinking water problem of the region (*Krishi Jeevan*, 1996: 2). By 1993-4, the well-recharge movement had begun spreading to other parts of Gujarat, notably Sabarkantha, Mehsana, and Panch Mahal districts.

In the meanwhile, a large volume of anecdotal evidence has begun doing the rounds informally and through the mass media to highlight the exceedingly beneficial economic impact of the well-recharge programmes. Then there are assertions based on back-of-the-envelope calculations. For example, Vayak and Khanpara (1994) estimated that the 10,000 wells recharged in Kutch and Saurashtra during 1993-4 cost the farmers Rs. 50 lakh but raised net output by Rs. 80 crore. In a rare study based on interviews with 60 farmers of 6 villages from Junagadh, Mendarada, and Visavadar *talukas*, Munshi, Savalia and Kanani (1996) found significant differences in farming before and after the well-recharge work as outlined in Tables 4 and 5. Table 4 shows that access to stable, reliable groundwater irrigation increased significantly after the well-recharge movement. Similarly, Table 5 shows that the risk

TABLE 4: IRRIGATION IMPACT OF THE WELL-RECHARGE MOVEMENT

Practice	Number of farmers before well recharge	Number of farmers after well recharge
Transplant groundnut	8	17
Protective irrigation to groundnut crop	15	31
Grow a *rabi* crop	37	52
Raise a summer crop	15	31

TABLE 5: IMPACT OF WELL RECHARGE ON FARMING CHOICES

	Number of farmers before well recharge	Number of farmers after well recharge
Short-duration *kharif* crops	34	17
Long-duration *kharif* crops	26	43
Short-duration *rabi* crops	17	9
Long-duration *rabi* crops	20	53
Short-duration summer crops	9	19
Long-duration summer crops	6	12

of moisture stress—which forces farmers to choose short-duration and low-yielding crops—was considerably eased after the well-recharge work.

Improvisation and Propagation of Recharge Technologies

Two aspects about the well-recharge movement are significant. First, the dynamic of the movement, especially with respect to appropriate technological innovation in water harvesting, conservation, and recharge; and second, why did it succeed in attracting as broad people's participation as it seems to have. According to some observers, since 1992 several dozens of new methods have been designed for capturing rain water, conserving it, and using it for recharge. In terms of complexity, these are no big deal. Most of these are improvisations of old methods, but they have been devised by farmers experimenting, learning, improving, perfecting, and then propagating. As we outlined earlier, Swadhyaya Parivar has an ingenious communication machine that propagates information about new techniques widely and rapidly.

The basic technique of well recharge is simple and involves drawing channels to direct all the rain water in a sump (typically 4'×3'×3') made besides the well; a channel is made from the sump to the well 6" above the bottom of the sump so that dirt and soil in the water settles at the bottom and the water that flows into the well is free from them. A 3" diameter pipe is inserted into the wall of the well to direct the water into the well; and the outer end of the pipe is covered with a sieve; the inner portion of the pipe is made to extend to about 1' so that the flowing water does not damage the inner wall of the well. Where the farmer has access to a rivulet/stream by the side of the field, the basic technique was changed to provide a pipe or a channel that enabled water to flow from the river straight into the well. Another improvisation was of *khet talavadi* (farm ponds); this is a 20'×30'×3' tank to which all the channels capturing water from the tank's catchment area are linked; water is channelled from the tank into the well either by a PVC pipe of 9-12" diameter or by an open *pucca* channel; both are sieved at their tank-end.

These methods need some modifications for recharging through bore wells. The standard method is to build an underground *pucca* tank (of 6'×5'×5') in which all the rain water gets directed; to filter the water, the bed of the tank is lined with layers of stones and sand; a 4-6" diameter pipe fitted at the base of the tank (complete with a sieve) links the tank with the casing pipe of the bore well through which water stored in the tank flows underground. The method recommended for using a nearby river to recharge a bore well is somewhat difficult. That involves using a siphon; the bore is linked with the river through a pipe whose river-end is fitted with a sieve and the other end is linked with the casing pipe of the bore well. The delivery pipe of the bore well is closed to commence recharge. After this, the bore is pumped so that water from the bore well flows into the river. When the pumping is stopped, a vacuum is created inside the connecting pipe which now serves as a siphon as water begins flowing from the river into the bore well.

As the well-recharge movement spread, farmers began to carefully study the uses of pumped water. Many of them observed that around 20 per cent of the pumping was used not for irrigation but for bathing, washing clothes and utensils, and drinking by humans and animals. Moreover, a lot of water used for irrigation also tends to accumulate in low-lying areas and gets lost by evaporation. So the farmers began constructing soak pits of various types to recharge the waste water into the ground. This also resolved the problem of waterlogging in low-lying areas and made that land available for more productive use.

Soak pits are now also put to innovative uses. Saurashtra has hundreds of rivers and rivulets; modified soak pits are gradually becoming popular to use the rain water captured by these rivers to recharge groundwater. Soak pits of 8-10' diameter are made at intervals of 500 meters in a river bed. Inside the pits, 4-6" diameter bores are made. Of a 5' strainer pipe, 1 foot is lowered inside the bore; and the remaining 4 feet above the pit-base is covered with a sieve. The casing pipe is then filled with bigger stones up to 3-4 feet, then small pebbles for 2 feet and further with sand for another 2 feet. This contraption which will constantly recharge the aquifer with river water. Also used for recharge by river flows by social workers like Chunighai are techniques like *gupat bandh* built across the rivers in the river beds. After the monsoon flow is emptied, the sand is dug out of the river bed where a *gupat bandh* is to be sited to create a large cavity in the bed. Jute bags filled with clay are used to build embankments which are covered first by a polythene sheet and then by a thick layer of clay. The cavity is then filled with the sand which was dug out earlier. Because the flow of water is impeded by the *gupat bandh*, it augments groundwater recharge in the area.

A new contraption now being used for groundwater recharge in north Gujarat is the *holiya*. A depression of 5'×5'×5' or 6'×6'×5' is made in the lowest area of a field; and a bore is drilled to a depth of 80' to 100'. A strainer pipe is inserted inside a casing to the depth where underground 'river' is found; this is typically a layer of porous soil which at some point in the past was a good aquifer. During the monsoons, rain water from the catchment collects in the low-lying area and recharges the upper aquifer through the bore well. Some 35 *holiyas* were made by the Mehsana Agro-Foundation in Mehsana district during 1996-7. However, these cost more than the simple recharge techniques developed by farmers in Saurashtra. Therefore, there is a subsidy available from the DRDA for making *holiyas*.

The well-recharge movement has also produced an urban impact. In towns as well as cities, owners of *pucca* houses have begun experimenting with all manner of roof-water harvesting. The most common method is of linking roofs with an underground water tank which in turn is linked with a bore well made for a hand pump. In some houses, traditional *tanks* which were the standard fixtures on all old *pucca* houses in Saurashtra, Kutch, and Rajasthan have made a come back. Many builders devoted to Swadhyaya doctrines design roof-top water-harvesting systems into the new residential complexes they build. In 1996 itself, some 1,500 houses built a *danki* recharge system of roof water harvesting alluded to above.

In many parts of Saurashtra and north Gujarat, experiments were also undertaken to recharge groundwater through tanks. Many tanks commonly get filled up several times during a monsoon; their capacity to capture and store water from second or third showers becomes impaired because they get filled up during the first shower or two and stay that way throughout the monsoon. A popular recharge method recommended through tanks involves sinking a 100-50' deep borewell (of 4-6" diameter) in the centre of the tank bed. A vertical pipe is fitted into the bore at the top so that its upper end stays above the highest water level in the tank. There is a cement-concrete base provided around the upper pipe to keep it secure in the bore. At the top of the pipe two inlets are provided through a 3' pipe fitted horizontally inside the vertical pipe. This contraption is designed to capture the spillover water after the tank gets filled and to use it to recharge the groundwater.

More recent experiments in north Gujarat build very similar structures but with the idea of maximizing the tank recharge. In these, the inlets are provided close to the base of the tank bed so that water constantly flows into the bore well as long as there is water in the tank. The advantage with these structures is that they allow the tank to fill up several times in a monsoon. In north Gujarat, this can be a great strategy to fight fluoride contamination of groundwater. Fluoride is normally found in groundwater pumped from great depths, as in districts like Mehsana.

In such areas, it has been suggested that if tank recharge methods are used to capture rain water in the upper aquifers, and if this water tapped through community wells on tank-shores is reserved for drinking water uses of the community, the fluoride problem can be solved to a considerable extent. The residents of several villages in Mehsana have been intuitively following a variant of this practice. Unjha town, for example, has a large tank with several recharge bores inside; three community wells on the tank-shore are used to lift water exclusively for domestic use. Other smaller villages have also used such arrangements for some time.

Overall, then, the well-recharge movement has brought in its wake a veritable revolution in experimentation and improvisation in recharge techniques. Starting with wells, the movement began encompassing other recharge sources such as roof-tops, waterlogged lands, soak pits, rivers, and tanks. Also, starting with the Swadhyayees, and later with the Swaminarayan Sampradaya and other religious sects played a crucial role in capturing this continuous learning in print and propagating it across the countryside. What makes this a movement is that none of the participating organizations play a dominant role in supporting or spreading the movement. In most senses, the movement is thus self-orchestrating, self-coordinating and self-propagating. Somewhere along the way, a seed was sown in fertile ground, and it has begun to pay back ten-fold, hundred-fold, thousand-fold.

Understanding the Engine of Saurashtra's Well-Recharge Movement

Both for understanding Gujarat's water future as also as a social phenomenon, it is critical to explore why the well-recharge movement catalysed by the Swadhyaya Parivar grew into a movement. After all, the recharge movement was only one of the several—and later—of Athavale's experiments in redesigning the community.

The concepts of *Matsya Gandha* (common property fishing boat), of *Vriksha Mandir* (common property forest), of *Yogeshwar Krishi* (common farm) are also exciting. These too have fired the imagination of the people. These too have multiplied. But none of these has spread like wildfire quite like the well-recharge experiment has.

Several reasons can be advanced for this development. But the correct response is probably a combination of several of these. First, the strong allegiance of the core Swadhyayees to Athavale, and their readiness to give a serious try to his ideas catalysed the first generation of well-recharge experiments in Saurashtra. Shorn of all the frills, there is nothing new about recharge. Many notables have in the past exhorted farmers to conserve rain water and recharge the aquifers. What made the difference this time was that it was Athavale who exhorted them; and there were individuals and communities willing to do whatever he asked them to do.

Second, Athavale marketed the message of well recharge in the package of instrumental devotion; at no stage in the early years did the Swadhyayees ask farmers to recharge their wells because it was economically profitable; they untiringly cited Athavale's teachings that, if you quench Mother Earth's thirst, she will quench yours'. This helped to underplay the economics of well recharge in the making up of the individual minds about recharge; the early pioneers undertook recharge experiments as an act of devotion to God and to follow the path shown to them by Dada.

Third, the fact that Athavale's ideas about well recharge had to do with one of the most pressing, urgent, and critical problems facing the people of Saurashtra explains why the movement took off in Saurashtra rather than in districts like Kheda or Baroda which are also Swadhyaya strongholds.

Fourth, and most critically, the spread of the Swadhyaya movement is in the form of communities. In numerous cases, there are entire villages that have turned to Swadhyaya. Even otherwise, in the countryside, it is more common to find group allegiance to the Swadhyaya movement rather than allegiance by scattered individuals. This meant that in the early recharge experiments, either the entire village or a substantial proportion of the farmers agreed to participate, which helped the community to internalize the positive externality produced by each recharged well. If, instead, only isolated farmers had recharged their wells individually, it is doubtful if the early results would have been as strikingly beneficial as they were found to be. That the internalization of this positive externality of well recharge has produced powerful, snowballing effect on people's participation is evident from the experience of many villages.

Fifth, post-1994, however, the large-scale adoption of well recharge through promotional and extension effort of NGOs and other religious movements was facilitated greatly by widely shared reports about exceedingly beneficial productivities and income effects of well-recharge programmes on farming. It was at this stage that the driving force of the movement began to change gradually; well recharge as an act of instrumental devotion began to get replaced by well recharge as a technically rational economic act as the movement began spilling out of the Swadhyaya movement and the Swaminarayan Sampradaya.

Probably, even amongst the followers of these, there was an added economic impetus to do the devotional act. The government's intervention was awkward: first promising, but then withholding for a long period a subsidy on recharge of wells by private farmers. Enticed by the promise, many farmers who were otherwise game to recharge their wells on their own postponed their decision to benefit from the subsidy. Vigorous efforts by Swadhyayees, the Swaminarayan Sampradaya and well-meaning NGOs to divorce the well-recharge movement from the subsidy programmes—almost by discrediting the government initiative—finally prevented what might have spelled the doom of the movement.

Sixth, and finally, post-1995, the scale of participation and the resulting momentum that the movement has achieved spontaneously has been a powerful engine for the growth of the movement. In terms of the theory of externality, the reluctance of an individual farmer to invest in well recharge is explained by his inability to internalize the positive externality produced by his investment. However, if a substantial proportion of farmers take to well recharge, it progressively makes more and more sense for the farmer on the margin to recharge his own well (Runge, 1985: 31-55).

CONCLUSIONS

Normal people, driven by their narrow technical rationality, have their own responses to the growing scarcity of groundwater. Compared to many other regions, the dominant Gujarati farming community of the Patidars—like the Marathas in Maharashtra—are dynamic as entrepreneurs, and have been way ahead of the rest of the country in terms of tapping opportunities for economic advancement. Since groundwater is amenable to private development, the Gujarati farmer took an early initiative to promote and propagate technological and institutional innovations that broadened the scope of groundwater irrigation. Thus, while the government of Uttar Pradesh initiated its public tube well programme to popularize modern pumping technologies amongst farmers in 1930s, the farmers of central Gujarat had already been using diesel pumps for well over three decades. While it was only with the onset of the 1990s that the World Bank discovered the value of water markets in bringing greater efficiency in production, allocation and use of water, many parts of Gujarat had thriving groundwater markets as far back as in the 1930. Similarly, for over a decade, governments and NGOs have worked over time but with uncertain results to create water users' associations (WUAs) to manage irrigation assets better than agency bureaucracies.

In large areas of north Gujarat, small WUAs—called farmers' irrigation companies—have existed and operated in the thousands without any assistance from the government or NGOs. Finally, created entirely by the Gujarati farmers are some of the most efficient water conveyance and distribution systems based on underground pipeline networks. All in all, then, Gujarat has been the fertile ground for irrigation innovation in technology as well as institutions.

To a great extent, these institutional and technological innovations—which are otherwise beneficial in promoting efficiency and equity in irrigation—have also contributed to the problem of groundwater depletion in many parts of Gujarat. In coastal Saurashtra's Junagadh and Amreli districts, the problems of salinity ingress would have been far less acute but for the enthusiasm with which farmers took to diesel and electric pumps, and pervasive groundwater markets had not made irrigation available to small and large farmers alike.

Similarly, in the north Gujarat districts of Sabarkantha, Mehsana, and Banaskantha, groundwater irrigation would have declined long ago had the farmers not devised irrigation companies that enabled them to share the extraordinarily high capital costs and risks involved in sinking tube-wells in that region. Indeed, the problems of water salinity and fluoride contamination—that has hit the lives of large numbers of people in north Gujarat and Saurashtra—would have been far less acute but for the rapid growth of irrigation through these institutional innovations.

Groundwater institutions have responded to the growing water scarcity but far too slowly and sluggishly to produce the desired effect. A lot—between half and two-third—of groundwater irrigation in Gujarat occurs through farmers buying pump irrigation service. As a result, total groundwater withdrawal is determined by terms and conditions prevailing in these pump irrigation markets. Studies from different parts of the state show that pump irrigation rates have risen faster where groundwater depletion has been rapid. Field studies show that water prices rose nearly 1.5 times faster in north Gujarat villages than in central Gujarat. To what extent have these rising prices been able to restrict groundwater irrigation is a matter of conjecture. In all likelihood, total groundwater withdrawal has been checked more effectively by the declining supply of power for pumping than by the rising prices of groundwater.

In conclusion, then, there are diverse responses to groundwater depletion in the north Gujarat and Saurashtra regions where the water crisis is most critical. But most of these are peripheral or ineffectual. There are Big Thinkers' whose grandiose ideas of a Gujarati Garland Canal are fascinating. They seem feasible in the distant future when the state will achieve a higher economic and system management capability, but it is of no relevance now. Work by Gujarat's vibrant NGO movement is imaginative and inspired. However, it is essentially experimental, as it has to be. NGO work is expected to generate alternative solutions; but it is futile and unfair to expect NGOs to solve the problems at the level of the state. NGO work, moreover, is as local and situation-specific as the Big Thinkers' work is global and grandiose. Most disappointing has been the response of the state: its different organs have worked at cross-purposes, often nullifying the impact of each other's work. In recent months, Gujarat's courts have shown exemplary astuteness on issues of water pollution; but the executive has been hell-bent on undoing what the courts do. The only important way in which the government has helped restrict groundwater withdrawal is by involuntarily failing to keep up the supply of power to farmers, much as it would have liked to.

The most appropriate response to groundwater depletion has been provided by the people of Kutch and Saurashtra under the inspiration of social formations such as the Swadhyaya Parivar and the Swaminarayan Sampradaya. The rate at which this movement has grown since 1992 in terms of people's participation, in terms of technological innovation, and in terms of regional spread suggests that the recharge movement might well equal the enormity of Gujarat's groundwater challenge. So far the recharge movement appears completely autonomous of government programmes and policies. It has emerged on its own; it has found its own sources of energy and dynamism; it has devised its own expansion plans. It is difficult to figure out what the public agencies and policies can do to support and nudge the movement along. At the least, what it can do is not compromise it by the kind of subsidy programme it had announced; at the best, the government can make different policies, programmes, and organizations to work for the shared purpose of promoting environmentally sustainable management of ground and surface water resources.

NOTES

1. Mehsana's limited groundwater was located much deeper underground than Junagadh's. As a result, Mehsana has to use eWEMs and is thus somewhat more amenable to public regulation through the electricity supply route. Junagadh is not, despite being declared mostly 'dark' or 'gray' and denied electricity connections on that ground. Groundwater pumping in Junagadh carries on unabated, albeit with dWEMs.
2. In fact, the state's Groundwater Development Corporation that manages its 3,500 STWs has been losing so much money that during the last decade, the central priority of the government has been how to wind up the corporation. This proposed step is justified on all imaginable grounds. However, the corporation's 6,500 employees would not let this happen.

REFERENCES

Bhatt, Devendra B., 'Prabhu Prem Ni Pehchan', in *Krishi Jeevan*, vol. 26, no. 10, May 1994.

Dubash, Navroz K., 'Pumping for Power and Profit: A Study of Groundwater Markets in Gujarat, India', A Dissertation Prospectus for the University of California, Berkeley, 1995.

Jal Sanchaya Visheshank, *Krishi Jeevan*, Vadodara: Gujarat State Fertiliser Company, 1996.

Khanpara, G.M., 'Asha Nu Pahelu Kiran', in *Krishi Jeevan*, vol. 26, no. 10, May.

Moench, Marcus, *When Good Water Becomes Scarce*, Ahmedabad: VIKSAT, 1993.

———, *Debating the Options: Groundwater Management in the Face of Scarcity, Gujarat, India*, Ahmedabad: VIKSAT, n.d.

Moradia, R.P. and P.D. Kalaria, 'Khetare Khetare Narmada', in *Krishi Jeevan,* vol. 26, no. 10.

Munshi, M.A., V.J. Savalia and P.R. Kanani, 'Development of Agriculture through Well Recharge (in Gujarati)', in *Krishi Jeevan*, vol. 28, no. 11, June 1996, pp. 24-5.

Nanavati, Reema, 'Water as Regenerative Input: SEWA's Experience', Ahmedabad: Self Employed Women's Association (unpublished note), n.d.

Patel, Arvind S., 'Swadhyaya Experience in Well Recharging', Vadodara: M.S. University (unpublished document), 1995.

Patel, P.P., 'Water Problem of Gujarat: Resource Development and Management', Vadodara: M.S. University, n.d.

Patel, Vithubhai, 'Samtol Vikas Mate Gujarat ma Sarvagrahi Jalsanchalan (Integrated Water Management in Gujarat for Balanced Development)', in Rohit Shukla, et al. (eds.), *Gujarat Ni Jal Samasya*. Ahmedabad: Sardar Patel Institute of Economic and Social Research, 1995.

Phadtare, P.N., *Geo-hydrology of Gujarat State* (*with status of work carried out by the Central Ground Water Board up to August 1988*), Gandhinagar: Government of Gujarat, n.d.

Prabhakar, R., Vina Vani, Ketan Tatu and Sweta Patel, *Case Studies on Fresh Water Management in the Arid Region*, Anand: Institute of Rural Management, 1997.

Runge, C. Ford, 'Common Property and Collective Action in Economic Development', in *Proceedings of the Conference on Common Property Resource Management*, Washington, D.C.: National Academy Press, 1985.

Shukla, Rohit, V.K. Shah, Rohit Desai, Ramesh B. Shah, Manjula Naik, *Gujarat Ni Jal Samasya*, Ahmedabad: Sardar Patel Institute of Economic and Social Research, 1995.

Vaid, Chunibhai, *Hidden Dam: Special Features*, Ahmedabad: Gujarat Lok Samiti, n.d.

Vani, Vina, V. Ballabh, Tushaar Shah and R. Prabhakar, *Typology of Competition for Water in Gujarat*, Anand: Institute of Rural Management, *c.* 1993.

CHAPTER 10

Water Pollution Control: Role of Community and Public Action

HIMANSHU THAKKAR

INTRODUCTION

Rising, and increasingly widespread, incidence and levels of pollution of surface and groundwater sources in different parts of the country is a matter of grave public concern. Pollution of the Ganga, Yamuna, Sabarmati, Lower Bhavani, and Palar have attracted nation-wide attention. But the problem is far more widespread. According to the National Environmental Engineering Research Institute (NEERI), a staggering 70 per cent of available water in India is polluted. 'From the Dal lake in the North to the Periyar and Chaliyar rivers in the South, from the Damodar and Hooghly in the East to the Thana creek in the West, the picture of water pollution is uniformily gloomy. Even our large perennial rivers like the Ganga are today heavily polluted' (CSE, 1982; CSE, 1984-5).

Groundwater pollution, which is more dangerous because it is more difficult to reverse, is also becoming a major problem. Its seriousness is illustrated by the prevalence of arsenic poisoning in Bengal, the growing salinity problem in Gujarat, and the prevalence of high levels of pesticide residues in groundwater in several areas.

1. Scientists from Jadhavpur University estimate that 38 million people in 8 districts of the state of West Bengal are exposed to health risks from the high arsenic content of groundwater, an

*This paper was written in mid-1997 and last updated in late 1997. The main conclusions of the paper remain valid.

important source of drinking water. The increase in arsenic content is attributed to excessive pumping of groundwater for irrigation following the advent of high-yielding paddy varieties.
2. In Gujarat, the area affected by salinization in 19 *taluks* around the Gulf of Khambat is estimated to have increased from about 7-8 per cent in 1960 to 64 per cent in 1993. About a fifth of the state's lands are currently afflicted by this problem and, in the absence of effective corrective action, is projected to reach 30 per cent by 2000.
3. In Delhi, one-third of the groundwater samples examined were found to have nitrate concentrations above the permissible limit; the incidence of TDT in the Yamuna waters at Delhi and that of T-HCH and TDDT in the Cauvery was found to be very high.

These are but illustrative examples. The available data and institutional arrangements for measurement and monitoring are inadequate to get a comprehensive picture of water quality. Nevertheless, the fact that the main sources of pollutants—namely agricultural chemicals, untreated industrial effluents, and urban sewage—are growing rapidly, with little being done to contain the process, strongly points to a progressive aggravation of the problem. The consequences to the health of the people and to the sustainable use of water resources over the long run are very serious indeed.

THE GOVERNMENT'S RESPONSE

The government has declared its good intentions about pollution control through numerous policy statements and formal legislations.

- The National Water Policy (Ministry of Water Resources, 1987) declared: 'Economic development and activities including agricultural, industrial and urban development, should be planned with due regard to the constraints imposed by the configuration of water availability.'
- The National Conservation Strategy and the Policy Statement on Environment and Development was adopted by the Government of India in June 1992 to provide the basis for the integration and internalization of environmental considerations in the various policies and programmes of the government.
- In 1994, a far-reaching notification by the Ministry of Environ-

ment and Forests (MEF) made it obligatory for almost all developmental and industrial concerns to conduct Environment Impact Assessment (EIA) studies.

- In a decision intended to make industry more accountable to the people, the Cabinet Secretariat, in January 1997, made public hearings mandatory before giving environmental clearance to any industry project or activity, whether private or governmental.
- In April 1997, by an extraordinary gazette notification, MEF established the National Environmental Appellate Authority (NEAA), in exercise of its powers under a 1997 Act. The NEAA is to deal with complaints against developmental projects that fall under the EIA of 1994. 'The aim is to enable people directly affected by developmental and industrial projects, as well as NGOs, to present their point of view and ensure transparency in the process of EIA.'

But these steps do not seem to have made much of an impact. Pollution continues to spread and increase in intensity in spite of these pronouncements. There are several reasons for this. For one, when the government acts both as a promoter and a regulator of economic activity, often there is a conflict between these two roles. In practice, the government has acted more as a promoter than as a regulator. It seems in the name of development everything is expendable, including the health and future of the people. Worse, when responses of the government are looked at in the courts, one finds that the government is only trying to defend the interests of the polluters. For effective pollution control, the pollution control bodies will have to be independent of the government.

Pollution control agencies being part of the government do not have the independence and autonomy to initiate action or enforce the provisions of the law. Monitoring mechanisms are weak, and often absent; where they exist, the information they collect is not available to the public; and there is little coordination between the concerned departments. Nor do the elected representatives and legislatures show any interest in pollution-related issues, much less in exerting pressures for remedial action. On the contrary, they act more often in ways which favour, if not actively support, the interests of the polluters.

Since the affected communities have a direct stake in controlling pollution for the health and well-being of their members, it is essential that they have access to information regarding the state of their water sources and the right not only to voice their concerns but also to seek effective redress for damage already done and to prevent potential damage in the future from development projects. The laws do not provide for consultation, not to speak of the involvement, of the affected communities in government processes concerning pollution monitoring and control. Attempts by the affected communities to seek redress through petitions and appeals to the government rarely yield a positive response.

Frustrations with the ways in which the official regulatory framework functions have led communities severely affected by water and other pollution to take recourse to public protests, mass agitations, and the courts. The responsive attitude of the judiciary to public interest litigation has many times facilitated the raising of pollution complaints in the courts. There has been a significant growth of such community-NGO initiatives. The purpose of this chapter is to focus on the experiences in certain specific cases drawn from various states in which affected communities or citizens' groups have played an active role. The strategies they followed and the role and response of other actors (namely, polluters, the government, and the courts) is also examined briefly in some cases.

These are not meant to be exhaustive accounts, but essentially to give a feel of the kind of issues involved, the attitudes of the concerned parties, and the difficulties of securing full and effective redress, to indicate the potentials as well as the limitations of such public activism by communities and NGOs. We argue that the active involvement and participation of local communities is essential to bring the problems onto public platforms, to promote informed discussions of solutions, and to force the concerned public agencies (pollution control authorities, courts, and the government) to take appropriate remedial action.

CASE STUDIES OF PUBLIC ACTION

Gujarat: Campaign to Save the Narmada River

In Gujarat, one of the most industrialized states in the country (also earning the sobriquet 'most polluted state', as noted by the Gujarat Ecology Commission (May 2001)), vying to be the most industrialized, more and more instances of severe water pollution are coming to light, particularly in what is rather proudly called the golden corridor between Valsad and Mehsana. A number of cases were filed in the Gujarat High Court against polluting industries in and around 1995. These were mostly at the instance of concerned citizens; the affected communities and NGOs were not involved. The court did act and passed a number of orders for closing down hundreds of industrial units. These orders made headlines in newspapers in 1995. However, possibly due to the economic clout of the industries and the political support of the government, these units restarted operations without much change within months. The weaknesses of the cases, filed without the necessary homework, were also evident (Chakraborty, 1995).

The courts' intervention resulted in some remedial action in terms of the installation of treatment facilities. But these were limited in both scale and impact. Effective pollution control remains a distant goal. Even today the industries are merrily operating without much change in the situation. However, the cases helped to generate greater awareness among the public about the polluting nature of industries and stimulated greater community and NGO involvement in these issues.

One such initiative was taken by the people of the villages around Mehgam in the Amod *taluka* of Bharuch district in central Gujarat. In 1993 they approached Manviya Technology Forum (MTF), a voluntary organization active in the area, to say that the Gujarat Industrial Development Corporation (GIDC) was constructing a pipeline to release effluents from the proposed Vagra GIDC industrial estate into the Narmada river and was destroying their land and standing crops in the process. Soon an organization called the Narmada Pradushan Nivaran Samity was formed, and letters were written to the relevant officials asking if any environment impact study of the proposed Vagra industrial estate had been done. According to the GIDC officials only an engineering study was done and no environment impact study was done regarding the release of effluents. The Samity also discovered that the proposal did not have the sanction of the Gujarat Pollution Control Board (GPCB).

The people had no option but to stop the work of the pipeline. Soon the fishing community of the villages on the banks of the Narmada, fearing the destruction of the rich estuarine fisheries if the effluents were released into the river, also joined the struggle. Their point was that if the upcoming Indian Petrochemicals

Corporation Limited complex in the same area was required to release its effluents into the sea and not into the river or the estuary, then why was the GIDC or any other agency being allowed to release the effluents into the river? The Bharuch Fisheries Directorate also opposed the release of any effluents into the river.

In October 1995, the collector told the GIDC to give up its plan to release effluents into the river or the estuary and hand over the study to the NEERI to find out where the effluents could be released into the sea. The GIDC has yet to take a decision on this issue. But in the meanwhile, both the work of the pipeline and the setting up of the industrial estate remained stalled. For the people this was a victory. This was to be the first step in the struggle of the Samity to control the pollution of the Narmada river.

More Struggles Against the Pollution of the Narmada

As the news of this breakthrough spread, the people of Piramal village in the Bharuch *taluka* approached the Samity, saying that they were facing a similar problem. The GIDC was constructing a channel from its upcoming Jhagadia industrial estate to release effluents into the Narmada at Haripura village, which is just upstream of the point where Amlakhadi had been releasing untreated pollutants from the Ankleshwar and Panoli GIDC estates for over two decades now. People on the banks of this 28-km.-long Amlakhadi were living with this poisonous channel which had killed the fish in the rivulet, killed or maimed the domestic animals who drank water from it, flooded their lands, and destroyed their groundwater. For them, the proposed new channel was like rubbing salt in their wounds.

When the activists mobilized the people and stopped work on the channel in an attempt to pressurize the authorities, the response from the various quarters was rather quick. The GIDC, the industrialists, and the administration immediately called a meeting. Fresh from their victory at Vagra, the people put up the same grounds for not allowing the release of any effluents into the river. It was soon discovered that no EIA of this new industrial estate had been done. Strangely, the GPCB had given its consent for releasing the Jhagadia effluents into the river, contrary to the statement of the GPCB member secretary that 'We are not giving consent to anyone to release effluents into a freshwater body, including the river.' The district collector, who called the meeting to resolve the issue, could not find fault with the agitating people.

But in this case the GIDC was desperate. Four industries at the Jhagadia estate were ready to start production, and if more industries were to be attracted, the problem could not be allowed to linger. In a meeting called by the collector on 18 January 1996, the GIDC agreed that a study to find out where in the sea and through what kind of channel the effluents of the Jhagadia, Ankleshwar, and Panoli estates should be released would be given to the NEERI; that till the effluent channel was ready, the existing four industries would not be allowed to start operations at the Jhagadia industrial estate; and that the last 8 km. of Amlakhadi would be cleaned up before the 1996 monsoon at the estimated expense of Rs. 65 lakh to be shared by the industries, the GIDC, and the Members of Parliament from the area.

This success was a major boost for the people. The channel was desilted before the monsoon. Though only a temporary measure to reduce the flooding of the area, for the people of the village on the banks of the channel, it was a great boon indeed. However, the other clauses of the agreement of January 1996 are proving to be much more difficult to implement.

First, even as the rapid survey report by NEERI was awaited, officials are raising the finances for the effluent channel to the sea, estimated to cost over Rs. 350 crore. The industries which have for decades released untreated effluents into the river for free would not be willing to spend that kind of money. When last heard, the World Bank was considering funding this project.

Second, the GIDC is already wavering about not allowing any more than the existing four industries at its Jhagadia estate to start operations. In an interview with me, the deputy chief engineer of GIDC asked: 'What is the logic in such an understanding? We are not allowing any industry to come up in Jhagadia without effluent treatment. And you know that the effluents of both the Ankleshwar and Panoli estates have been flowing into the Narmada without any treatment for decades. Which is more polluting? Anyway, that understanding was reached only by our local officer.' Implying thereby that they were unlikely to honour the agreement reached with the protesting people in January 1996.

While this struggle was going on, Rajshree Polyfils Ltd., a plant of the Birlas, was constructing an effluent channel through Umarva-Sir-Bamalla and other villages to release effluents into the Narmada at village Navra. Strangely, the GPCB had given its consent for this release at Navra village just upstream of the point from where the Gujarat Water Supply and Sewage Board (GWSSB), GIDC, IPCL, Bharuch city, and a number of other units were lifting water from the river for domestic and industrial uses. When I asked GWSSB and GIDC officials if they knew of such a move, they were shocked to 'know' that this was indeed happening.

People, of course, stopped the construction of the effluent pipeline in March 1996. The GPCB said in June 1996 that they were reviewing the consent for the release of the effluents of the Birla unit. This review led to the cancellation of the consent. When it was pointed out that in a similar way the industries in the Valia *taluka* of Bharuch district were operating in a landlocked area where the effluents from the industries were entering the Amaravati river—a tributary of the Narmada—from where GWSSB was lifting water for villages of Valia *taluka* and Ankleshwar town, the GPCB member secretary said that the consent of these Valia industries too was being reviewed. While this was another round of victories for the people, it also showed the utter lack of consistent policy for pollution control in the Gujarat administration.

Bichri Village in Rajasthan

Bichri in Rajasthan is another case where in spite of several scientific and official reports showing that an industrial establishment, Silver Chemicals, was responsible for making the groundwater—the only source of water for the villages around—toxic and non-potable, no action was taken. The case was taken to the Supreme Court. Strong protests by the local people backed by a number of NGOs led to the closure of the factory in 1988, but the guilty could not be punished. Not surprisingly, the victims were never participants in the decision-making process after 1988 when the factory stopped working.

The Supreme Court ruled in favour of the petitioners and upheld the principle of absolute liability of the polluter, that is, when an enterprise is engaged in a hazardous or inherently dangerous industry posing a potential threat to the health and safety of persons either working in or residing in the surrounding areas of the factory, it owes an absolute and non-delegatable duty to the community to ensure that no harm results to anyone as a result of this activity. If any harm does occur, the enterprise must be held liable to compensate for it.

This was indeed a landmark judgement. However, some people argue that exemplary damages should have been awarded to the people of Bichri, and that the Rajasthan government and the MEF should have been held guilty for allowing the polluting industries to come up in the first place. Till date, not a paisa has been paid as compensation. Mannarma, a Bichri farmer, pointed out that the nation's three main acts concerning the environment are defective as they put the onus on the ill-functioning PCBs, which lack adequate and popular representation (Shankar, 1993; Sharma, 1996).

Some seven years after the closure of the main errant units, in February 1996, the Supreme Court ordered that five units of a particular company producing toxic chemicals in Bichri village in Rajasthan's Udaipur district be closed down. In the meantime, the same company has reportedly shifted base, and is merrily doing an encore in Vapi, Gujarat, arrogantly confident that no one can touch it. Events over time have vindicated this confidence.

Madhya Pradesh: Pollution of the Betwa River

Over the years, the Betwa river, lifeline of the obscure and sleepy town of Vidisha to the east of Bhopal, had degenerated into a sewer because of the effluents from a distillery. A movement to restore this lifeline took shape gradually, cutting across social strata and political ideologies, and bloomed into an un-ignorable protest against the pollution of the river. The incidence of dead fish in the Betwa in 1993 did lead to some turmoil, and the distillery was then ordered to shut down, but the distillers obtained a stay order from the Gwalior High Court. Subsequently, the government was not able to vacate the stay.

The local administration, however, played a crucial role of keeping open the information channel to its citizens. All political parties decided to set aside their differences to work together on this issue. The entire business community also threw in their lot. The deputy chief minister of the state agreed that the studies carried out by the Madhya Pradesh PCB had proved that Som distillery,

Sehatganj in Raisen district, was responsible for the extreme pollution loads in the river. And yet the system failed to act. In fact, he was quoted in the press as saying that the state government and the official machinery had sold out to the liquor mafia, referring obviously to the infamous Som distillery. That led to an uproar in the assembly, forcing the chief minister to call a meeting and directing the MPPCB to order the closure of Som distillery. But even then the chief minister said that his government was not responsible for the mess and that only the MPPCB was responsible.

To gather mass support for the movement, for three days in November 1996 the people of the town were contacted on the street corners and requested to sign letters of pledge. In the process, they were individually educated on the issue, and their future support pledged. Over twenty thousand letters were signed. Paryavaran Vahini, the citizens' forum for action on the local environment led the movement. Madanlal Sharma, affectionately called *Betwa ke baba*, has been trying to educate the local people about the necessity of keeping the river clean for the last 22 years.

The mass death of fish in the upstream *anicut* in July 1996 after such incidents had earlier led to the final ignition of the movement against pollution. Ultimately, in December 1996, a distillery owned by a powerful industrial house, identified as the major source of the pollution, was closed down. The quiet exit of an environment minister who had staunchly defended the polluters and, on the face of it, a 'concerned' government valiantly rushing to the rescue of the harassed people *after* all the fish had died were some other notable gains of this struggle (Banerji, 1997).

Goa: The Nylon Plant Struggle

The struggle against the setting up of a nylon plant in Goa is a good example of how a movement of people can prevent a potentially polluting industry from being established. The government formed by the Progressive Democratic Front was forced to appoint a Legislative Assembly House Committee, which took the unusual and unprecedented step of going in for a public hearing to decide on the matter. The Committee recommended that the project should be shelved on account of the likely threat of pollution: '. . . it would be an ill-advised move to allow large chemical industries discharging even their treated effluents into our eco-rich and virgin rivers. . . . We have to safeguard our rivers against any conceivable environmental and pollution threat. . . .'

The report also indicted the government and government agencies for bypassing the well-laid down procedures for grant of licenses, selection and grant of sites, and other related matters. Among the lapses noticed was the failure to get an EIA for the project, which is mandatory before any clearance is given. The Goa State Pollution Control Board was pulled up by the Committee for its failure to uphold the mandate given to it under the provisions of the Water (Prevention and Control of Pollution) Act, 1974 and the Air Act of 1981 as the documents pertaining to the no objection certificate were found to be very superficial. Despite such a scathing report, the project went ahead under a new government that followed the PDF-ruled government.

The decade-long people's agitation in Goa led by the Anti-Nylon Citizens' Action Committee also launched an awareness campaign about the harmful effects of the project, about the outdated technologies being used, and the track record of the company. The issue made such an impact that it figured in the panchayat elections held in 1991 and those who opposed the project were elected. The local panchayat refused to give permission for construction activities at the project site but the government authorities invoked some deemed provisions in the law to give the go-ahead to the project's promoters (Mohan, 1995). It was only after a prolonged agitation by the people in 1994-5, resulting in police repression including arrests, firings, and deaths, that the project was finally shelved and was forced to go to another state, Tamil Nadu.

The Zuari Agro Chemicals Factory

In another case, Zuari Agro Chemicals, a Birla concern supported by US Steel, the International Finance Corporation, and the Bank of America, started operations in 1973. Within three months fish began to die in the water bodies around the factory; within a year cattle were also dying in the area; and in another year, coconut trees began to wither away. The factory was shut down by an order under Section 133 Cr.P.C. in 1974, six months after it had formally opened. This was the beginning of a legal battle whose eventual outcome shows how even dogged perseverance may not be rewarded by adequate judicial relief.

By 1975 protests began. A Citizens Anti-Pollution Committee was formed. The Goa Government gave an ultimatum to the factory in April 1975, but without effect. The district magistrate ordered the immediate closure of the factory in the same month. At different stages of protests, different political parties supported the campaign of the villagers. Ultimately, the company had to pay compensation, provide drinking water to the affected villages, and install pollution control equipment under the control of the then Central Pollution Board.

In 1983 a case was filed for a permanent injunction against the ZAC in the district and sessions court of South Goa. This was rejected on technical grounds. After the enactment of the Environment Protection Act (EPA) in 1986, it was decided to file a case under the EPA. However, in order to prosecute under the EPA, it is necessary that air and water samples be taken and analysed by statutory agencies. Even after a notice was given of the intention to prosecute the ZAC under Section 19(b) of the EPA, neither of the authorized agencies, i.e. the state PCB or the Inspectorate of Factories, would take the required samples. Without the analysis of the samples by one of these agencies, a criminal suit cannot be entertained under the EPA or under the Air and Water Acts.

A petition was filed by the Goa Foundation and citizens praying that it was incumbent upon the two authorities appointed under Section 11(1) of the EPA to take samples of air and water within a specified period on the requests from citizens and also to disclose to the parties the results of the analysis and the inquiries carried out as a result of the complaints filed under the EPA. The Inspectorate of Factories refused to take the samples since the alleged pollution was occurring outside the factory premises and the Inspectorate of Factories had no jurisdiction outside the factory. The state PCB stated that it was not bound to take any samples under any provision of the statute! The petition drags on. The villagers of Sancoale, where the offending unit is located, have had an ongoing battle with the ZAC for over two decades now (Alvares, 1994; CSE, 1982).

Maharashtra: Sterlite Industries

The story of why and how Sterlite Industries decided to cut its losses and leave Ratnagiri in Maharashtra is one of the more significant events that shows what a well-informed movement against a potentially polluting industry can achieve. The people of Ratnagiri, a coastal district in southern Maharashtra, successfully stopped a copper smelter unit from being put up in their area in 1994 as they feared that the unit would lead to the pollution of their land, air, and water. The 60,000-tonnes-per-annum capacity copper smelter plant costing Rs. 700 crore was to be put up by Sterlite Industries. A report by an expert committee instituted by the state government vindicated the opposition of the people, saying that such an industry would endanger the region's fragile ecosystem. Sterlite nevertheless went ahead.

Unfortunately for Sterlite, the *sarpanch* of the Zadgaon gram panchayat, within which the factory premises fell, happened to be a former civil servant. He understood the law and knew that the company should have sought the permission of the gram panchayat before beginning construction. When he sent a *gram sevak* to serve notice on the factory, the man was thrown out by the general manager.

This led the gram panchayat to send a demolition notice. In May 1993, the *sarpanch* called a press conference to declare that the gram panchayat planned to rent bulldozers and demolish the plant. The response of the state government, quite significantly, was to promulgate a new ordinance saying that the approval of the Maharashtra Industrial Development Corporation was mandatory for such construction (Sharma, 1995; Shrivastava, 1994).

It took years of strident and sustained protest by the local people to stop this unit. The struggle included direct action by some 35,000 people who marched to the factory site and demolished the sheds of the factory. A local newspaper, *Ratnagiri Times*, played a very active role in the campaign. It set about finding out from independent environmental experts the consequences of setting up such a large copper smelter plant within a short distance of a heavily populated urban area. Once it was known from environment experts that the plant could not be located so close to the city, a variety of lay people and professionals came together and formed the Ratnagiri Bachav Sangharsh Samiti. A social boycott was declared on all the employees of the company. The employees were asked to vacate the houses by their landlords, and even autorickshaw drivers would not transport people employed by Sterlite.

As a result, Sterlite was forced to abandon its Zadgaon site. It is now putting up the plant in Tamil Nadu, where too it is facing some agitation, though not as effective as the campaign in Ratnagiri. Two copper smelter plants are also being set up in Bharuch, Gujarat despite some local resistance.

Delhi: Pollution of the Yamuna

In early 1997, after some months of study, the Centre for Science and Environment, an NGO, came out with a report on pesticide pollution of the Yamuna river. The report was deliberately released at a public meeting by a former prime minister. The media gave the event prominent coverage as the report focused on the presence of pesticides in the drinking water supplied to Delhi.

A debate was raised in the Delhi legislative assembly following the release of the report. The Delhi Water Supply and Sewarage Undertaking (DWSSU) came out with blanket statements asserting that these were myth-spreading reports and that there were no pesticides even in the raw water. With time these statements were changed to the effect that the pesticide levels were within the prescribed limits, and then further that efforts were being made to meet the standards. A citizen picked up the report and filed a public interest petition in the Delhi High Court, alleging that the right to life of the citizens had been violated. The CPCB agreed with the CSE's report in the court. The Haryana PCB, however, which was made party in the case, as Haryana was the pollutant state in this case, blamed Delhi for the pollution of the Yamuna river upstream of Delhi! Dozens of reports have been filed in the newspapers. The debate and the court case go on.

Some significant gains have been achieved. Some awareness has been generated about the importance of the quality of drinking water. The DWSSU has agreed that they had not been monitoring the pesticide levels in the Yamuna waters. They have now decided to buy equipment to constantly monitor the same. They have also started talking about acquiring the appropriate technology to treat water to remove the pesticides. However, as the experience of other countries shows, technological solutions are, if not impossible to implement, prohibitively costly. The authorities will have to seek management solutions to see that pesticides do not enter the water. For once pesticides enter the water it is difficult to remove them (Agarwal, 1996a).

According to a study by some scientists of the Nuclear Research Laboratory of the Indian Agricultural Research Institute, 33 per cent of the groundwater shows nitrate contents exceeding the general acceptable limits (20 ppm), and 15 per cent crossed the maximum permissible limit (45 ppm). Health effects of long-term exposure to nitrates is potentially serious.

A recent report by the Industrial Toxicology Research Centre, Lucknow, says that wherever surface waters were polluted, practically all pesticides were found in higher concentrations in groundwater compared to surface water, some of them higher by over 100 per cent. Once the groundwater is contaminated, it is practically impossible to remove the pesticides and more people depend on groundwater thinking it is safer than the surface waters. The other significant observation of the report is that the range of pesticide residues in drinking water in rural areas was comparable to the concentrations in the major urban centres like Delhi, Kanpur, and Kolkata. Thus, rural areas are as vulnerable to the adverse impact of such non-point sources of pollution, perhaps even more because, unlike urban areas, people in rural areas generally drink untreated water (CPCB, 1996; Kumar, 1995; Anon., 1996).

Karnataka: The Mangalore Refinery

In Karnataka, strong pressure from affected constituents forced the Mangalore Refinery and Petrochemicals Limited to install a Rs. 200 crore effluent treatment plant and set up measures to recycle the treated effluents. In December 1995, thousands of fisherfolk assembled to stage a *rasta roko* on the national highway. The Karnataka State PCB had given a clean chit to the project on the water pollution aspect. But the data and analysis done by the National Institute of Oceanography did not correspond with the survey conducted by the state, the centre, or the MRPL. The government's stand since the beginning had been skewed in favour of the company. Finally, in December 1995, the government had to issue an order whereby the MRPL was directed to 'install suitable equipment and facilities for the recycling of waste and removal/ disposal of sludge, if any; and maximum extent of recycling will be achieved to conserve water and at the same time to minimise discharge of effluents into the sea' (Sitaraman, 1996).

However, the refinery started partial production without the effluent discharge pipeline in place. Within three months, by July

1996, a large amount of oil-contaminated water leaked out of MRPL, damaged crops on over 200 ha. land, and polluted practically all the sources of water. Activists of the movement immediately identified the culprits. Faced with a furore from the opposition in the legislative assembly, the state government, otherwise an apologist for the unit, had to warn the company that if it refused to pay compensation to the farmers and install a recycling plant, the company would not be allowed to restart production. The state PCB, invoking Section 32(1)(c) of the Water Act, prohibited the company from any discharge outside the plant's premises. None of these actors would have acted the way they did in this instance but for the expression of strong protest by the people. And the people could muster their courage and strength to protest only because they had knowledge and experience (Krishnaswamy, 1996).

The people of Bidar in northern Karnataka launched an agitation against some drug companies because they seemed to have brought in more misery in the form of widespread pollution than prosperity to the region. In their enthusiasm to develop Bidar, the official machinery totally ignored the need to regulate the disposal of toxic substances by these factories. Soon the region's air, water, and land were in a state of deterioration. For over three years, the Karnataka Vimochana Ranga has been organizing the villagers against the polluters. The government has been compelled to constitute a committee, which came out with a very unfavourable report with respect to the industrialists. Arguing that the very purpose of developing industries is being defeated as it is becoming dangerous to human life, the report recommended urgent measures and ordered the temporary closure of five polluting factories until remedial measures were undertaken.

But some factories procured a stay order from the Karnataka High Court and continued to pollute the area. In June 1995, an expert committee was constituted whose report, known as the Kudakavi report, reiterated the need for a treatment plant. When neither the government nor the Karnataka PCB took any action to bring this into effect, an intensified agitation was launched. When even after the vacation of the stay order by the High Court, on 19 September 1995, the KPCB did not take action to close down the factory, on 11 October 1995, the people decided to march to the factories to lock the four polluting drug factories in the Kolar industrial area. The action invited violent encounters with the Bidar police over the next few days. Finally, five days later, unable to control the mounting pressure from people from all over the state, the chairperson of the KPCB was compelled to invoke Section 33A of the Water Act, and order the shutdown of the four units (Sitaraman, 1996).

Andhra Pradesh: Patancheru

An industrial estate in Patancheru, just off the Hyderabad-Pune highway in Andhra Pradesh, lets out untreated industrial effluents from the estate into the Nakkavagu drain. As a result, the groundwater was poisoned, affecting hundreds of acres of agricultural land and also the drinking water of about 15 hamlets.

According to the Status of Environment in Andhra Pradesh, Citizen's Report, 1990, published by the Academy of Gandhian Studies, Hyderabad, 'Since its inception, the work of the Andhra Pradesh Pollution Control Board (APPCB) has been more on the lines of pollution-promotion rather than controlling it. Most of the time, it (the APPCB) pursues loosely structured, sham legal cases.' The Campaign Against Pollution was launched by some NGOs and citizens, which led to a petition being filed in the Supreme Court, among other outcomes.

Following the filing of a PIL in 1990, the Supreme Court directed the district judge of Medak to inspect the location and submit a report. This report, submitted in November 1995, concluded that over 100 industries were responsible for causing groundwater pollution in the region. In May 1996, the Supreme Court directed that the guilty industries reimburse Rs. 1.5 crore in compensation to the people, set up a proper effluent treatment plant, and the government supply treated municipality water to the villages.

The long struggle had led to some improvement in the situation, which would not have come about if the industry and the government were left to themselves. But the industries continue to operate and release untreated effluents as before. The struggle continues.

Tannery pollution in Tamil Nadu

The North Arcot district of Tamil Nadu is a good example of how an industry can ruin the environment. Nearly 70 per cent of India's total exports of leather and leather products come from

700 odd tanneries in Tamil Nadu. Ambur is one of the worst affected towns. Untreated effluents from about 250 tanneries, most of which are located in and around the towns of Ambur and Ranipet, have almost turned the land around these towns into wasteland. Crop production is down by more that 70 per cent. The water is undrinkable. Respiratory and skin diseases are widespread.

IIT Madras and the King Institute in Madras found in the early 1980s that the water supply in Ambur was unfit for human consumption. The factories were then given a year's time to install effluent treatment plants. But nothing was done.

Over half a million people have been affected following water pollution by the tanneries. Besides laying waste over one lakh hectares of farm land, they have contaminated ground and surface water, making it unusable. An estimated 40 million litres of untreated effluents are discharged daily into the rivers, irrigation ponds, and wells. The worst hit is North Arcot-Ambedkar, which, with the largest number of tanning units, accounts for over 80 per cent of the foreign exchange earned from leather processing in the state. Over 35,000 ha. is now unfit for cultivation and the area under paddy has been reduced by half.

The problem, as the industry sources say, began in the mid-1970s with the Centre bringing pressure on this sector to raise exports by fixing targets for the states. This led to a scramble to produce more in the shortest possible time, and the tanners switched to the faster chromium process route from the eco-friendly vegetable process that uses barks and plant extracts. The new process reduced the time and space required for tanning, but it meant the entry of highly toxic waste, which was—for want of ETPs or other pollution control measures—let out into the open. The TNPCB, set up in 1982, let the problem fester by not doing anything beyond issuing notices to the tanneries (in one case, a single unit was warned 25 times) to set up ETPs (Krishnakumar, 1996).

An NGO, Community Action for Development, has been active in mobilizing people against the pollution. The secretary of the Vellore Citizens Welfare Forum, on whose public interest petition the case was taken up by the Supreme Court, explains: 'Besides seeking the restoration of the ecological balance, this time we have also insisted on providing adequate compensation to the affected people.' Following their petition, 54 tanneries were closed in the

North Arcot-Ambedkar district in May 1995 by the Supreme Court.

In another landmark judgement, the Supreme Court, in April 1996, ordered that more than 200 tanneries in Tamil Nadu be shut down for failing to install effluent treatment facilities as required by law. The century-old industry in the state has played havoc with the lives and livelihood of the people, irreversibly fouling up fields, crops, flora and fauna, and destroying living spaces by releasing toxic wastes. The managements of the tanneries were prompt in appealing to the Supreme Court to reconsider its order to close down the units. The court rejected this plea, and when the tannery owners said that the installation of pollution-control systems would cost over Rs. 100 crore, the judges said: 'These units must know that they cannot operate like this.'

In arriving at its decision, the Supreme Court was aided by the National Environment Engineering Research Institute's February 1996 report on the status of CETPs. According to this, of the 30 ETPs identified in the tannery cluster of the five districts of North Arcot-Ambedkar, Erode-Periyar, Dindigul-Anna, Tiruchi, and Chengai-MGR, only seven were operational. In fact, none of these operational ETPs met the standards laid down by the Union Ministry of Environment and Forests or the TNPCB.

In August 1996, on the basis of a petition from the Vellore Citizens Welfare Forum, the Supreme Court imposed fines of Rs. 10,000 on each of the 700-odd tanneries in Tamil Nadu, and asked them to pay compensation for polluting the environment and also to install pollution treatment plants in four months (Banerji, 1997; CSE, 1984-5; CSE, 1996-7).

SOME LESSONS

State Policy

The above-mentioned cases clearly show that determined and sustained action by the affected communities, public action groups, and non-government organizations, helped by a sympathetic judiciary, have in several cases succeeded in placing the pollution problems prominently on the public agenda, getting their complaints heard and getting some redress.

But they also bring out the lukewarm attitude of the government in monitoring pollution and the reluctance even to entertain complaints under the relevant law, gather and make available relevant information. Reluctance to invoke legal sanctions and enforce court judgements is also evident in several cases. There is also reluctance to take action sanctioned by the law. Many of the court judgements ordering the closure of plants or compelling them to install treatment facilities have not been implemented or have been implemented partially and only after protracted delays.

The efficacy of such treatment plants as have been set up is also in some doubt. For example, a study by Greenpeace on toxic contamination in India found that combined effluent treatment plants have been of little use in reducing the extent of pollutants present in water. It feels that CEPTs are non-functional, and that the money being spent on such plants by the World Bank would be better spent instead to control and stop pollution at the source. This also applies to the governments and the courts, which tend to see CEPTs as the panacea for all pollution ills.

The problem lies partly in the lacunae in the law. The laws are many, scattered, and sometimes conflicting in goals and standards. They are often not based on the ground realities. They are not effectively communicated to the stakeholders. They are written in a language and in a style which an average Indian would find difficult to understand. There are also large gaps, like the absence of any legislation regarding the treatment of municipal waste and pollution control in the small industry sector and in agriculture. Standard fixing is inadequate and largely incomplete with reference to many toxic materials.

Pollution control agencies being part of the government do not have the independence and autonomy to initiate action or enforce the provisions of the law. Monitoring mechanisms are weak, and often absent. Where they exist, the information they collect is not available to the public. There is little coordination between the concerned departments. Nor do the elected representatives and legislatures show any interest in pollution-related issues, much less in exerting pressures for remedial action. On the contrary, they act more often in ways which favour, if not actively support, the interests of polluters. The laws do not provide for consultation, not to speak of the involvement, of the affected communities in government processes concerning pollution monitoring and control. Attempts of the affected communities to seek redress through petitions and appeals to the government rarely yield a positive response.

The implementation of laws is difficult in a context where several departments and agencies with different, and often inconsistent, mandates are involved. The department of industries, for instance, being concerned with attracting new industries, considers anti-pollution laws a hindrance. The agencies for monitoring and enforcing pollution laws are typically understaffed, ill-equipped, too weak relative to the development departments, and are liable to be opposed by vested interests. In some states, for example in Maharashtra, the government has authorized private consultants and laboratories to issue no-objection certificates, giving rise to manipulated results and grant of no-objection certificates even to polluting units. These difficulties are compounded because the government acts both as a promoter and regulator of economic activity and because pollution control boards lack the requisite authority and autonomy to effectively act as regulators.

A deeper problem is the perceived conflict between economic growth and the need to safeguard the environment. Insistence on meeting emission and effluent standards involves additional (often large) investments and increased operating costs and lower returns to investment. This is resisted on the ground that it would make new industrial investment less attractive to entrepreneurs and would result in a dampening of investment and growth of output and employment. Since strict enforcement of pollution standards on the existing plants also raises costs and reduces profits, they are reluctant to incur the extra costs. Instead, they successfully pressure the government, on threat of closure, to bear the costs directly or through subsidies. That closure of the plants may throw workers out of their jobs is a contingency which governments are unwilling to face.

The government currently relies almost entirely on anti-pollution laws, which lay down standards and threaten violators with civil and criminal consequences. Self-control on the part of polluters in the face of laws and even as part of business ethics and social responsibility can help but these steps are hardly effective. Strong economic disincentives are one possible measure that may help. But currently state policies work in the contrary direction by grossly under-pricing water and not charging industries for the release of polluted effluents. Moreover, there is no reason why the total expense of the government for pollution-control activities, including the staff and administrative expenses, should not be borne by the industry and other polluting actors.

Costs for pollution abatement can be reduced by more economical use of water. Among industrial units using similar technology and product mix, the specific water consumption varies over a wide range. For example, in India the paper and pulp factories consumed, during 1987-8, between 230 cu.m. to 450 cu.m. per tonne of output; the fertilizer industry used between 9 to 40 cu.m. per tonne of urea produced (during 1988-9). Overall, water use per unit of output of Indian industries is considerably more than that of their counterparts in the developed countries. The fact that even the lowest water consumption figures of the Indian plants are significantly higher than the international levels in most cases highlight the scope for saving costs by reducing water consumption.

Since water charges in India are very low, industries are not charged anything for the release of effluents and even nominal penalties for violations are not enforced. Factories and manufacturing centres throughout the country seem to have a *carte blanche* approach to water—they consume it, waste it, pollute it, all without worrying about the costs. In the pollution legislation, most of the penal provisions relate to source-specific standards. And these source standards are not really married with ambient standards. There is a need to link source-specific pollution through the total volume of pollution-causing activity. That has not been done in India. Nor is there an agency to monitor overall water quality.

In sharp contrast, many industrialized countries charge heavy levies for both the use of freshwater and for the release of effluents, thereby creating a strong incentive for industry to enhance water use efficiency. Policies that made water consumption, wastage, and pollution more expensive provided a boost to water recycling and conservation. Faced with levies of this kind, industry does look for enhanced water use efficiency. In 1965, Japan's industrial output was worth US $21 per cubic metre of water. In 25 years this value had increased to US $77 in real terms. In the former West Germany, water use stabilized between 1975-90 even as industrial output increased by 44 per cent. In the US, industrial water consumption in 2000 is expected to be only one-third of what it was in 1977. All this was made possible by policies that provided a boost to water recycling and conservation and that made water consumption, wastage, and pollution more expensive (Thakur, 1995).

In Sao Paulo, Brazil, water use in manufacturing dropped between 42 and 62 per cent, depending on the industry, from 1980 to 1982, after effluent charges were introduced. Consumption of water in the GDR dropped from 400 to 120 litres per capita per day, or 70 per cent, between 1989 and 1994, primarily as a result of cost-based pricing (Serageldin, 1995).

THE COURTS

Among the roles of various other actors, that of the courts has been the most significant in recent times and hence it has been in the limelight in the arena of pollution control activities. However, the courts can play a role only when an individual or organization or community approaches it; there are but few instances where the courts have taken *suo mottu* action based on a letter or a media report. By allowing pollution-related complaints under the fundamental rights violations, the courts have greatly increased the chances of complaints getting heard and decided speedily. However, the effectiveness of the actions of the courts is directly dependent on the quality of the petitions and the commitment and dedication of the lawyers. Hence, even as the courts have acted in a number of cases, they have not been able to set examples or trends for long-term sustained control of pollution.

For example, the courts, even when taking note of the destructive nature of the impacts of water pollution, have generally failed to either quantify the impacts, or penalize the polluters to compensate the losses. This failure—as the concrete cases discussed earlier in this chapter show—have not been as much of a deterrent to the polluters as one would like and expect. The complexity of environmental problems and the rationale behind laws and regulations, taken together with the quality of the evidence, difficulty of proof, and calculation of damages pose formidable challenges to the courts. This naturally affects the quality of the judgements, leading to a situation where in a number of cases the judgements are challenged or remain unimplemented.

Also, it is only the High Courts or the Supreme Court that have been active in this sphere. They have not taken cognisance of the fact that when an industry violates water pollution norms, the PCBs file cases in the lower courts which are notorious for inaction and delay. Also, even when the failure of the pollution control is evident, the courts have failed to pin-point the responsibility and penalize the guilty. Thus, the actions of the higher courts have also failed to achieve effective pollution control. It is clear that access to the courts is an option exercised by citizens because the executive has abdicated its responsibility. There is little attempt to identify the key problems and reverse the trend.

One other notable problem with judicial action has been that while handing down orders for the closure or shifting of industries from certain areas, they have not ensured that the workers do not suffer due to these actions. Moreover, they also have not ensured that at the new places, the industries follow the minimum pollution norms.

THE MEDIA

In almost every instance of successful struggle by the communities to control or remove polluting activities, the media have played a significant role. More effective action has come about when the communities, awakened by some voluntary organization, individuals, or the media, have risen to take action. It is also seen that such cases, publicized by the media, have the potential of inspiring other communities to also act to curb the destructive impacts of pollution. The role of the media in both informing the people about the pollution and in publicizing the communities' and courts' actions has been generally exemplary, though one thing lacking in this is that the media have not taken the issue to such levels where the government has been forced to take institutional steps to control the pollution effectively. Also, the media have many times failed to effectively analyse the situation and to follow up on court or other actions to show how useful or harmful such actions have been. (In the late 1990s, the role of the media in this regard was less encouraging.)

COMMUNITIES AND NGOs

There is an urgent need to learn lessons from such instances and to institutionalize the role of communities in pollution control, which is non-existent today. This institutionalization can be achieved, for example, by stipulating that every large industry or group of industries should have a citizens' committee from the surrounding area, with the right to inspect the plants and records of any company, whenever necessary. Second, each industry should be asked to make periodic reports to the community about its pollution-control activities and records. Third, the local pollution control officers must be in some way accountable and answerable to the local communities on a regular basis.

The resources, energy, and expertise which have to be mobilized for pollution control are often beyond the means of the communities and even the NGOs. They have to contend with a paucity of relevant information and struggle to remove the veil of secrecy which shrouds such data as are available. Some of the needed information can be collected by training local people and by utilizing the resources in local colleges and research institutions. But organizing it takes effort and far more responsive and active institutions in civil society than is currently in evidence. In any case, a great deal of data analysis requires facilities and skills, which only the government and higher-level research laboratories have. The pollution-control agencies must obviously be strengthened and made transparent and accountable. Greater involvement, on a sustained basis, of universities and of research institutions outside of government would help increase the credibility as well as the accessibility of the data. In either case, full and prompt publication of the information and a legal right for interested members of the public to gain free access to such information is imperative.

ACADEMIC AND RESEARCH ORGANIZATIONS

As we go through the cases described in this chapter, we see that academic and research institutions have a very important role to play when communities are agitated and mobilized over water pollution. Their role is not only to help the communities and NGOs assess the levels and extent of pollution and its constituents, but also to play a more proactive role of monitoring the state of affairs. They also need to make their reports public in such a way that the reports reach the people who are affected most. While there have been instances where someone or the other agency has helped the affected communities to increase the general level of awareness and understanding of pollution problems and their solutions, one cannot escape the conclusion that the scientific community has largely failed to conserve scarce water resources or come to the aid of the affected communities.

What is important after coming out with reports which have a significant bearing on national resources is to at least see to it that the conclusions and recommendations get widespread publicity, raise awareness, and build up some pressure so that the authorities are made to change faulty policies and programmes. However, unfortunately, most research institutions have singularly failed to even publicize their findings of a serious nature, leave aside the question of lobbying for policy changes. It is generally left to the media, NGOs and concerned citizens to do that (Chaudhuri, 1996; GEC, 1997; Kumar, 1995).

Moreover, learned professionals and institutes of higher learning do not seem to have got into the act despite their better access to information and their greater clout in decision making. If local communities are to be informed, it will be necessary to cultivate the social responsibilities of the professions in building civil society and in organizing good governance.

CONCLUSION

There is a need to institutionalize the right to information of the communities regarding the activities that affect their lives. Various actors including the government, industry, and the aid agencies, must be made to make public reports about their activities. The communities must also have a right to inspect and a right to participate in the decision-making process about all activities that may affect their lives. The pollution-control bodies have to be independent of the government in every sense of the term. Third, every user of water must be made to pay the full costs that have been incurred in providing the water to the user. This is true even for community resources like groundwater. Similarly, pollutants must be made to pay for the release of effluents even when if they are treated. There is no reason why the total expenditure incurred by the government for pollution-control activities, including the staff and administrative expenses, should not be borne by the industry and other polluting actors.

The research and academic institutions have got to play a much more significant role in identifying relevant issues and in informing all the concerned parties about the status of pollution and its impacts. They must also push for changes in policies that they have found to be necessary for reducing and checking the pollution of resources. They need to work hard to increase the water resource literacy and the availability of the necessary knowledge about the state of pollution of our water resources.

REFERENCES

Academy of Gandhian Studies, *Status of Environment in Andhra Pradesh*, Citizens Report, Hyderabad, 1990.

Agarwal, Anil, *Homicide by Pesticides: What pollution does to our bodies*, New Delhi: Centre for Science and Environment, 1996.

———, 'Pay-offs to progress', *Down to Earth*, 15 October 1996.

Alvares, Norma, 'Pollution Battles: Limited Legal Redress', *The Hindu Survey of the Environment*, 1994.

Anon., 'Water alarm', *Down to Earth*, 30 November 1996.

Banerji, Rajat, 'Dead fish tell tales . . .', *Down to Earth*, 31 January 1997.

Banerji, Rajat and Max Martin, 'Courting green', *Down to Earth*, 15 May 1997.

Central Pollution Control Board, *Groundwater Quality in Flood Affected Areas of Delhi*, CPCB, March 1996.

Centre for Science and Environment, *The State of India's Environment: The First Citizen's Report*, New Delhi, 1982.

———, *The State of India's Environment: The Second Citizens' Report*, New Delhi, 1984-5.

———, various issues of *Water Newsletter*, 1996-7.

Chakraborty, Sujit, 'Green justice: Up in smoke?', *Down to Earth*, 30 September 1995.

Gujarat Ecology Commission, *Ecological Degradation around the Gulf of Khambhat,Gujarat: A Status Report*, February 1997.

———, *State of the Environment: Gujarat*, May 2001.

Krishnakumar, Asha, 'Facing the music', *Frontline*, 31 May 1996.

Krishnaswamy, Chetan, 'Paying for a spill', *Frontline,* 23 August 1996.

Kumar, Sanjay et al., *Organochlorine residues in rural drinking water sources of northern and northeastern India*, Industrial Toxicology Centre, 1995.

Madan, M. Mohan, 'Nylon 6,6 Project: Advantage Goa', *The Hindu Survey of the Environment*, 1995.

Ministry of Water Resources, *National Water Policy*, New Delhi: Government of India, 1987.

Serageldin, Ismail, *Towards Sustainable Management of Water Resources*, Washington, D.C.: World Bank, 1995.

Shankar, Uday, 'Purifying the Ganga', *Down to Earth*, 30 September 1992.

———, 'Groundwater: Disappearing act', *Down to Earth*, 15 July 1993.
Sharma, Anju et al., 'The blind Court,' *Down to Earth*, 30 April 1996.
Sharma, Kalpana, 'Konkan Smelter Project: What people's power means', *The Hindu Survey of Environment*, 1995.
Shrivastava, Rahul, et al., 'Polluting industries: Just shifting the issue', *Down to Earth*, 31 May 1994.
———, 'Copper unit comes a cropper', *Down to Earth*, 15 May 1994.
Sitaraman, Sudha, 'By hook or by crook . . .', *Down to Earth*, 15 April 1996.
Sitaraman, Sudha et al., 'Beyond backwardness', *Down to Earth*, 30 April 1996.
Thakur, K.B., 'That sinking feeling' *Down to Earth*, 30 June 1995.
The World Bank, *Water Resources Management: A World Bank Policy Paper,* Washington, D.C., 1993.

CHAPTER 11

The International Water Tribunal

ARTHUR VAN NORDEN

INTRODUCTION

In 1981, the International Water Tribunal Foundation (IWT) was created by Dutch NGOs involved in the struggle against the pollution of the river Rhine and the North Sea. The Foundation is an independent organization set up as a remedy for the widely recognized inadequacy of the earlier and some existing mechanisms for water management. These mechanisms have proven to be especially inadequate for the resolution of conflicts between, on the one hand, the interests of the local population whose livelihood depends on a water source, and, on the other hand, the interests of industry and public authorities in the use of large quantities of water.

The recognition that the interests of many individuals and collectives were (and still are) not adequately protected by the existing mechanisms resulted in the organization of the First International Water Tribunal (IWT 1), in 1983. IWT 1 dealt with cases of water pollution in Western Europe. In 1992, the Second International Water Tribunal (IWT 2) was organized. IWT 2 focused its attention mainly on water management projects in Africa, Asia, and Latin America and the relationship of such projects with activities from the industrialized countries.

The IWT has no juridical power, and hence it cannot impose legal sanctions. The verdicts of the jury appointed by it are of a moral/ethical nature. The IWT enables plaintiffs to bring cases of water (mis)management to the attention of the public worldwide and to start a discussion with defendants in front of a jury which gives a judgement. The discussions are therefore not without consequences.

This chapter explains the general principles, objectives, and mechanisms of the IWT and gives an overview of the proceedings, results, and conclusions of the Second International Water Tribunal. It also discusses the IWT as a possible instrument to tackle water-related problems in India, as discussed during the seminar at which papers in this volume were discussed.

IWT FOUNDATION

The objectives of the IWT Foundation are several fold. They include:

- to offer an international platform to citizens concerned about (water-related) environmental problems;
- to stimulate research and provide support to the development of sustainable livelihood;
- to improve communication between those responsible and those suffering;
- to raise awareness of the transnational character of environmental problems, especially in developing countries;
- to stimulate the sensible and rational use of (inter)national environmental legislation;
- to influence public opinion;
- to share knowledge.

The IWT Foundation is a not-for-profit organization located in Amsterdam, the Netherlands. It was formed with the aim of providing a world forum to focus attention on the ethical and moral responsibilities of individuals, governments, and corporations to care for the management of the environment, particularly the water aspect of the environment.

Its highest body is the Board, which consists of a minimum of five persons. The Board is responsible for:

- the organization of a tribunal;
- the formulation of a declaration as the basis for the jury's pronouncements;
- the establishment and surveillance of the rules of procedure before and during the tribunal;

- the appointment of the members of the jury and the committee of experts;
- the contacts with the press before and during the tribunal;
- the coordination of the scientific programmes of plaintiffs during the preparation of the cases;
- the evaluation of the tribunal.

The activities are carried out by a small secretariat under the supervision of the Board.

The IWT Foundation operates on a project-by-project basis and is therefore not active on a permanent basis. Each time a tribunal has to be organized, the Board recruits a number of staff members for the required period.

HOW DOES THE IWT FUNCTION?

Potential plaintiffs (for example, NGOs) hear about the IWT through newspapers, magazines, the radio, or other contacts. Any NGO wishing to invite the IWT to examine the dispute in any case is required to submit a short description of the case, which is then checked by the IWT staff to assess whether it meets the formal acceptance criteria. After the initial approval has been granted, the NGO is invited to submit more details of the case proposed. If financial support is required, a fund is set up. The cases are then examined by external experts.

If a proposal is approved, the NGO receives a manual with guidelines indicating how to prepare and present the case, and, if required and approved, it also receives a first funding payment. After the completion of independent research work and the receipt of all case documents by the IWT, a decision is made as to which and how many cases can and will be heard by the jury. At least two months before the tribunal starts, the defendants are informed and are sent all the relevant documentation. As soon as the cases, defendants, jury, dates, and locations, etc., are known, the media are informed of all these through press conferences and/or press releases.

The final stage is to give a concrete form to the actual tribunal. After the completion of the tribunal, all the case documents, jury pronouncements, evaluations, and immediate results are presented in the IWT Case Books. If possible the IWT Foundation also tries to give follow-up details in those cases where both parties, plaintiffs and defendants, have expressed their willingness to continue the discussion.

Jury

The jury is composed of people from various disciplines and they may be from any continent. The credibility and the impact of the tribunal depend on the acceptance of the jury by the plaintiffs, defendants, and media representatives. The jury must consist of persons who are widely accepted and respected, independent and unbiased. The final selection of the jury members is made by the Board of the IWT Foundation through personal interviews with all the candidates.

The jury is assisted by a Committee of Experts which provides advice on the more specialized aspects of the case. To guide the jury in its deliberations, a declaration is prepared, adapted to the specific objectives of the tribunal. The jury is assisted by a Registry that prepares the jury's pronouncements. The decisions, or rather the pronouncements, of the jury are ethical/moral judgements lacking juridical power.

Plaintiffs

Plaintiffs consist of NGOs and public interest groups such as citizens' action committees, environmental groups, consumer groups, or professional associations of fishermen or farmers, and so forth. Plaintiffs can use the IWT as a neutral and international platform to express their concerns, and criticisms about (trans-boundary) water management problems in their country/region. Plaintiffs are fully responsible for the content, the preparation, and the presentation of their cases in accordance with the criteria for acceptance and selection of cases and the general quality standards as formulated by the IWT.

Defendants

Because of the generally large-scale nature of the environmental deterioration which often has a national or transnational origin, the defendants are government agencies, international financial institutions (such as the World Bank), multinational corporations, and large domestic industries. Defendants are informed of their status as early as possible in the case-preparation process and at least two months before the tribunal. They are free to participate in the tribunal and to submit information and evidence to counter the plaintiffs' arguments.

Media

The IWT is not a court, but comprises a series of public debates. Because of the lack of juridical power, the IWT is highly dependent for its success on the attention of the media in the North as well as in the South. The communication objective of the IWT is to create greater awareness among defined target groups about the importance of water.

Legal Aspects

The IWT demonstrates the need for international mechanisms through which cases can be brought against those responsible for environmentally harmful activities. Water is an important area where the linkage between environment and development is manifest. However, water, being an essential element for the survival of both present and future generations, is one of the areas where the issue of sustainable use emerges most prominently.

The IWT has been constituted to protect the interests of those dependent upon water resources. For each tribunal a declaration is prepared, adapted to the specific objectives of the Tribunal: for IWT 1 the Declaration of Rotterdam and for IWT 2 the Declaration of Amsterdam. The declarations formulate both the rights of those whose interests may be affected and the duties of those who may affect those interests and thereby infringe upon the fundamental rights protected by the declarations.

The declarations are built on earlier documents adopted in international forums. They do not contain radically new ideas, but rather they conform to what many individuals and entities have already expressed as being their ideas regarding the rights and duties relevant for environmental protection. These declarations are important instruments to guide the IWT jury in their deliberations. The experience and outcome of the IWT can be used to develop effective mechanisms for the resolution of conflicts concerning water resources in particular but also for environmental conflicts in general.

THE SECOND INTERNATIONAL WATER TRIBUNAL

The Second International Water Tribunal focused on the issues of water management in developing countries. Plaintiffs had the opportunity to present to an independent jury cases of environmentally unsound water management practices such as, for instance, ecologically disastrous projects involving large dams, water pollution, mining, cases on inadequate or non-applied environmental laws, and the relationship of such projects with activities from the industrialized countries. It gave a chance to the defendants (who ever was considered responsible for the environmental damage caused by such projects) to defend themselves before the same independent jury. The IWT 2 was held in Amsterdam from the 17–21 February 1992. A total of 12 cases concerning the (mis)use of water were submitted to the international jury by NGOs from different regions of the world. (See the appendix for an overview of these cases.)

Jury

The 1992 jury consisted of Ms Keri Hulme, author and chairperson of the Maitahi Maori Committee; Ms Devaki Jain, economist and director of the Institute of Social Studies Trust, New Delhi and Bangalore; Professor A. Kiss, chairperson, president of the European Council for Environmental Law; Mr Mochtar Lubis, director general of the Press Foundation of Asia; Ms Mary Okelo, banker, vice-president of Women's World Banking and board member of the Hunger Project and the International Centre for Research on Women; Mr Vicente Sanchez, Chile's ambassador in Kenya and representative to UNEP; Professor Oleg F. Vasiliev, civil engineer and director of the Institute for Water and Environmental Problems, Novosibirsk; and Professor Gerd Winter, director of the Centre for European Legal Studies. The public hearings of the jury were chaired by Mr Michel van Hulten, special adviser to the Dutch Foreign Ministry.

Plaintiffs

Three years before IWT 2 took place, more than one hundred NGOs, especially in the developing countries, were directly approached to consider the preparation of a case. Out of the numerous reactions received and after careful deliberations by the IWT, 22 NGOs were invited to prepare a case. Ultimately, 12 cases were selected for the jury hearings. The remaining cases were presented to the press, and all participants, plaintiffs and defendants alike, had the opportunity to present themselves in an exhibition hall.

During the preparatory period, the involved NGOs were assisted by especially appointed case coordinators, one for Africa, one for Latin America, and one for Asia. These local case coordinators provided scientific and organizational support to the NGOs and acted as the IWT's liaison person for the region in question. (See the appendix for an overview of the plaintiffs.)

Defendants

For IWT 2, most defendants submitted written information to the tribunal, but only in three cases did the defendants participate in the actual hearings. These were: (1) the Canadian multinational Hydro-Quebec (Canada) in the case 'The James Bay Hydroelectric Project, Quebec, Canada'; (2) the Israeli Government in the case 'Denial of clean water provision to un-recognized Arab villages in Israel'; and (3) Petroecuador, Ecuador's national petroleum company, in the case 'Petroleum in the Ecuadorian Amazon, water pollution due to petroleum exploitation'. (See the appendix for an overview of the defendants.)

Media

The communication objective for IWT 2 was to create greater awareness among the defined target groups with respect to the importance of water in the developing countries (and related problems) and the role of the industrialized world in these issues.

In 1989 a communication plan was developed by a professional media specialist. The general line followed was to send out press releases and hold press conferences only when there was real news to present, with activities being stepped up towards the start of the tribunal event.

By and large, the interest of the Western media was adequate. The media in the South hardly reacted at all, except in the Philippines. To tackle this problem the IWT decided to invite 'local' journalists from relatively unknown newspapers in the region for the cases to be presented at the tribunal.

The result was that 17 journalists from as many developing countries attended the event and extensively reported in their own country, in their own language, not just about IWT 2 as an event, but more importantly about the details of the specific case in their region. During the week of the tribunal, as a special service to the local journalists, a media workshop was organized in cooperation with the Media and Press faculty of the University of Amsterdam. During the workshop the journalists from the South had an opportunity to exchange information amongst themselves and to receive guidelines on how to present 'environmental news' from colleagues from the North who specialized in covering environmental issues.

More than 100 journalists from all over the world were present during the week of the tribunal. A special press room with full facilities was at the disposal of all those accredited, varying from representatives of large media corporations such as CNN (USA), CBC (Canada), and AFP (France) to journalists from small newspapers such as the *Times of Papua New Guinea*. Dutch newspapers reported daily on the press conferences where the jury's pronouncements were made public. Plaintiffs and defendants were given an opportunity to hold their own personal press conferences. Several months after the tribunal, IWT 2 was still being reported on in the press, not only as an event, but also as a good instrument for dealing with international environmental conflicts.

Legal Aspects

IWT has identified the need for strengthening national legislation and its enforcement. The IWT Foundation, therefore, initiated a legal project in 1990. The project consisted of two parts: (*a*) the Declaration of Amsterdam and (*b*) preparation of an overview and a comparative analysis of environmental and water laws in the South.

The Declaration of Amsterdam was prepared specially for IWT 2 for guiding the jury in its deliberations. It was prepared by a group of lawyers from different parts of the world, both South and North. The final version of the Declaration was approved by the Board of the IWT Foundation in 1991.

The major objective of the overview of environmental and war laws in the South was the dissemination of information on the instruments available for the legal protection of water resources in Africa, Asia, and Latin America, and on the legal actions that can be taken within the country concerned. The aim was to prepare a reference work accessible and comprehensible to a wide readership, whether or not legally trained, and in particular NGOs from developing countries. The main reason for undertaking such an extensive overview was simply that nothing comparable existed. Information about this kind of legislation is scattered all over the world in specialized institutes which are highly inaccessible. The study offers valuable material with which to compare how various countries in the South have shaped their water legislation.

An enormous amount of information was gathered from various sources, including the Library of Congress in Washington D.C. and the Environmental Law Centre in Bonn. It proved very difficult to obtain up to date information about environmental legislation in African countries. By approaching various African lawyers on a personal basis, a great deal of information was gathered. In 1991 two workshops were organized in Colombia and Sri Lanka to consult the legal experts from the NGOs involved in IWT 2 from Latin America and Asia. The African NGOs were consulted during the week of the Tribunal.

The overview was made by Mr José Borrero Navia, jurist in Cali, Colombia. He was assisted by Ms Liesbeth Bijvoet, international jurist in Amsterdam, the Netherlands.

As the task was too large to handle in the relatively short period of time, the overview is not yet complete. It is hoped that the gathering of information can be continued in order to render the overview as complete as possible.

EVALUATION AND RESULTS OF IWT 2

Evaluation

Overall, IWT 2 has had good results. A number of the defendants did reply seriously in writing and three defendants appeared before the jury in their defence. Through the exercise of performing scientific research, writing a case document, and presenting their case before the jury, the NGOs have gained a great deal of self-confidence and experience. As a result they felt and feel much stronger in their attempt to openly question the environmental consequences of water management projects. The attention of the press in the North and the South has influenced the general public's attitude towards the various water management problems. As a result, and since the jury's pronouncements were made public, the defendants concerned as well as the authorities could no longer deny the existence of a conflict. The NGOs in question consequently became serious discussion partners—an entirely new situation relative to the period before the tribunal, when many NGOs were either ignored completely or were regarded as agitators endangering the safety of the state.

Not all NGOs met the expectations and not all documents were scientifically watertight, but all documented the immediate urgency of the problem concerned; and that was picked up by the journalists who publicly questioned the defendants' acts. The conclusion can be drawn that a public platform such as IWT 2 is not only needed but has also been accepted as an instrument for publicly debating environmental conflicts. In a number of cases IWT 2 has certainly narrowed the gap between the conflicting parties.

Results

In a number of cases it is not possible to indicate whether the observed actions are due partly or fully to IWT 2, viz., in those cases where the IWT case document was part of an existing campaign and when defendants did not react at all. What follows are some concrete results observed before, during, and up to one year after the Tribunal.

ARGENTINA: POLLUTION

Just prior to the start of the Tribunal, a written statement from Pelikan S.A. was received stating that the company had been unaware of the severity of the pollution caused by its industrial activities and that measures would be taken to prevent further pollution. These measures were to include a reduction of the pH and solid organic matter content of effluents, a halt to sewer discharges of substances soluble in ethylic ethers, and the complete elimination of sulphur from the production process. A reporter from Inter Press Service visited the plant a short while after the Tribunal and was able to ascertain that Pelikan had altered its manufacturing process in order to comply with local legislation.

CHILE: MINING

Before the beginning of the Tribunal, CODEFF (a Chilean NGO) and El Indio (a gold-mining corporation in Chile) had already entered into negotiations. The plaintiff prepared a document which was sent to the defendant, who expressed a willingness to establish a dialogue on the condition that the case document would not be made public. At first, the plaintiff did not wish to cooperate, one reason being the refusal of the defendant to fully accept a set of demands presented by the plaintiff. Through intermediation by IWT, the two parties came to an understanding, the defendant accepting most of the plaintiff's demands, and the plaintiff agreeing not to publish the case document. Both parties expressed the will to continue the dialogue.

PERU: MINING

Just before the hearing of this case, IWT received a letter signed by 22 members of the Peruvian parliament in Lima. The letter not only endorsed the accusations directed towards SPCC, a North American copper mining corporation operating near Ilo, southern Peru, but it also strongly supported the concept and implementation of the Tribunal. The plaintiffs were supported by a government official, Dra. L. Aurea Saenz, General Controller of the Republic of Peru, who presented before the jury à titre personnel, a testimony of the negative effects of the copper-mining operations.

In December 1991, the defendant announced its intention to invest several million dollars to improve environmental conditions in the area in question. Although a direct link with the IWT cannot be proven, it appears more than coincidental that this announcement was made public two months after the defendant had been informed about the complaint. The plaintiffs have announced that they are considering a court case in the USA against the defendant.

In September 1993, the IWT received the following notices from the plaintiffs:

- Since one year ago SPCC has undoubtedly changed its attitude from being intolerant and secretive into expressing the will to create a dialogue and to open the road for cooperation with governmental institutions and local agencies.
- During a conversation between the vice-president of SPCC and a member of the National Congress of Peru, the first mentioned that the company had made a mistake in not attending the IWT 2 and that the company's environmental policy has been wrong.
- SPCC has taken several steps to improve its image in connection with environmental problems.
- This change in attitude of SPCC is undoubtedly a reaction to the new global context towards the environment and in particular a reaction to the pronouncement of the jury of IWT 2.

PAPUA NEW GUINEA: MINING

As a result of the hearing of this case, the National Government of PNG organized conferences on the subject both in PNG and Australia. When the head of the government of PNG's Western Province threatened to close the Ok Tedi mine he was put in jail, but was later bailed out. The national government reacted by sending a mobile police squad to protect the mine.

The Land Owners' Association in the region where the mine is located has called upon the government to implement the IWT jury pronouncement. The association is seriously considering suing the mining company. One of their lawyers visited the IWT office in Amsterdam for further information and support.

On 14 January 1993 a motion was accepted in the Federal Parliament of Germany. The motion asked the German Federal Government to influence the German shareholders of the mine and the Government of Papua New Guinea to reduce the environmental harm within the limits of Western standards, to find better solutions for the retention of the mine's tailings, and 'to take care that the regulations for compensation and for participation are checked, with the aim of including all people affected, compensating them adequately for the economic, ecological and social costs and distributing the profits of the project as widely as possible'.

In August 1993, the PNG Government admitted in Parliament that there was considerable damage to the environment and the Fly River system as a result of the operations of the Ok Tedi gold and copper mines in the Western Province, but that it is unable to repair the damage because there is no technology available to neutralize the pollution caused. The Government was doing everything possible to minimize the effects of the mine's operations on people's lives and the ecosystem along the Fly River. Furthermore, the Government stated that in the past it had denied accusations made in some reports by international environmental groups that the mine was causing dangerous pollution.

COSTA RICA: POLLUTION

Prior to the Tribunal the problems on banana plantations had hardly been debated in the national press. After the Tribunal this situation changed radically, and as a result the national government and the defendant, Standard Fruit, had to accept the plaintiff as a serious discussion partner. The defendant announced its intention to undertake a study of drinking water quality in the plantations; whether that happened is not known. Through this case international attention was drawn to the problems in question.

ISRAEL: MANAGEMENT

Through the publication of the IWT jury pronouncement and the media coverage of this case, issues related to the country's unrecognized communities were for the first time recognized for their political value. Prior to the Israeli elections of 1992 the Interior Ministry agreed to officially recognize three hitherto unrecognized Arab villages. After the elections the number of villages which came to be recognized was extended to five.

REACTIONS FROM PARTICIPANTS OF IWT 2

Plaintiffs

During the preparations for IWT 2 it became clear that individuals and groups often do not have access to the technical and legal expertise, or the financial means, required for the preparation of their claims. The IWT Foundation acted as an intermediary without which the NGOs would not have been able to prepare their complaints, gather the relevant experience, etc. For instance, in those cases dealing with water pollution due to mining activities, industrial waste, and so forth, it appeared that the NGOs involved had hardly any knowledge of and/or experience with the water-sampling techniques that were a prime necessity to prove the pollution. The NGOs recognized this fact and gladly accepted the offer of the IWT to send out a professional to assist them.

An overall fact recognized by the NGOs was that with this external support they were able to conduct a multidisciplinary analysis of the environmental effects of, for instance, mining and metallurgical operations, a task that would have otherwise been impossible as Southern NGOs do not themselves have the resources to confront big transnational companies.

The NGOs also remarked that the IWT created an opportunity to generate international publicity for their respective cases. This was considered an important development, as all the channels for solving the problem in their own countries had been exhausted and international pressure and attention were necessary, especially from and in those countries that ultimately benefit from the activities of the transnationals.

Another positive effect of participating in the Tribunal was the opportunity to establish numerous contacts with people involved in the ecology movement, legislation of water management and so forth, and with the press. Later, that allowed a number of NGOs to carry out campaigns and to establish channels of communication in Europe, the USA, and Australia in order to initiate lawsuits against the companies involved. And last but not the least, the NGOs made progress in describing the problem in a scientific way and in developing alternative solutions to the problem.

Defendants

In view of the spirit of the time (February 1992, six months before UNCED), with considerable international attention on and awareness of global environmental problems, it was hoped and expected that the majority of the accused parties would make an appearance at the Tribunal—or at least react in writing—with the objective of making a constructive contribution towards solving the problems. It is disappointing, however, to note that some important parties, like the Brazilian Government and the World Bank, accused in cases internationally recognized as very serious, did not react at all. Other defendants sent a simple note stating that they would not be accepting the invitation because they had doubts about the tribunal's sincerity, as it was biased, incompetent, and/or of no juridical value. While not weighing heavily on the IWT, such statements do affect the integrity of the jury members and have a negative influence on the much-needed public debate on how to solve environmental problems.

Fortunately, a number of defendants did reply seriously in writing before and after the tribunal and three defendants appeared before the jury in their defence. In particular, one defendant, Hydro-Quebec from Canada, took the matter very seriously. Its European representative visited the IWT office in Amsterdam, accompanied by some company lawyers to investigate the credibility and professionalism, but above all the unbiased status of the IWT Foundation. Hydro-Quebec requested that it should be given more time during the hearings, otherwise they would not appear. After long negotiations, the IWT fulfilled their wish on the condition that their opponents, the Cree Indians, would have equal time. After coming to an agreement, the IWT could announce to the press the upcoming appearance of an international giant. That fact alone generated an enormous interest from the media, in particular from North America, a development from which the other cases profited.

After the tribunal, Hydro-Quebec evaluated in writing the Tribunal hearing and procedures and made valuable comments for the future. The main points that they raised were:

- It is foolhardy to ask a multidisciplinary panel whose members come from around the world to rule within a period of five days on the number of files that IWT 2 had on its agenda;
- The legal framework of reference, the Declaration of Amsterdam, was considered too broad in its approach. It should have been restricted to water rights only and should not have included human rights;
- The procedural rules were not clear enough, in particular the role of the technical chairman and the chairman of the jury;

- All information should be made accessible to the various parties involved with more rapidity;
- The attitude of a few members of the jury and the committee of experts gave the impression of partiality, not to say hostility, toward Hydro-Quebec;
- The nomination process of the members of the jury and the committee of experts should be refined and clarified by asking all plaintiffs and defendants to make recommendations in advance;
- The tribunal was, after all, an enriching experience for Hydro-Quebec.

Observations by the Jury/Committee of Experts

Mr Alexandre Kiss, chairman of the jury considers it very important that an independent international body composed of experts examine cases brought by NGOs. This body should be created by governments and should be able to examine also cases which would be submitted to it by state authorities. For the future, it should be convenient to designate in advance one or two rapporteurs for each case among the members of the jury and one or two experts who should help them with scientific and technical advice. Missing, mainly at the beginning, at our sessions in February, was a comprehensive presentation of the case, before the audience, both from a scientific/technical and legal point of view, so that the members of the jury could have asked questions based on a better knowledge base.

Mr Gerd Winter, jury member points out that the IWT 2 was not an international conference. Quite consciously a court-like procedure was adopted, implying that cases were discussed, not general information and ideas. The case method makes causes and consequences more easily understandable and it inspires the search for solutions. An amazing richness of data, explanations, evaluations, and reformatory proposals were obtained about water issues in North-South transactions. The IWT demonstrated the relationship between the human rights of local people and the nature preservation goals of Northern protectionists.

The jury felt that it is not an intrusion into a state's sovereignty if one requires big international companies to apply the standards of their home country elsewhere also where they conduct business, because everyone has an original obligation to protect the environment, notwithstanding the legislative framework within which he or she is acting. The outcome of IWT 1 and IWT 2 is encouraging enough to continue, but not by persuading the international communities of the states to conclude a convention setting up an international adjudicatory commission with the states wanting to control the proceedings. It should build on the IWT approach to appeal to the international public, bring cases to its attention, and provide a fair procedure for the handling of the cases. The selection of the jury and the committee of experts could be improved in a way which would allow both sides to trust in the independence and wisdom of the selected persons.

The main problems in his view relate to (*i*) finding funds for financing the organization and assisting indigent parties in the preparation of their cases; (*ii*) procedural questions concerning the role of the committee of experts; (*iii*) the preparation and exchange of documents and the style of the oral proceedings; (*iv*) deciding whether the normative frame of reference should be based on the national and international law in force, or whether international morality or on a trust that the normative criteria emerge from the cases themselves. He also mentions that in some languages 'tribunal' has a connotation of blaming someone without allowing a proper defence. Therefore, it may be advisable to choose a label which avoids this misunderstanding.

Michel van Hulten, technical chairman felt that experience had justified the assumption that the IWT has no judicial powers and cannot enforce its conclusions, that the pronouncements of the IWT would carry only a moral or ethical weight, and that any practical results would have to derive from a public disclosure of the facts, placing pressure on companies and governments to correct their behaviour with regard to the use and pollution of water.

He noted that in several cases the companies acted in contravention of the existing national or local laws and regulations. Strangely enough, these accused parties showed an unwillingness to attend, defend, and justify their behaviour, because the IWT lacks a judicial basis. In several instances, the jury observed that the laws protecting the environment are not inadequate at all but that the problem is rather the observance of the rules. Willingly or unwillingly, governments take very insufficient care of the environment; laws have been established but the implementation of these laws is undercut; there is uneven distribution of power in the countries concerned; the economy overrules the environmental needs and protection. Verdicts on environmental degradation cannot be based solely on environmental considerations.

In general, the cost of repairing environmental damage is rarely or never included in the calculations. Therefore, it will be impossible to weigh environmental considerations against production costs. In the case of the construction of dams in the 'Three Gorges' in China, it was stated explicitly that if all costs were to be included in the calculations, in many cases the costs would prove to exceed the benefits, thus undercutting the economic justification of the project in question.

It was most disturbing to hear during the sessions that practically everywhere information is concealed, consciously, by governments, by companies, and by those responsible for activities, thus having a negative impact on water quality and quantity. In most cases, the accused companies did not provide the jury with information that could have been helpful in formulating correct conclusions as to whether water quality is harmed or not. Transparency of decision making with regard to the quality of the environment is of prime importance. A basic requirement for such transparency is the availability of good-quality information about the current situation and the proposed changes—and such information has to be provided by the companies initiating new or altered production methods and activities.

With responsibilities transparent, those concerned about likely or actual damage are in a better position to take the necessary action in defence of their interests—but this is probably precisely what certain decision makers want to prevent, when information is not gathered or, if gathered, is not made public. If this assumption is correct, then there is even more reason to continue with activities like this International Water Tribunal in Amsterdam.

Mr Vijay Paranjpye, Committee of Experts observed that in some of the cases, for example, the tanneries in Chennai, in India and the Sao Paulo case in Brazil, there appeared to be a lot of important and legally relevant information which was available but which was not included or which was drafted ineffectively in the final case document. It may therefore be necessary in the future to appoint a professional advocate/copywriter/editor to assist each organization or a coordinator during the last two months of the case-preparation period so that the final plaint becomes precise and comprehensive without losing its legal edge.

Follow-up and the Future

In the early 1980s and early 1990s, it appeared that individuals and groups found it difficult to gain access to existing technical and legal expertise, necessary to question the acts and policies of national governments, international financial institutions, transnational companies, and the like. In the late 1990s that is still the case. International assistance and cooperation in this field is essential.

The IWT Foundation has tried and not without success to provide assistance especially to the NGOs of the South to overcome this handicap. As one of the activities the IWT Legal Project was initiated. It is hoped that this project can be continued to make it as complete and up to date as possible. So far this goal has not been realized because of the lack of funds.

Another initiative was to develop a manual for water-sampling campaigns focusing on situations in the South, and to investigate the possibilities of creating small-scale, mobile laboratories where the local population could have their (drinking) water samples analysed for indicative purposes. The manual appeared in 1994 in English, French, and Spanish. Hundreds of copies have been distributed among NGOs, libraries, development organizations, and so forth, meeting the need for a concise, comprehensible, and surveyable publication in this field. Lack of funds have stood in the way of investigating the creation of mobile laboratories despite the enthusiastic reactions of the target groups.

Environment, development, and especially their interrelationship in respect to the proper use of water resources were and will continue to be the focus of the IWT. However, it was felt that the formula of the Tribunal as an instrument together with the expertise of the IWT could and should be used to raise public attention in regard to other global environmental problems. In 1992, the IWT was approached by several Dutch environmental and development organizations to investigate the possibilities of organizing an international tribunal on world wide forestry problems. This resulted in a feasibility study. The reactions of the consulted NGOs and forestry experts who were consulted demonstrated the need for such a tribunal. Unfortunately, the potential donors, such as ODAs, who were approached reacted negatively, not because they questioned the tribunal as a proper instrument, but because of possible political conflicts between the states importing and exporting timber. The economic power of the timber (related) industry and consequently its political influence certainly played a role in forming this negative reaction.

Since then nothing substantial has occurred at the state level to effectively deal with the world wide forestry problems. In particular, the role of the local forest communities in the exporting countries, the protection of endangered species, the promotion of sustainable alternatives, and the adaptation and development of proper legislation have not and are hardly ever taken into consideration.

Another initiative made by the IWT had a similar fate. In 1993, a proposal for the development and instalment of an International Arbitration Commission, an international mechanism for the settlement of environmental disputes of the international order, a mechanism falling beyond the traditional inter-state mechanisms to be addressed in particular by non-state actors.

The proposal was the result of an inquiry made among many NGOs all over the world. It was to ascertain whether there was a need for such an arbitration commission. Representatives of a number of corporations and governments were also asked to give their views. The idea was also discussed with some international organizations such as the Earth Council, the Environmental Court Foundation, and the International Green Cross. The need for such a mechanism was confirmed. Many particularly applauded the idea of easy accessibility for non-state actors, the permanent character of the proposed commission, and the possibility of examining a conflict with speed (as opposed to the long preparatory periods for IWT 1 and IWT 2). The proposal was introduced to the UNDP, the EC, and the Dutch Government for initial financial support. Although all the parties approached reacted positively to the need and the general idea of the proposed mechanism, no financial support was provided.

Since 1994, after the completion of the IWT 2 Case Books, the water manual, the International Forest Tribunal feasibility study, and the International Arbitration Commission proposal, the IWT Foundation's activities have been limited to answering the numerous replies for information regarding its functioning, its experiences, and tribunals planned for the future.

CONCLUSIONS

Despite the Earth Summits 1 (1992) and 2 (1997) and the other inter-state conferences on various subjects related to the environment and development, no real progress has been made. Economic interests and political power have prevailed over social needs, particularly the needs of the local population in rural and urban areas in the southern hemisphere.

The availability of clean water for everybody has been recognized as a basic human right. In view of the scarcity of this vital natural resource, it can no longer be accepted that the protection of its quality and availability are overruled and endangered for the sake of economic growth and because of political interests.

States have demonstrated their inability to tackle the problem, big (inter)national corporations are merely interested in making more profits, and the general public seems powerless and unable to intervene. They seem so but being the consumers of industrial products and being the electorate, having the power to install or sack their political leaders, the general public has a huge potential to influence the powers that rule them. A primary condition for developing and using this potential is to have access to information and to be properly informed. NGOs can play an important role here as the mouthpiece of public interests.

Initiatives such as those undertaken and proposed by the IWT can make a valuable contribution to this end. A systematic approach to national and international environmental disputes, in particular over water, should be encouraged. The installation of a 'permanent' IWT would put into effect a mechanism of practical value and create a means for the implementation of international law, making a difference at the practical level.

As a result of the IWT's easy accessibility, it will provide a standing, that is to say, a means of protecting the interests of those whose environmental and socio-economic interests are at stake. It will permit them to participate in the management and development of the environment and in the achievement of a sustainable livelihood.

COMMENTS DURING THE SEMINAR

A number of points regarding the utility of a IWT-type mechanism were raised and discussed during the Amersfoort seminar. While the idea of an IWT sounds attractive and appropriate, several questions require attention if the IWT is to become a credible alternative to the existing mechanisms, especially within the Indian context. For instance, there was an impression that the IWT selectively identifies the jury, and empowers and finances it. All of this is likely to hamper cooperation from (state) governments and industries as potential defendants and make the participating NGOs suspect as plaintiffs. The results could be counterproductive, especially in India which is characterized by a highly competitive environment and extreme inequalities among its people.

In response, to the raising of this concern it was pointed out that the jury-selection process is a difficult one. It could be improved by involving plaintiffs and defendants right from the beginning by seeking recommendations for specific candidates. But the final decision should still remain with the IWT. An option could be to develop a jury member profile approved by both defendants and plaintiffs, thus enabling the IWT to look for suitable candidates. Another option could be to install a more or less permanent jury, supported by different committees of experts depending on the subject of the tribunal. In any case, the jury is not controlled by the IWT. The jury members are independent and are not paid for their services. They are only asked to take into consideration the IWT guidelines and to accept the tribunal's procedures.

A second source of concern related to the possibility of incorrect interpretations of facts and principles, possibly causing a variety of problems. It was recognized that this is of course always possible in the 'real' courts. But, unlike the 'real' courts, the IWT has developed procedures which are public, easy to understand and, if absolutely necessary, they can be adapted without having to change the law.

The IWT has also developed strict rules and conditions regarding the scientific quality of the case documents prepared by plaintiffs. If false or wrong accusations are made based on supposedly right facts and principles, it is the jury's responsibility to come to a balanced conclusion. Besides, the jury is not alone in coming to a final conclusion. The members are assisted by external experts on the scientific side and by a registry for the legal side. If both parties, defendants and plaintiffs alike, have accepted the procedures as well as the jury and have publicly stated so, there is little chance that the final conclusion of the jury will cause major problems.

A third concern was that the small number of defendants involved in the cases of IWT 2 limited its effectiveness in resolving the problems. In response, it was pointed out that since the IWT is not a legal court, it does not aim to convict defendants *per se*. The overall goal is to raise public attention and awareness regarding water-related environmental, social, and economic problems and to improve communication between the conflicting parties. It is true that due to the non-legal status of the IWT, defendants are reluctant to participate. The relative obscurity of the functioning and purpose of the IWT is another reason for their reluctance.

But the experiences of IWT 2 proved that most defendants reacted in one way or another under the pressure of the expected media attention and in several cases because of the outspoken and proven evidence brought forward by the plaintiffs. The IWT as a new mechanism is still in an experimental phase. It must certainly be promoted in a better manner and its imperfections must be remedied. Organizing more tribunals is the only way to solve these teething troubles. It is questionable if the IWT should seek the patronage of governments and/or the UN, as has been suggested. That kind of patronage could weaken its position and status of being a neutral, unbiased instrument and could lower its credibility in the eyes of the general public. An option is to get official recognition from the international community. That could be enforced by creating massive support from the (international) NGO community and likewise from that section of the (international) business community which recognizes the value of the IWT.

Another problem raised related to the name 'tribunal' which seems to evoke different meanings and generate different feelings in different places and contexts. For instance, there are environmental tribunals in India which are part of the court system. If the IWT format were to be used as an instrument to deal with water scarcity problems in India only, one could look for a different, more appropriate name. But if the format is to be used to establish an internationally recognized platform with a permanent character, the word 'tribunal' should be kept. It is hoped that the initiative of the IDPAD will be continued amongst others by investigating the possibilities of organizing an event based on the format of the IWT dealing with the problem of water scarcity in India as one major issue based on three or four cases. The focus could be on the Indian water legislations, their functioning, and their implementation.

APPENDIX

An Overview of Cases, Plaintiffs, and Defendants of IWT 2

It would be too long to fully describe all the 22 cases submitted to the IWT 2. However, all the cases are briefly mentioned. The 12 cases heard by the jury are marked with an asterisk (*). The cases are subdivided into 4 categories: mining, dams, pollution, and management. The full documents and pronouncements of the jury are published in the *IWT 2: Case Books* in 4 volumes.

ARGENTINA

Category: Pollution

Pollution of the Matanza-Riachuelo River Basin, Argentina

Plaintiff

- Centro de Estudios Ambientales, Buenos Aires, Argentina

Accused

- Obras Sanitarias De La Nacion (OSN), Argentina
- Lever Y Asociados, Argentina (Unilever, UK/Netherlands)
- Pelikan Argentina (Gunther Wagner, Germany)
- Suchard Argentina (Phillip Morris, USA)

BANGLADESH

Category: Management

Action Plan for Flood Control

Plaintiff

- Bhuapur Development Project (BDP), Bhuapur, Tangail

Accused

- World Bank, *French Government, *French Engineering Consortium, *Asian Development Bank, *UNDP, *German Government, *Japanese Government, *USAID, *and the consulting agencies: Mott MacDonald, BCEOM, Halcrow, ISPAN, Louis Berger, Pacific Consultants, and any construction firms under their coordination and operations.

BRAZIL (1) (*)

Category: Dams

Hydroelectric power plants of the Amazon

Plaintiffs

- Pro-Indian Commission of Sao Paulo (CPI), Sao Paulo, Brazil
- Coordinating Committee of Those Affected by Dams in the Amazon, Altamira (Pará), Brazil

Accused

- Brazilian Federal Government as represented by: Mr Fernando Collor de Melo, President

BRAZIL (2) (*)

Category: Mining

Mercury pollution in the Tapajos Valley, Santarem, Brazil

Plaintiff

- Group for the Defence of the Amazonian Region (GDA), Santarem (Pará), Brazil

Accused

- Brazilian Federal Government as represented by Mr Fernando Collor de Melo, President
- Pará State Government as represented by Mr Jader Fontenelle Barbalho, Governor

CHINA/CANADA (*)

Category: Dams

(A) The Three Gorges dam project, China ('Three Gorges'), and
(B) The James Bay Hydroelectric project, Quebec, Canada ('James Bay')

Plaintiffs

(A)

- Probe International, Toronto
- Mr Guo Yufang, Chairman, Federation for a Democratic China, Dutch Branch, The Hague

(B)

The Grand Council of the Crees (of Quebec), Nemaska, Quebec

Accused

(A)

- The Government of China, Beijing

- The Government of Canada, Ottawa as represented by:
  Canadian International Development Agency (CIDA), Ottawa

  CIPM Yangtze Joint Venture (CYJV), a consortium composed of:
  - Hydro-Quebec International, Montreal
  - British Colombia Hydro International, Vancouver, BC, Canada
  - CIPM Canadian International Project Managers Ltd.

[owned one-third each by SNC Holdings Canada Inc., Lavalin Inc. (now merged as SNC-Lavalin Inc.), and Acres International Ltd.]

(B)

- The Government of Canada, Ottawa as represented by:
  - Mr Jean Charest, Minister of the Environment
  - Mr Tom Siddon, Minister of Indian Affairs and Northern Development
- The Government of Quebec, Quebec, Canada as represented by:
  - M. Robert Bourassa, Premier
  - M. Pierre Paradis, Minister of the Environment
  - Ms Lise Bacon, Minister of Energy and Resources
- Hydro-Quebec, Montreal, Canada
- SNC-Lavalin Inc., Montreal, Quebec, Canada

CHILE

Category: Mining

Water pollution of the Elqui River, Chile, because of the mining of gold

Plaintiff

- National Committee for the Defence of the Environment (CODEFF), Santiago, Chile

Accused

- Compañia Minera El Indio, Santiago, Chile (Subsidiary of Lac Minerals Ltd., Canada)

COSTA RICA (*)

Category: Pollution

Claim against the "Standard Fruit Company", (Estrella Valley, Limón, Costa Rica)

Plaintiff

- Association for the defense of the hydrographic valleys of Costa Rica and for clean water, San José, Costa Rica

Accused

- Standard Fruit Company, USA Branch/Costa Rica Branch

ECUADOR (*)

Category: Pollution

Oil in the Ecuadorian Amazon: Water contamination by oil activities

Plaintiff

- Corporation of Legal-ecological Investigation and Defense of Life (CORDAVI), Quito, Ecuador

Accused

- Petroecuador, Ecuador
- Texaco Petroleum Company, USA
- City Investing Company, Bahamas

GHANA

Category: Management

Negligence of coastal developing countries by their rich trading partners of the North on curbing pollution from ships

Plaintiff

- Friends of the Earth—Ghana, Accra, Ghana

Accused

- The industrialized countries of the North (the international donor community)

INDIA (1)

Category: Dams

Narmada Project—India's largest, planned, environmental and human tragedy

Plaintiff

- Mr Ramesh Billorey (Narmada Bachao Andolan), India

Accused

- The Chief Secretary of the Government of India, New Delhi
- Mr Lewis Preston, President, World Bank, Washington

INDIA (2) (*)

Category: Pollution

Water pollution due to leather industries in Tamil Nadu, India

Plaintiff

- Trust-Help, Madras

Accused

- The Government of Tamil Nadu
- The Secretary of the Ministry of Environment and Forestry

INDIA (3)

Category: Pollution

Radionuclide concentration in the marine ecosystem due to activities of the Indian Rare Earths, Kochi Plant, Cochin, Kerala, India

Plaintiff

- Centre for Industrial Safety and Environmental Concerns (CISEC)

Accused

- Indian Rare Earths Ltd., Tiruvananthapuram, India

ISRAEL (1) (*)

Category: Management

Denial of clean water provision to unrecognized Arab villages in Israel

Plaintiffs

- The Galilee Society for Health Research and Services, Rama
- The Follow Up Committee on Health in the Arab Sector, Rama

Accused

- The Government of the State of Israel via its representatives: Mr Rafael Itan, Minister of Agriculture, Mr Arieh Der'i, Minister of the Interior

ISRAEL (2)

Category: Management

The legal status and the legal provisions applicable to international groundwater resources geographically located between Israel and the West Bank of the Jordan

Plaintiff

- Palestinian Hydrology Group, Jerusalem, Israel
- Palestinian Advocates Group, Jerusalem, Israel

Accused

The State of Israel, via its representative, the Ministry of Agriculture, Jerusalem, Israel

NIGERIA

Category: Mining

Tin Mining and Watershed Degradation on the Jos Plateau, Nigeria

Plaintiff

- Earth Search, Enugu, Nigeria

Accused

- Consolidated Tin Mines Ltd., Bukuru & Partners, namely:
  Gold and Base Metals PLC, Jos, Plateau State, Nigeria
  Amalgamated Tin Mines PLC, Jos, Plateau State, Nigeria
  Exlands PLC, Jos, Plateau State, Nigeria
  Bisichi-Jantar PLC, Jos, Plateau State, Nigeria

PAPUA NEW GUINEA (*)

Category: Mining

Ecological Damage Caused by the Discharges from the Ok Tedi Copper and Gold Mines, Papua New Guinea

Plaintiff

- Wau Ecology Institute (WEI), Wau

Accused

- Ok Tedi Mining Ltd. (OTML), a consortium formed by:
  State of Papua New Guinea (20 per cent); AMOCO, USA (30 per cent); BHP, Australia (30 per cent); Metallgesellschaft, Germany (7.5 per cent); Degussa, Germany (7.5 per cent); DEG (Bank), Germany (5 per cent).
- National Government of Papua New Guinea as represented by: The Secretary of the Department of Minerals and Energy

PERU (*)

Category: Mining

The irrational water management by the mining industry in the south of Peru

Plaintiffs

- LABOR (Civil Association), Ilo, Peru
- The Provincial Municipality of Ilo, Peru
- The Permanent Multisectoral Commission on the Environment (CMPMA), Ilo, Peru

Accused

Southern Peru Copper Corporation (SPCC), USA

PHILIPPINES (1) (*)

Category: Mining

The operations of Benguet Corporation, its Antamok Gold Project, and the consequences for the people of Itogon, province of Benguet, Philippines

Plaintiff

- The Mining Communities Development Centre Inc. (MCDC), Baguio City

Accused

- Benguet Corporation, Benguet Centre

PHILIPPINES (2)

Category: Management

Laguna Lake Crisis: A case of government neglect and mismanagement

Plaintiff

- Solidarity for People's Power, Southern Tagalog (SPP-ST), Los Baños

Accused

- The Government of the Philippines, Manila
- Laguna Lake Development Authority (LLDA), Manila
- Asian Development Bank, Manila
- The Government of Japan
- Overseas Economic Cooperation Fund (OECF) - Philippines
- The Government of France

- Societé Grenofloise d'Etudes et d'Applications Hydrauliques (SOGREAH)

SOUTH AFRICA (*)

Category: Pollution

THOR Chemicals: pollution contamination of water resources around a mercury plant

Plaintiff

- Earthlife Africa (Pietermaritzburg Branch), South Africa

Accused

- THOR Chemical Sa (PTY) Ltd., Cato Ridge, South Africa

TANZANIA

Category: Pollution

Lake Tanganyika: Water Management Problem

Plaintiff

- Mr S.G.M. Ndaro, Department of Zoology and Marine Biology, University of Dar es Salaam, Tanzania

Accused

- AMOCO Petroleum Co., Houston, Texas, USA

CHAPTER 12

The Increasing Role of Public Policy in Integrated Water Management

WIM VAN LEUSSEN

INTRODUCTION

In the Netherlands, as in many other countries, the subject of public policy making, particularly the implementation aspects of policy plans, is currently a much discussed subject. It is realized that central governance is not quite effective, and so policy making should be the result of an interactive process between many actors of which only some should be governmental agencies. In this discussion are also included various aspects of the implementation of plans in the area of integrated water management. The role of public policy in the implementation of water management plans is the main topic of this chapter. Developments in their relative importance are investigated both in the Netherlands and the United States. Notwithstanding major cultural differences between the two countries, a strong similarity is observed concerning the approach and planning towards the main topics in integrated water management.

The chapter begins with a concise summary of relevant developments in the area of integrated water management in both countries. The overview of these developments in the United States is based on the results of a two-month study tour in 1995, during which I visited a large number of stakeholders in water management. Among these were agencies responsible for the formulation of water policy in specific areas and those working on the development of water management plans at the federal, state, and local levels. I also visited some non-governmental organizations, citizens' groups, businesses, industries, and universities. My research was concentrated on rivers, lakes, and estuaries with special attention to the implementation of management plans for the *restoration of aquatic ecosystems*. More details of this study can be found in Van Leussen (1996).

The focus of this chapter is on the role of public policy in integrated water management. On the basis of developments in the USA and the Netherlands, it illustrates how the role of public policy is of increasing importance because of the increasing complexity of the decision-making process in public policy. Such developments need new instruments for their success. Therefore, attention will be given to on going studies searching for adequate concepts and approaches to arrive at improved methods for analysing complex policy processes and for reaching planned goals of projects in good time and with sufficient public support. Attempts to arrive at such new approaches are then discussed with special attention to innovative partnerships. The chapter concludes with three representative case studies from the Netherlands and the United States.

DEVELOPMENTS IN INTEGRATED WATER MANAGEMENT

The Netherlands

Water management has a long tradition and history in the Netherlands. This is not surprising since more than half the country is situated below sea level. In fact, the very existence of the Netherlands has been based on a continuous struggle with water. Therefore, sometimes the Netherlands is called 'Waterland'. Verwey (1980) shows how Dutch culture is interwoven with water. For a recent overview of the history of water management and land reclamation in the Netherlands see Van de Ven (1993).

In the 3rd National Water Policy Document (Ministry of Transport, Public Works and Water Management, 1990), *integrated water management* is defined as a coherent policy and management strategy carried out by the various governmental agencies with strategic tasks on water management following the *water systems approach*, which means that both internal relations (that is, relations between quantitative and qualitative aspects of surface water and groundwater) and external relations (that is, relations between water management and other policy areas, such as environment, physical and nature conservancy) are taken into account. Due to the changing needs of a society, integrated water management also becomes a dynamic policy field, in which the main topics change over time.

Flood control and navigation have a history of many centuries and are still, of course, very active. They comprise safety against river flooding as well as defence against the sea. In the beginning of 1995 high floods resulted in inundation along the rivers Rhine and Meuse, and about 250,000 inhabitants were evacuated within two days.

Water quality is of more recent concern. In the 1960s much attention was given to arresting the intrusion of salinity caused by the deepening of the Rotterdam Waterway. In the 1970s, degradation of the water quality due to pollution became evident in several rivers, lakes, and estuaries such as the Rhine, the Meuse, and the Western Scheldt. The problem still exists but important improvements have been effected through adequate legislation, construction of waste water purification installations, and (international) action programmes.

Nature conservancy and landscape ecology are aspects of integrated water management, which are in their infancy at this moment. However, they are receiving much attention from the Nature Policy Plan (Ministry of Agriculture, Nature Conservancy and Fisheries, 1990), Ministry of Transport, Public Works and Water Management (Ministry of Agriculture, Nature Conservancy and Fisheries, 1994). It must be stated that at the moment most of the attention is going to nature development along the rivers, lakes, and estuaries (cf. Meire et al., 1995). Developments on a landscape scale seem to be the next step and may be expected in the coming years.

There is long experience of *system analysis*, which is the study of the water system in relation to other policy fields. Here it is usual to try to arrive at rational solutions to complex policy problems by applying computer simulations. An example of this is the PAWM (Policy Analysis for the Water Management of the Netherlands) study in the period 1976-83 (Pulles, 1985). Using this experience as a starting point, now decision support systems (in Dutch, Beslissings Ondersteunend System or BOS) are being developed for solving actual water policy problems. This includes a flexible system of models, database management, and GIS applications.

For the implementation of policy plans in real practice, *public policy* is assuming increasing importance. The main difficulty in implementing a policy is 'the complexity of joint action'. Generally, the route to implementing a plan in practice needs the cooperation or approval of a number of institutions, groups, committees, etc. Often working out such a route takes several years, even decades, during which circumstances and responsible persons keep changing. Moving forward along such a route, many competing interests are met. Therefore, in the process what apparently seems to be a simple programme may turn out to be a very complex one, involving numerous participants, a great number of differing perspectives, and a long and tortuous path with decision points that have to be negotiated. It illustrates that modern integrated water management is no longer just a typical technological question, but also includes public policy elements of increasing importance. This aspect will be discussed in greater detail later in the chapter.

The United States

The USA has long experience of water management and handling of water resources. It is not the intention here to give a general overview of these developments, but rather to concentrate on the public policy aspects, especially to focus attention on those points which are important for understanding the present-day developments. For information and discussion on such points, I am greatly indebted to Dr Stakhiv and Dr Hansen of the Institute for Water Resources of the US Army Corps of Engineers in Alexandria, VA and Professor Shabman of the Virginia Polytechnic and State University in Blacksburg, VA. Two important written sources for this section, namely, the report on 'Environmental Activities in Corps of Engineers Water Resources Programs: Charting a New Direction' (Shabman, 1993) and 'Restoration of Aquatic Ecosystems' (NRC, 1992) are also gratefully acknowledged.

It has for a long time been realized that water resources management is closely related to economic development. In 1950, President Truman's Water Policy Commission stated that integrated land and water management could lead to the development of the nation's economy: '. . . the American people are awakening to the new concept that the river basins are economic units; that many problems center around the use and control of the water resources. . . .'

Many water control projects were planned and constructed for multiple purposes of flood control, hydroelectric power generation, and navigation particularly for regional economic development, and many hydrological regimes in watersheds changed tremendously as the result of the construction of dams, levees, and channels. In this way, it was expected that watershed alteration would promote human welfare, and policy makers saw opportunities for using water development as the engine of economic prosperity.

Through the 1960s, the USA built a large capital stock of dams and water delivery systems, mainly through federal government investments. The achievements were impressive. Irrigated agriculture was established in the West, hydroelectric/power production potential was harnessed, inland navigation on the nation's rivers expanded with the improvement of channels, and the construction of locks and dams and flood control projects resulted in a higher level of safety against flooding, while making the original flood plains available for agricultural and commercial uses.

In the 1970s, these developments, looking for possibilities in the remaining free-flowing rivers, met increasing public opposition. In the early 1970s, immediately after the passage of the National Environmental Policy Act (NEPA), many lawsuits were brought against federal water development agencies for their failure to adequately report the loss of environmental values that could occur from allegedly questionable expenditures to alter river flows.

The 1980s were characterized by a policy of reducing the size of and spending by the federal government. For water resources this meant a shift of cost responsibility for water projects to non-federal interests through significant intergovernmental cost-sharing reform. The Reagan Administration commissioned the Water Resource Council to publish the so-called *P&G: Economic and Environmental Principles and Guidelines for Water Related Land Resources Implementation Studies* (WRC, 1983). These guidelines establish standards and procedures for use by federal agencies in formulating and evaluating alternative plans for water and related land resources implementation studies.

One of the subjects in the P&G is 'general public participation'. It is stated that:

Interested and affected agencies, groups and individuals should be provided opportunities to participate throughout the planning process. The responsible federal planning agency should contact and solicit the participation of: other federal agencies; appropriate regional, state and local agencies; national, regional and local groups; other appropriate groups such as affected Indian tribes; and individuals. A co-ordinated public participation program should be established with willing agencies and groups.

Efforts to secure public participation should be pursued through appropriate means such as public hearings, public meetings, workshops, information programmes, and citizens' committees. From this new conception, today, more than ever, groups directly affected by resource-use decisions, or those having an interest in such decisions, increasingly engage in bargaining and negotiation as a means of making choices on the use of water resources.

The increasing public interest in environmental issues continued into the 1990s. Investments in maintaining and restoring the health of the environment were seen as responsible investments for the future. Starting from the principle of 'sustainable development', this care for the future was also adopted by the Clinton Administration. Another important development under the Clinton Administration were the recurrent attempts to 'reinvent government' (cf. Osborn and Gaebler, 1992; Gore, 1993).

This new policy shift gave much importance to the role of cost benefit analysis in the environmental decision-making process, and attempts were made to force public agencies to function on a competitive level with non-government agencies, between offices with other agencies, and with the private sector. It was thought that this would create a responsive, innovative, and entrepreneurial government, where the market mechanism would create a new dynamic throughout the federal government. It is not surprising that these new developments generated a lot of uncertainty among the various agencies in looking for new missions.

The above-mentioned increased requirements for cost sharing with state and local governments, as well as the increased concern of the public with environmental resources, included in the more recent regulations, means that decisions are to be made in consultation with state and local officials and also with various interest groups. This also means that in the new world of public decision making the *negotiation model* has been accepted. This model involves gathering local (and perhaps not so local) 'stakeholders' or their representatives, providing the resulting group with information as requested and perhaps with mediation and help, and waiting for a consensus project to emerge. The conse-quences of this new approach are discussed below.

THE INCREASING ROLE OF PUBLIC POLICY

It is clear from the foregoing discussion that at present both in the USA and in the Netherlands centralized decision making by the government for water management programmes does not generally work. Many actors in society feel that they should be connected with such projects and programmes, and that their opinions and interests are often in conflict. The main groups of stakeholders in such environmental disputes are: federal, state, and local governments; non-governmental organizations (NGOs); private industry; and citizens or groups of citizens. The problem is how to work out a programme which is more or less acceptable to all these groups.

Although public recognition of the importance of an environmental resource is significant in the decision-making process, it may lead to controversy, conflict, or opposition. This means that the formation of cooperative partnerships is not always an easy task. For example, within the framework of the Little River Adaptive Management Plan in Oregon, meetings were organized to form a partnership. The local community members were not interested in participating in these meetings because they perceived this as an attempt by federal agencies to increase their control over private lands. This incident illustrates the importance of careful and clear communication and the value of working on a joint vision and the building of mutual trust at a very early stage of any process.

In this new setting of the public decision-making process, the role of the governmental agencies is changing dramatically. Management through dialogue and cooperation at local and regional levels will be quite different from traditional bureaucratic management. Grumbine (1994) presents several fundamental questions: How might managers establish a common base of information among people with widely varying levels of knowledge and different values? Are 'experts' capable of playing less powerful roles in decision-making? If everyone has an equal voice in decision making, who takes responsibility for the implementation of any decision? How can time-consuming decision making be facilitated in an increasingly multicultural society? It seems that a number of agencies are unaware of the radical implications of creating these partnerships.

The main point in the bargaining process is *consensus building*. An example from the USA of such a process is the water policy in California, with a large number of stakeholders and strong conflicting interests from the agriculture, environment, urban growth, and recreation sectors. Both through the greater public awareness of environmental issues and the very high complexity of present-day regulations, presenting many possibilities for vetoing a proposal, the consensus-building process here seems to be the only possibility for making progress (Kennedy and Barnes, 1992). Continuing with step-by-step activities based on negotiation and accommodation among the various interests is indicated as the most promising approach. Fraser (1992) shows how crisis situations of severe drought contribute essentially to the bargaining process, setting this subject on the top of the agenda of public issues. Nevertheless, it is shown how divergent interests and competing agendas can slow or even paralyse problem resolution.

It is recognized that the building of a consensus must begin within the framework of individual groups before the process can advance to the next level. This is where many efforts are stalled, as consensus at the grass-roots level is not achieved. Proposed solutions tend to polarize the participants even more than the issues themselves. Therefore, a step-by-step process with several levels of increasing detailing of the project is advised.

Consensus building is by no way a perfect process; it has its weaknesses. It is time consuming and often, while consensus efforts are being made, problems can become even more critical. Appropriate deadlines obviously must be established to prevent potentially devastating situations from worsening. Consensus-building efforts require that the parties involved should be willing to compromise and to give something up (Fraser, 1992).

It is recognized that *leadership* by the elected officials is an essential factor in providing incentives for the various interest groups to reach a consensus. Fraser (1992) concludes that the types of decisions that face the present-day leaders are distinct from those faced by visionary leaders of decades ago. Sharply divided publics such as those found today do not lend themselves to similar 'visionary' decision-making. What is needed are leaders who are willing to assume the role of risk taker, forging ahead through sometimes murky waters to facilitate consensus (Fraser, 1992).

From the experience of achieving consensus on water policy in California, it can be concluded that in concept, the path to consensus building is relatively straightforward. However, a simple concept should not be confused with an easy process. Stakeholders must be prepared to give up a little to gain a lot. They must negotiate in all sincerity to assure a 'win' for all sectors and regions of the state. The vital catalyst for achieving consensus is courageous and skilful leadership. If ever there was an issue in need of leadership, it is this (McPeak, 1992).

Innovative Partnerships

In the previous section, the increasing complexity of the present-day decision-making process has been demonstrated. The interconnectedness of the various policy fields, and the recognition that traditional planning methods are no longer adequate resulted in working within networks with a variety of stakeholders. This is also true in the case of the management of aquatic ecosystems, particularly since the strong interconnection between land and water has been recognized, and environmental, health, and economic prosperity are handled simultaneously within the framework of sustainable development.

Due to these developments, the realization of restoration plans is only possible when partnerships are formed through which various groups of stakeholders work together on the basis of joint responsibility. Partnership formation is based on the following factors:

- gathering of the various disciplines
- prevention of doublures
- breaking through the fragmentation characteristic of various governmental agencies
- efficiency: optimizing the use of governmental finances
- cost sharing due to limited budgets
- optimization and fine tuning of various policy fields; integrated objectives
- fine tuning of legislation in the various policy fields

- obtaining of a sufficiently strong basis of support within the larger society
- increasing public involvement

Formation of partnerships depends essentially on the nature of the problem to be solved, the circumstances, and the concerned stakeholders. To meet the problem of the increasing number of governmental agencies that have a stake in programmes for the restoration of aquatic ecosystems, first of all new inter-agency partnerships should be built to ensure that the concerned governmental agencies work together. Second, the role of the governmental agencies is essentially changing from being a centralized decision maker to being a negotiator, working in dialogue and cooperation with the various stakeholders at local and regional levels. Therefore, an increasing number of *public/private partnerships* have been initiated.

Given the increasing awareness of the strong interconnection between land and water, watershed management has assumed increasing importance. However, watersheds are often very large, with a variety of different properties in different regions, from the hydrological, environmental, and social points of view. Therefore, the restoration plans should be tuned to the circumstances of the smaller scale operation, preferably on the scales of sub-watersheds where people feel connected with the local landscape and its rivers, estuaries, and lakes. This calls for *local/state partnerships,* where locally-based strategies are central. These innovative approaches are discussed in the following paragraphs.

Inter-agency Partnerships

In the 1990s an increasing number of departments related to projects for the restoration of aquatic ecosystems came into existence. One of the reasons for this development was the discovery that the health of a water system is strongly related to that of the surrounding land, particularly the watershed, as well as the air, through which substances are transported to rivers, estuaries, or lakes. Therefore, there is an increasing need to look for more collaboration with the concerned governmental agencies. Tippie et al. (1991) argue that the following developments stress the need for inter-agency partnerships:

- management issues have become more controversial
- technical needs have grown more complex
- financial resources have become more limited.

In February 1991 President George P. Bush announced the inter-agency partnership called 'Coastal America', which is based on a coordinated multi-agency approach to address the environmental problems along the US coasts. At that time four federal agencies with principal stewardship for coastal resources participated. Since that time the number of partners has increased to 12 departments. This illustrates how many policy fields are engaged in the process as well as the complexity of the present-day decision-making process.

The strength of the Coastal America partnership is its operational structure, working on three levels (Coastal America, 1993, 1994, 1995):

- national level: policy formulation
- regional level: planning process
- local level: project implementation

The main task of this inter-agency partnership is to stimulate and organize the process of collaboration. At the national level, this means to identify policy issues and conflicts and to ensure resolution in a timely manner. At the regional level, its task is to ensure that environmental protection and restoration objectives are incorporated in regional efforts. The concept of sustainable development is the central starting point. At the local level, care is taken to pool financial resources, technical expertise, and legislative authorities to accomplish the tasks that no agency can accomplish alone. In this way, efforts are made to contribute to cost-effective and innovative solutions (Tippie, personal communication).

Public/Private Partnerships

An increasingly popular way to develop and implement programmes for the restoration of aquatic ecosystems is to foster *public/private partnerships*. In the USA, this is strongly advocated by the President's Council on Sustainable Development (PCSD, 1995). The background is a 'shared responsibility'. It is stated that programmes should be characterized by innovative and integrated approaches to overcome the fragmented historical structures within the government. Of course, the reduction of polarization and the prevention of litigation will also be one of the reasons. Such partnerships will include all relevant groups of stakeholders, such as federal, state, and local governments, non-governmental organizations, private industry, and citizens or groups of citizens.

An example is the San Francisco Bay Estuary Project. The management plan was signed by a large number of representatives from the public and private sectors who felt that the restoration of the Bay was a joint responsibility. It was concluded that as the project moves into the implementation phase, it would be essential to continue this unique public/private partnership (SFBE, 1994). On the basis of this partnership, the implementation committee of the project will have 20-5 members comprising

- key local, state, and federal governmental representatives
- three environmental/conservation representatives
- three business/water/discharger representatives
- one fishery representative.

More examples will be presented in the next section.

Local/State Partnerships

Even in the case of a public/private partnership, giving a good representation to all relevant groups of stakeholders under the agreed upon management plan is not easy. One of the main reasons is that a one-size-fits-all solution does not work because generally the circumstances vary enormously within a watershed. These varying circumstances include differences in land use, industries, population densities, sources of nutrients, heavy metals, organic micro-pollutants or radio-nuclides, as well as differences in economic circumstances and cultural variations, as well as the distance to the river, lake, or estuary. Such a local-level approach also meets the requirement to develop a joint vision in areas where people feel connected to and responsible for their place of residence and/or work.

Moreover, all the representative groups have their own priorities, expectations, and demands, and a joint vision for managing the development of the area must evolve taking all these into consideration. This is of course not easy but it has been found that unless relevant local groups of stakeholders are involved right from the beginning, from the planning and then at the implementation stage, success is almost impossible. Recently, this kind of local-level approach has been widely adopted.

In the USA such an approach has been proposed in the Potomac basin. The Potomac river is one of the three largest river systems draining into the Chesapeake Bay (watershed 14,670 square miles). The geography of the basin is extremely varied. Its landscape is characterized by mountains, valleys, regions of low hills and ridges, low-lying plain regions, and steep ravines close to some of the streams. The main problem is the reduction of nutrient input, both from 'point sources' and 'non-point sources'. Looking at the Potomac river basin from a land-use perspective, it is predominantly rural, dominated by forests, pastures, and other agricultural land uses. However, urban areas, which account for a smaller percentage of the basin, make a significant contribution to the load of nutrients into the Potomac river.

It was realized that for a successful strategy public involvement is a key element. Through public meetings and a questionnaire, the government of the State of Virginia has been directed by citizens in the Potomac Basin to solve the nutrient, reduction problem through partnerships with local governments and locally-based citizens' groups and interest groups. The state organized meetings between officials of local government and representative local interest groups with a view to promoting local initiatives and incorporating local perspectives into Virginia's Potomac tributary strategy. Localities were encouraged to form local or regional 'tributary teams' consisting of representatives of local government, business, industry, environmental groups, agriculture, and others as appropriate. The 'local strategies' could include a number of programmes such as: local ordinances to reduce the nutrient impacts of future land use conversion or development; funds for regional storm-water treatment facilities; upgrades of waste water treatment plants; promotion of voluntary nutrient reductions among local businesses and industries; and additional funds for the implementation of agricultural BMPs within their jurisdiction (Commonwealth of Virginia, 1995).

From this local/state partnership we learn that an integrated water management plan, for example a restoration programme, should take full account of the economic, social, and geographic variations within the watershed, and should also take cognizance of the feelings of various people in the concerned area, and try to convince them of the importance of the water system for their quality of life and of their joint responsibility for the success of the plan.

ILLUSTRATIVE CASES

Biesbosch

The Biesbosch is a freshwater tidal area comprising a complex network of rivers, creeks, and islands. Its surface is about 5,500 ha. In 1421, due to the St. Elizabeth Flood which destroyed 13 villages, this area was changed into an inland sea with strong tidal effects from the North Sea. Under these conditions a specific freshwater tidal landscape developed. It is the largest freshwater tidal wetland in Europe. In the nineteenth century the river 'Nieuwe Merwede' divided this area into three parts: the Sliedrechtse Biesbosch, the Dordtse Biesbosch, and the Brabantse Biesbosch.

Since 1970 the tidal range has been reduced from 2.00 m. down to 0.2-0.3 m. and in some areas to 0.6-0.7 m. As a result, the intertidal area decreased drastically and a large amount of (polluted) fine-grained sediment has been deposited. In 1985 the first proposal was made to develop the Biesbosch into a national park, restoring the tidal regime as much as possible. In 1991 the 'Management Plan Biesbosch' became available, and in 1993 this area received the status of a national park. Under this heading the management plan has been accepted. A covenant was signed in March 1994, by which 4 ministries, 2 provinces, 5 cities, 2 recreation agencies, and 2 local water agencies agreed with the objectives of the management plan (Overlegorgaan Nationaal Park De Biesbosch, 1994).

From a public policy point of view the main points are:

- *Multidisciplinary approach.* Under the project a number of institutions dealing with a variety of disciplines are working together. These include the physical sciences and chemistry, aquatic ecology, spatial planning and landscape ecology, cultural history, sociology, recreation, and economics.

- *Cooperation of various policy fields*. The project is an agreement between 4 ministries, which means that the following policy fields are engaged: agriculture, nature, and fisheries; transport, public works, and water management; physical planning and environment; finances.
- *Network with many actors*. The covenant illustrates the co-operation of a number of actors from the national, regional, and local governments.
- *Consensus building*. Agreements, as indicated in the management plan, were made on the basis of consensus. The total period of negotiations took about 10 years. Within the management council, decisions are taken by a majority of votes.
- *Education programme*. Education is a special point of attention in the covenant. Two visitors' centres and one museum are dedicated to the Biesbosch. Biesbosch leaflets have been published with information on this area and daily sailing cruises are planned.

The River Meuse

The river Meuse is about 850 km. long and its source is 450 m. above sea level, on the Plateau de Langres in France. Its watershed has a total area of 33,000 km^2 of which 10,000 km^2 is situated in France, 13,000 km^2 in Belgium, 4,000 km^2 in Germany, and 6,000 km^2 in the Netherlands. The water of the Meuse is largely supplied by precipitation in its catchment area and not by melting snow or ice. The discharge regime is extremely irregular, because the river reacts on precipitation within a few days. The river discharge varies from values lower than 10 m^3/s up to more than 3,000 m^3/s.

Due to the extremely varying discharges, in the past there were many problems concerning discharge capacity and navigation. Therefore, during the past two centuries a large number of engineering works were carried out to normalize the river. Such works included straightening some of the river bends, channelization of the river, reinforcement of some stretches of the river banks, and the construction of a number of sluices and weirs. Such interventions changed the river enormously; some natural woodlands disappeared, a number of islands were removed, and some land adjacent to the river has been taken over for farming or building.

Such interventions in the watershed had implications for the discharge regime of the river. After a heavy rainfall the water rushes in quickly from the steeper and higher peaks. The consequences of this were clear in 1993 and 1995, when high floods resulted in inundations along the river. After the flood of 1993, plans were made to prevent such occurrences (Commission Boertien, 1994). These plans gained momentum during the floods of 1995, and found a firm basis in public opinion and in the parliament. Shortly after the 1995 flood, a special law was passed by the Dutch parliament, the so-called 'Deltawet Grote rivieren' (Special Law for Large Rivers), on the basis of which the procedures for taking public decisions were shortened drastically. Local embankments were legalized in the same year to guarantee these areas a minimum safety against flooding of once in 50 years; it was also agreed that higher safety conditions (1/250) should be obtained before 2005. To reach that stage, a number of strategies were worked out, such as deepening and widening the summer bed, digging of the winter bed, cutting of adjacent branches and/or high water channels.

These were among the activities of the 'Meuse Works', aimed at the enlargement of safety against flooding; improvement of possibilities for shipping; and nature development. To realize these ends in good time and within the estimated costs, a special institution was founded, in which two national and one regional governmental agencies cooperate: Ministry of Transport, Public Works and Water Management; Ministry of Agriculture, Nature Conservancy and Fisheries; and the Province of Limburg. The responsibilities of each of these agencies were laid down in a covenant. Indeed this is a clear example of inter-agency partnership.

In 1997 the available budget was not sufficient, and the Dutch government proposed to postpone the date of higher safety from 2005 to 2015. This caused much opposition in the local region of the river Meuse. Groups of representatives of industry (gravel dredgers) and nature conservancy organizations (NGOs) have made alternative plans and have asked for cooperation from the government. So it is possible that under these changed conditions other partnerships might be formed to realize the Meuse Project.

Mono Lake

The Mono Lake Project is an example from the USA, where local initiatives ultimately resulted in the restoration of a damaged aquatic ecosystem. The Mono Lake is situated in California. The lake occupies about 65 square miles of the 700 square miles Mono basin. The basin is recognized as an extensive, fragile visual resource, with a landscape character typical of the Great basin, but much enhanced by the presence of Mono Lake. The Sierra Nevada, seen as a backdrop, vastly increases the lake's visual value. The surface of the lake is highly reflective, mirroring the surrounding elevated landforms. In calm summer conditions, the water is clear and usually reflects the vivid blue of the sky; in winter, the lake may appear green. Tufa formations, scattered around the lake's current shoreline, are a unique scenic resource of the basin.

However, the quality of Mono Lake was gravely threatened when the city of Los Angeles started diverting water from the freshwater streams which feed the lake. This resulted in a major decline in the water level in Mono Lake. Concerned citizens formed the Mono Lake Committee in 1978 in order to fight for the lake's survival. After numerous rounds of legal hearings, the California Supreme Court unanimously declared, in 1983, that Mono Lake must be protected. Then, in 1991, a California court took emergency action, ruling that Mono Lake must remain above the 6,377 foot elevation.

In 1994, after a 16-year battle and 59 years of the degradation of the lake, the Water Board issued an order to raise the water level of Mono Lake. This is an example of how a small group of citizens, supported by scientists and lawyers, can initiate the restoration of an aquatic ecosystem. During the celebration of this outcome, Martha Davis, executive director of the Mono Lake Committee, said: 'Today's victory is an affirmation that we all can make a difference—if we care enough to follow our hearts'.

The case is not a typical example of an innovative partnership, but it does show how in a specific situation people feel connected to their surroundings and feel responsible for the quality of their place of living. Such relations between the local people and the water bodies on which they depend for survival are expected to be of increasing importance for the successful implementation of restoration plans.

CONCLUSION

We have seen how in both the Netherlands and the USA, integrated water management has shifted from a policy area with impressive investments in closure works or dams, levees, water delivery systems, reclamation works, navigation channels, etc., and to just having faith in the natural sciences to a situation characterized by increased public opposition, particularly due to loss of environmental values.

This trend is still increasing due to the discovery of the social relevance of specific water systems, changes in social values, and stronger relations of water management with other policy fields. Presently, projects of integrated water management can only be realized amidst a number of stakeholders with generally a diversity of ideas and interests. These developments make public policy elements of increasing importance within modern integrated water management.

In these changed circumstances, traditional planning methods are seen to be no longer adequate. Partnerships with relevant stakeholders through working in networks seem to be critical in arriving at a successful integrated water management. Only in this way can the required public support be achieved.

However, the formation of cooperative partnerships is not an easy task. A consensus-building approach seems to be promising, but it does not give any guarantee for realizing water management projects in good time or at reasonable costs. Even the best choice from a number of alternatives is not guaranteed. It shows that the consensus-building process is by no means a perfect process; it is full of pitfalls and risks. Nevertheless, it is expected that the consensus-building process will contribute to a stronger basis within society, and will lead to a decision-making process of higher quality.

Courageous and skilful leadership with strong perseverance is essential to achieve consensus. In this new setting of the public policy decision-making process, the role of the governmental agencies is changing dramatically. Management through dialogue and cooperation at federal, regional, and local levels between a variety of stakeholders is quite different from traditional bureaucratic management procedures. In the new situation, social and economic skills are as important as scientific and engineering expertise.

REFERENCES

California State Water Resources Control Board (CSWRCB), *Draft Environmental Impact Report for the Review of the Mono Basin Water Rights of the City of Los Angeles*, vols. 1, 2, and 3, California State Water Resources Control Board, Division of Water Rights, Sacramento, California, 1993.

Coastal America, *Forging Partnerships to Restore Coastal Environments, 1993*, Coastal America Progress Report, Washington, D.C., 1994.

———, *Building Alliances to Restore Coastal Environments, 1992*, Coastal America Progress Report, Washington, D.C., 1993.

———, *Sustaining the Commitment to Our Coastal Environment, 1994*, Coastal America Progress Report, Washington, D.C., 1995.

Commonwealth of Virginia, *Virginia's Potomac Basin Tributary Nutrient Reduction Strategy*, Virginia Department of Environmental Quality, Virginia Department of Conservation and Recreation, Virginia Chesapeake Bay Local Assistance Department, 1995.

Fraser, J.P., 'Consensus on Water Policy in California', in *Achieving Consensus on Water Policy in California*, Water Education Foundation, Sacramento, CA, and Edmund G 'Pat' Brown Institute of Public Affairs, California State University, Los Angeles, 1992, pp. 37-248.

Gore, A., *From Red Tape to Results: Creating a Government that Works Better and Costs Less: The Report of the National Performance Review*, Plume, 1993.

Interagency Ecosystem Management Task Force, *The Ecosystem Approach: Healthy Ecosystems and Sustainable Economies, Volume I Overview*, 1995.

Kennedy, D.N. and S.M. Barnes, 'State water planning: Finding common ground', in *Achieving Consensus on Water Policy in California*, Water Education Foundation, Sacramento, CA, and Edmund G. 'Pat' Brown Institute of Public Affairs, California State University, Los Angeles, 1992, pp. 19-28.

McPeak, S.W., 'Achieving consensus on water policy', in *Achieving Consensus on Water Policy in California*, Water Education Foundation, Sacramento, CA, and Edmund G. 'Pat' Brown Institute of Public Affairs, California State University, Los Angeles, 1992, pp. 181-93.

Meire, P., M. Hoffman and T. Ysebaert (eds.), *De Schelde: een stroom natuurtalent. Gezamenlijk project van het Miniterie van de Vlaamse Gemeenschap*, het Ministerie voor Instituut voor Natuurbehoud, Hasselt: Rapport, 1995.

Ministry of Transport, Public Works and Water Management and Ministry of Agriculture, Nature Conservancy and Fisheries, *Natuur aan het werk. Een verkenning van mogelijkheden voor grootschalige natuurontwikkeling langs rijkswateren en rijkswegen*, Studierapport maart, 1994.

Ministry of Transport, Public Works and Water Management, *Third National Policy Document on Water Management* (in Dutch: 'Derde Nota Waterhuishouding'), SDU Uitgeverij, 's-Gravenhage, 1989.

National Research Council (NRC), *Restoration of Aquatic Ecosystems: Science, Technology and Public Policy*. Committee on Restoration of Aquatic Ecosystems: Science, Technology and Public Policy, Water Science and Technology Board, Commission on Geosciences, Environment and Resources, National Research Council, National Academic Press, Washington, D.C., 1992.

Osborne, D. and T. Gaebler, *Reinventing Government: How the entrepreneurial sprit is transforming the public sector*, Addison-Wesley Publishing Company Inc., Reading, MA 1992.

Overlegorgaan Nationaal Park de Biesbosch, *Beheers- en inrichtingsplan Nationaal Park de Biesbosch 1993-2004*, Overlegorgaan Nationaal Park de Biesbosch, Eindhoven, 1994.

President's Council on Sustainable Development (PCSD), *Status Update*, April 1995.

Pulles, J.W., *Beleidsanalyse voor de waterhuishouding in Nederland/PAWN. Hoofddirectie van de Waterstaat*, 's-Gravenhage, 1985.

San Francisco Bay Estuary Project (SFBE), *Comprehensive Conservation and Management Plan*, San Francisco Bay Regional Water Quality Control Board, Oakland, California, 1994.

Shabman, L., *Environmental Activities in Corps of Engineers Water Resources Programs: Charting a New Direction*, U.S. Army Corps of Engineers, Water Resources Support Center, Institute for Water Resources, Alexandria, Virginia, 1993.

Tippie, V.K., H.K. Sopher and F.P. Kineon, 'Coastal crisis calls for new directions in environmental policy: How do we address water quality degradation, habitat losses, and living marine resource declines?' *Sea Technology*, August 1991, pp. 10-20.

Van de Ven, G.P., *Man-made Lowlands: History of water management and land reclamation in the Netherlands*, International Commission on Irrigation and Drainage (ICID); Royal Institute of Water Management, Utrecht, Matrij, 1993.

Van Leussen, W., *Public Policy Aspects of Integrated Water Management: Implementation of management plans for the restoration of aquatic ecosystems, particularly estuaries*, Virginia Polytechnic Institute and State University; Netherlands School of Government, and Ministry of Transport, Public Works and Water Management, The Hague, 1996.

Verwey, G., *Searching for the fundamentals in Dutch history* (in Dutch: 'Op zoek naar het wezenlijke in de Nederlandse geschiedenis'), Elsevier, Amsterdam, Brussels, 1980.

United States Water Resources Council (WRC), *Economic and Environmental Principles and Guidelines for Water and Related Land Resources Implementation Studies*, Governmental Printing Office, Washington, D.C., 1983.

CHAPTER 13

The Rhine Research Project: The Port of Rotterdam Scour Control Programme

TIEDO VELLINGA

INTRODUCTION

Rotterdam is a port with open links to the sea, and the tributaries of the Rhine and the Meuse empty into it. Consequently, a huge amount of sediment enters this hydrologic system and settles in the port. Dredging is necessary to ensure that ships can pass freely through the Port of Rotterdam. If there is no dredging, the port would slowly silt up. Since it is vitally important to maintain a specific depth for ships to float, dredging has to be a continuous process for which the Rotterdam Municipal Port Management is responsible.

The dredging itself is not a difficult task, but the disposal of the dredged matter (in large quantities) is a problem. This matter was formerly used for raising low-lying polder areas. But once it was known that the dredged material contained contaminants, it was considered as undesirable waste matter. There are strict legislative regulations for the disposal of such matter; it cannot be disposed of just anywhere. The disposal of this dredged material is a problem with which Rotterdam has been struggling since the late 1970s.

Yearly about 20 million m^3 of sediment has to be dredged to maintain the necessary depth for shipping. Of this, 15 million m^3, being mainly of marine origin, is only slightly contaminated. It is put back into the sea; this is called sustainable relocation.

The remaining 5 million m^3 of the dredged material is

contaminated to such a degree that it has to be disposed of under strict environmental regulations. The full extent of the problem of the contaminated dredged matter became clear in Rotterdam in the 1970s. In 1972, samples were taken, albeit sporadically, for the first time from the waterbed. When it gradually became evident that the contamination of waterbed was a structural problem, a systematic approach to sampling sludge was developed. From 1982 onwards, the waterbed of the entire Port of Rotterdam was sampled every two years. From 1990 onwards, samples were taken annually. An entire range of some 40 contaminants was examined. These included heavy metals, polycyclic hydrocarbons (PCGs), polychlorinated biphenylene (PCBs), organochlorine compounds, and mineral oils.

An increasing amount of information concerning the degree of contamination was becoming available. Now the Port of Rotterdam had to seek alternative ways of relocating its dredged material. This was now regarded as waste material, and its disposal was regulated through legislation. When the Municipality of Rotterdam was faced with the problem of contaminated dredged material, it took the following action. At immense cost and great effort, two unique depots were constructed to store the contaminated dredged material: the Slufter for moderately contaminated sludge and the Papegaaiebek for heavily contaminated sludge.

The completion of the Slufter in September 1987 solved, for the time being at least, the problem of where to store the contaminated sludge from the Port of Rotterdam. The construction of the sludge depot had cost NGL 200 million and it had a capacity for storing 50 million m^3 of sludge. It was expected to get full in the year 2002.

The Papegaaiebek sludge depot was also completed in 1987. This depot, intended for heavily contaminated class IV dredged material, had a capacity for storing 1.2 million m^3 of sludge. Class IV dredged material occurred only at a limited number of locations.

Rotterdam had no wish to build another of these expensive depots and hoped that there would be no need to do so. If the sources of contamination could be successfully cleaned up, it was expected that all dredged matter would be so clean by the year 2002 that it could be either dumped into the sea or could be used on land without any danger. The final solution could only be found by cleaning up the sources of the pollutants once and, for all.

Only when the causes of the contamination of the sediments have been removed will it be possible to once again use the dredged sediment beneficially as before. Since the majority of the contaminant sources was located in the Rhine estuary, in 1985 the Rhine Research Project was initiated to bring this about.

THE RHINE RESEARCH PROJECT

Within the framework of its Rhine Research Project, Rotterdam focused its attention from 1985 onwards on the sources of contamination upstream of Rotterdam. The Rhine Research Project mainly concentrated on cleaning up specific discharge points upstream of Rotterdam. This project would ultimately cost the Rotterdam Municipal Authority a total of NGL 14 million. The aim of the Rhine Research Project was first to identify the sources and points of discharges into the Rhine and then to try to reduce or stop these through negotiations. The project was set up as a multidisciplinary project. Engineers, lawyers, and public relations people worked closely as a team.

The activities of the technical experts were focused on tracing the specific discharge points, quantifying them, and determining their relative share in the overall contamination of the Rhine. Legal research established the legal position of the Rotterdam Municipal Authority. The civil law approach played a central role here. The basic starting point was that the costs of the damage suffered by Rotterdam as the result of contaminated discharges could be claimed from those causing the contamination. Public relations activities made sure that the dischargers, the government authorities, and the public at large were informed about the goals set by the Rhine Research Project.

An investigation was carried out into the legal situation of the Municipality of Rotterdam *vis-a-vis* the polluters. On the basis of the legal position, a model contract was drawn up which could be used as a basis for signing reduction agreements with polluters. In a technical study, Rotterdam identified and quantified the most important specific sources of pollution upstream. The entire pollution situation of the Rhine was mapped out. Public relations activities were used to publicize the Rotterdam problem, particularly in Germany, Switzerland, and France. Based on the information it was strongly recommended to the polluters that they should negotiate with the Port Authority. After a slow start this process did result in the signing of a satisfying number of agreements on source reduction.

Right from the start, the Rhine Research Project found itself rowing against the current. It was not generally accepted that a municipal authority could take action to combat the problem of trans-boundary river contamination. A number of dischargers even considered the action taken by Rotterdam as reprehensible environmental activism, however, thoroughness and professionalism with which the technical experts, lawyers, and public relations specialists set about their work in the Rhine Research Project ensured that Rotterdam was eventually recognized as an equal partner in the negotiations by the discharging industries along the Rhine. The situation was undeniably facilitated by the fact that a change in social values came about in the social field of influence within which the Rhine Research Project operated with its critical and incentive standpoints. Dischargers and the various national governments along the Rhine have learnt to appreciate that people have a responsibility for the environment which extends beyond the borders of their own factory site or their own country,

The viewpoint propounded within the framework of the Rhine Research Project—that a port can operate best in a clean environment—is nowadays accepted almost as a matter of course. Now, after more than a decade of its inception this approach can be evaluated and it is clear that it has been very effective. The sources of pollutants are steadily diminishing and the sediment in the river Rhine today is of a much better quality. Whereas in the past, yearly 10 million m^3 of dredged matter needed special care, over the years this amount has been reduced to less than half.

Dischargers have signed agreements with the Rotterdam Municipal Authority concerning a reduction in their discharges. By doing this, they have in a certain sense recognized that the legal standpoints propounded in the Rhine Research Project and initially fiercely opposed are in fact valid. In exchange for a reduction in discharges, the Rotterdam Municipal Authority relinquished its right to seek compensation from the discharging industries.

Negotiations with the major dischargers identified by the Rhine Research Project achieved concrete results. One of the characteristics of the Rhine Research Project is that the different special activities had a stimulating and strengthening effect on each other. Although it is ultimately the total result that counts, which is more than the sum of the separate parts, this chapter will briefly highlight the different activities undertaken by the Rhine Research Project.

THE TECHNICAL STUDY

Looking back, it is quite amazing how little was known in 1985 about the sources of pollution of the river Rhine. The main purpose of the technical study was to map out the pollution situation as accurately as possible. Where did the pollution come from? What were its sources?

The Rhine Research Project is specifically concerned with the contamination of the dredged sediments in the river and the basins of the Port of Rotterdam. This contamination is assessed on the basis of some 40 substances: heavy metals, PAHs, PCBs, oil, and other organic micro-pollutants. It would be exceedingly complex and expensive to investigate the origin of all these substances. For pragmatic reasons it was therefore decided to restrict the technical study in the first instance to the five heavy metals, namely, cadmium, chromium, copper, lead, and zinc. This limited the scope of the problem to be investigated. Later on, during the course of the project, desk studies were carried out into the other pollutants.

The technical study began with an exploratory trip along the mainstream of the Rhine. Numerous discharge points were identified and samples were taken. On the basis of this identification trip, the sources of the pollution level (i.e. quantification carried out without the help of external data) were identified without too much difficulty. Points of discharge which released their pollutants suspected of playing a large role in the overall pollution of the Rhine were selected for quantification. In the following round of measurements, these sources were sampled in such a way as to assess a reasonable estimate of the yearly discharged load of contaminants.

An independent assessment of the pollution level (i.e. quantification carried out without the help of external data) appeared to be possible without too much difficulty for points of discharge which released their pollution into the river by means of a pipe or duct above or in the water. A sampling torpedo (the 'measuring fish') was especially designed for this purpose.

When these determined annual quantities proved to be substantial, they were discussed with those responsible for discharging the pollutants at specific points. These technical discussions, at which in most cases the accuracy of the measurements was endorsed, were followed later in the project by negotiations. In order to support these negotiations, once again a measuring trip was organized to measure the levels at the major discharge locations.

During the course of the Rhine Research Project, the area under study was extended. In 1987-8, the large Neckar, Main, and Moselle tributaries were studied in the same way as the Rhine mainstream. The identification of specific discharge points was followed by the quantification of discharge points which seemed to be important. In 1988-9, the basins of the Ill, Lahn, Sieg, Ruhr, and Lippe were investigated in an identical way. Balances were made for the rivers as well as overviews for the different regions.

The different sources were categorized as local and foreign industrial point discharges, domestic emissions, atmospheric contributions with the backgrounds in the river and the sea serving as baselines. This was used, on the one hand, to set priorities in the source control programme and, on the other hand to make a list of the most important dischargers. In total six measurement campaigns were undertaken between Rotterdam and Switzerland, and over 7,000 water samples were taken at 410 different discharge points and 105 tributary rivers. Some 38,000 chemical analyses were made.

The technical study ultimately produced a list of 10 discharge points whose contribution to the total pollution of the Rhine was more than 1 per cent for any one of the five parameters investigated. The city of Rotterdam conducted negotiations with those responsible for these discharges concerning the reduction of the discharges which were causing so much damage to the Port of Rotterdam.

THE LEGAL STUDY

The aim of the Rhine Research Project was to tackle the sources of pollution contaminating the silt dredged up in the Port of Rotterdam. In order to ensure that discussions with polluters were not entirely without obligation, the legal situation was carefully investigated. This task was carried out very constructively in collaboration with the Institute for Environmental Liability at the Erasmus University of Rotterdam.

The starting point of the legal study was civil law. The Port of Rotterdam suffers damage as a consequence of the pollution of its dredged material. The damage was assessed at around 200 million guilders: the cost of constructing a large-scale landfill for the safe disposal of the dredged matter. Those causing this damage could be held liable for this damage on the grounds of it constituted an unlawful act. An examination of the different national legal and jurisprudence systems produced standpoints which could be raised in relation to the major polluters. The legal study concluded that in the case of (environmental) damage, guilt already arises when a risk is created. Strict scientific evidence for the causal relationship between the discharge and the damage inflicted is not necessary. It is sufficient if it can be plausibly demonstrated in court that the prosecuting party is suffering damage due to the action of the accused party.

Since the pollution of an entire river is a very complicated affair, involving many complex factors, the study into the issue of multiple offenders and multi-causality in the different countries was of major importance. The pollution of Rotterdam's silt and sludge is caused by different parties, and faced with this problem it was best to act on the basis of 'group action'. On this basis, one polluter is held liable for the total damage. The advantage of this approach is that the complex task of assigning every individual's contribution to the total pollution is left to the polluters themselves. Once the court has passed the judgement, the prosecuting party is not involved in the responsibility actions between the different parties themselves.

The information, data, and distilled standpoints acquired from the legal study were brought to the attention of the academic and practising legal world via publications and conferences. Once the project was advanced far enough, it was possible to negotiate on the basis of the results of the discharges, and the lawyers came up with the standpoints they had developed. The aim of the negotiations was not, however, to bring about a legal trial, but rather to reach an agreement concerning the reduction of discharges.

During the process of negotiations, a polluter was offered the opportunity of having his liability rights waived or deferred in exchange for agreeing to undertake the reduction of his discharge. Depending on the reductions achieved, different types of covenants were consequently drawn up and signed.

THE PUBLIC RELATIONS CAMPAIGN

The Rhine Research Project was accompanied by public relations activities right from the start. The first objective was to get people to acknowledge the situation of the Port of Rotterdam. Most people in the Rhine basin were not aware that there was a problem of managing the dredged material at the end of the river Rhine. Most people thought that pollution of a river meant pollution of the water and in the end dilution into the sea. But the fact that most of the pollution is attached to the suspended sediments in the river and that most of these sediment particles flocculate, precipitate, and accumulate in the estuary, including the Port of Rotterdam, was unknown to most of them.

The aim of the public relations programme of the Rhine Research Project was to ensure that the Swiss, the French and the Germans in the Rhine basin were informed about the silt problem in Rotterdam. Journalists were invited to see the problem in Rotterdam with their own eyes. This led to publicity on television and radio and in newspapers and magazines. The public relations specialists were present at conferences and trade fairs to draw attention to the silt pollution problems in Rotterdam. Major cities in the Rhine basin were visited three times, in 1986, 1988, and 1992 during publicity campaigns. The problems of the Port of Rotterdam were illustrated with videos, slides, brochures, and exhibition walls. Lectures were given and discussions organized. Throughout these activities the policy was to mention no names of the polluters who had been identified through the technical study. This approach continued until a covenant was signed.

The publicity generated in this way was enormous. The problem of Rotterdam's silt was dealt with in innumerable television and radio broadcasts. Articles constantly appeared in leading newspapers in Germany, France, and Switzerland about the Rotterdam aspect of the Rhine problem.

During this entire publicity, the names of none of the dischargers identified by the Rhine Research Project were mentioned. This approach was dictated by the desire to create a favourable negotiation climate with the point dischargers with whom the Rhine Research Project was endeavouring to bring about an agreement for a reduction in the discharge. And a change in the climate between Rotterdam and the dischargers was desperately needed.

In the initial phase of the project, a number of dischargers took an exceedingly negative and sceptical attitude towards the activities of the Rotterdam Municipal Authority. Why was Rotterdam interfering with their effluent discharge and what right did they have to do so? Therefore, on every possible occasion, it was explained why Rotterdam was taking this action. In its public relations activities, Rotterdam explained that this was not simply an environmental problem. It was emphasized that the discharge of contaminated waste water was damaging Rotterdam's economic interests. As the party suffering the damage, Rotterdam wished to negotiate in a businesslike manner with those who were causing the damage. The public relations working party emphasized the thoroughness with which the technical study had been carried out, and the costs involved. The legal standpoints were also explained.

While this publicity campaign was in progress, a change began to take place in environmental awareness in the Rhine countries. People gradually began to see that environmental problems were problems to be taken seriously. The realization began to dawn on more and more of them that if society were to continue to function properly, it would have to pay serious attention to environmental problems. This shift in society's appreciation of the importance of dealing with environmental problems had repercussions with regard to the functioning of the Rhine Research Project. Rotterdam was at last accepted by the dischargers as a serious negotiating partner with whom business could be done.

THE COVENANTS

Talks were held with dischargers from 1986 onwards. First of all, the measurements taken in respect of the load assessments of the discharges were discussed. In addition, Rotterdam's silt problem was explained together with the way in which the city wished to solve this problem. At these talks, a discharger was also asked about his plans to further reduce his discharge in the future.

At the beginning of 1990, in the third phase of the Rhine Research Project, the negotiating team began talks with the principal dischargers identified in the first and second phases. The purpose of these negotiations was to reach an agreement concerning the reduction in discharges.

On the basis of the quality of the dredged material, it was calculated by what percentage a particular substance in the sludge needed to be reduced so that it could be dumped in the sea. A major discharger was requested to reduce his contaminated discharge before the year 2002 by this percentage (for copper 80 per cent, for example). If this were done, Rotterdam would relinquish its right to compensation in exchange for the reduction in the discharge.

When the first two stages of the Rhine Research Project were completed at the end of 1989, a start was made during the third stage, from 1990 onwards, on conducting the actual negotiations with the discharging industries. A list of ten specific discharge points, along the mainstream of the Rhine, was compiled on the basis of the technical study. These discharge points were chosen because each of them discharged to the extent of at least one parameter a quantity greater than 1 per cent of the total quantity in the river at the German-Dutch border (reference year: 1985). During the course of 1991 a further three discharge points along the tributaries were added to these original ten discharge points.

It was eventually possible to make agreements concerning a reduction in discharges with all the major discharging industries along the mainstream of the Rhine, either directly or with an umbrella organization representing them. Some of these agreements were recorded in contracts (Verbend der Chemischen Industrie, Duisburger Kupferhütte, Berzelius), and others in the form of an exchange of letters (Sandoz, Ara Pro Rheno, Rhone-Poulenc, Atochem). In some cases definite reductions were agreed upon until the year 2002. In other covenants interim agreements were made for 1995. Further negotiations were to be held concerning additional reductions. In return for an agreement the Port of Rotterdam will not sue the discharger for damages.

Agreements are made on all the substances that are discharged, but they differ from substance to substance. Also, the agreements contain paragraphs about the possibilities to monitor and control the load reduction. In the agreements concerning the reduction of discharges in general, the reference years 1985 and 1987 have been used in respect of the discharge level. The results are presented

TABLE 1: LOADS DISCHARGED WITH REDUCTIONS AGREED ON UNTIL 1995 WITHIN THE RHINE RESEARCH PROJECT

Parameter	Load 1986-7	Reduction till 1995	Reduction
Cadmium	2.4 tn	1.4 tn	58%
Chromium	64.8 tn	110.8 tn	67%
Copper	106.9 tn	60.9 tn	57%
Zinc	636.2 tn	269.6 tn	43%
Lead	35.4 tn	21.7 tn	61%
Mercury	0.7 tn	0.3 tn	43%
Nickel	70.2 tn	25.1 tn	36%
AOX	1,500.0 tn	600.0 tn	40%

in Table 1. The first column shows how much was discharged by the point dischargers, with whom agreements have been made, in 1986-7. The second column gives the total reductions agreed upon by the Rhine Research Project. The last column expresses the agreement made by the Project as a percentage of the total load jointly discharged by the Rhine Research Project dischargers.

THE DIFFUSE SOURCES

As the Rhine Research Project technical study progressed, it became increasingly clear that the diffuse sources of contamination played a significant role in the overall contamination. Diffuse sources of contamination occur everywhere in the Rhine basin. Contamination is caused at innumerable locations by domestic activity, road traffic, agriculture, and industry. The scale of this contamination is frequently not very large per individual source of pollution (e.g. one household, one car). But together they represent a huge quantity of contamination for a river such as the Rhine. The contaminants enter the water through the washing of streets, roads, and farmland, via the air, via drains and sewage purification plants. Since the start of the Rhine Research Project, the non-point sources of pollution have increased in relative significance. Table 2 presents the estimated non-point sources of pollution along the river Rhine.

It is more difficult to achieve results in the field of diffuse sources of contamination with the legal instruments developed within the framework of the Rhine Research Project. Contrary to the situation regarding specific discharge points, it is not so easy to hold a specific party liable for a substantial contribution to the overall

TABLE 2: SURVEY OF THE NON-POINT SOURCES OF CONTAMINATION OF THE RHINE (NO ACCOUNT HAS BEEN TAKEN OF THE NATURAL BACKGROUND EXCEPT FOR ARSENIC)

Parameter	Ind. non-point discharge	Discharge	Specification of non-point sources
Zinc	30%	70%	domestic 30%, small business 10%, deposition 25%, road transport 3%, agriculture 2%
Copper	35%	65%	domestic 60%, deposition 2%, shipping 2%, agriculture 1%
Chromium	60%	40%	small business 40%
Lead	30%	70%	domestic 35%, deposition (mainly road transport) 35%
Cadmium	75%	25%	domestic 15%, deposition 10%
Nickel	30%	70%	small business 35%, domestic 20%, deposition 10%, road transport 3%, agriculture 2%
Mercury	40%	60%	dental practice 25%, domestic 5%, deposition 2%, others 3%
Arsenic	10%	90%	natural 80%, agricultural 5%, deposition 2%, others 3%
PCHs	5%	95%	deposition 50%, shipping 20%, creosoted bank protection 25%

contamination of the river. Since the diffuse sources of contamination are difficult to identify, civil law actions are therefore not suitable for the time being. Nevertheless, the Rhine Research Project wished to cover the entire spectrum of the Rhine contamination problem with its activities. There was only one method left to tackle the diffuse sources of Rhine contamination: public relations.

On the public relations tour of 1992, the diffuse sources of contamination were dealt with extensively. Attention was drawn to the problems in seminars, slide presentations, and in a video film. It was pointed out to the authorities in the Rhine countries that they had a responsibility to reduce contamination from the diffuse sources by adopting and enforcing stricter legislation. The need was stressed for a reduction in this form of Rhine contamination, however difficult and complex it may be. In view of the complicated nature of the problems of contamination by diffuse sources, no immediate result was expected from these public relations activities. But due to the emphasis placed on the importance of the diffuse sources of Rhine contamination in these presentations, the awareness of the significance of this form of contamination is likely to have increased. In the long term this will probably lead to the adoption and enforcement of a more adequate form of legislation and the application of technical measures which will ultimately bring about a reduction in this form of contamination. These governments now keep a close track of developments with far more knowledge and insight than they possessed ten years ago. And this is in part due to the Rhine Research Project.

THE RHINE RESEARCH PROJECT IN A SOCIAL CONTEXT

The Municipality of Rotterdam has placed its Rhine Research Project in a wide social context. The problem of cross-border river pollution and the problem of contaminated dredged material are in the final instance, like all environmental problems, problems which, in the assessment of their gravity, are determined by social values. With its different activities, the Rhine Research Project has been operating in an area of tension characterized by continually changing social values. When considering the question as to whether the port management authority has developed a good instrument to solve a port problem with its approach, this wide social context should not be forgotten.

One difficulty in evaluating the Rhine Research Project is that for each of the three fronts on which the project operates, the changes brought about cannot be measured to an equal degree. It is clear, however, that developments in a legal/academic field have been stimulated in a specific direction due to the activities of the Rhine Research Project. Within the framework of the technical study, methods were developed to allow the independent determination of a discharge level. The techniques were later used in tackling other problems concerning cross-border river pollution (Danube, Meuse). Public relations activities made it clear that Rotterdam with its port activities was the direct victim of the pollution of the Rhine.

The Port of Rotterdam's presentation in the field of international cross-border river problems—dominated by national governments—has had a positive effect. The fact that a lower-level government body was the first to have a clear insight into the situation regarding the pollution of the Rhine undoubtedly encouraged the various national governments to carry out research studies themselves (Rhine Action Programme). The Rhine Research Project and the results of its technical study enabled critical and stimulating questions to be asked (for example, questions concerning the comparability of measuring station analysis results and the representativeness of samples at the measuring stations). The public relations activities of the Port of Rotterdam were not always welcome. But they did ensure that attention was drawn to the silt problem in the Rhine and also in states other than the Netherlands.

Many factors determine the activities of a government authority. It does not, however, seem too presumptuous to suggest that the methods and critical approach of the Rhine Research Project may have had a positive influence, either directly or indirectly, on the implementation of environment policy along the Rhine. The thoroughness, decisiveness, and drive with which the Municipality of Rotterdam tackled the Rhine Research Project has won it much respect.

CONCLUSIONS AND RECOMMENDATIONS

The approach of the Port of Rotterdam in combating the upstream sources of river contamination, in combination with the Rhine Action Programme of the Rhine States, has been very successful. The amount of contaminated sediment in the port has come down to half of what it was, the river has become notably cleaner, and consequently the deposits of contaminants from the river into the marine system have been reduced significantly.

When the Rhine Research Project started in 1985, there was a growing awareness of the contamination of the river water. But where that contamination came from exactly was not known in detail. The Rhine Research Project technical study precisely mapped out the contamination situation in relation to a number of substances of relevant to the assessment of the dredged material. An immense amount of information/data was gathered which allowed Rotterdam to play an active role and to critically follow developments.

Concerted efforts of professional experts under the project generated credible data on the nature and magnitude of pollution as well as the location of their sources and helped convince industries as well as the public about the legitimacy of the Rotterdam Municipal Authority seeking commitments from enterprises to reduce polluting effluents in an international river basin. Eventually the enterprises as well as the concerned national government accepted the legal standpoint of the RRP that the responsibility for protecting the environment transcends the boundaries of their factory sites or even their countries. The enterprises signed covenants agreeing to time-bound targets for reduction of hazardous effluents. The project thus contributed towards bringing about a change in the general social attitudes concerning the environment.

The change in environmental awareness which took place in the Rhine countries during the 1980s is naturally not entirely or exclusively due to the Rotterdam project. Like any other movement concerned with bringing about a change in social values, the Rhine Research Project also interacted with the society on which it was focused. The changing attitude towards environmental problems had positive repercussions on the Rhine Research Project. Acceptance of the project was greatly helped by this changing attitude towards the environment in the Rhine countries.

When evaluating the activities of the Rhine Research Project, one should not simply look at the measurable results. In addition to the concrete load reductions, the above-mentioned effects—which are more difficult to measure—should also be taken into consideration. When, at the conclusion of the fourth phase of the Rhine Research Project in 1994, the situation was reviewed, it was clear that a great deal had changed in the situation concerning the disposal of contaminated dredged material.

First of all, there is of course the improvement in the quality of the sludge. This improvement is due to the numerous initiatives designed to clean up the sources of contamination. Thanks to the activities undertaken within the framework of the Rhine Research Project, the Rotterdam Municipal Authority made an important contribution towards this operation to clean up the sources of contamination of the Rhine.

The amount of contaminated dredged material is now less than what had been predicted in the early 1980s. The period of time during which the Slufter can be used is consequently longer than the 15 years originally estimated. Given the current levels of dredged material to be stored in the Slufter, it looks as though this sludge depot will be able to store contaminated sludge until the year 2010, instead of 2002 as originally estimated.

If, however, the present quality of dredged material is compared with the standards which now apply to dredged material, it must be concluded that the Port of Rotterdam is not yet clean. And it does not look as though in the foreseeable future *all* the dredged material in the Port of Rotterdam will be of an adequate quality to allow it to be dumped into the sea. The areas of the Port of Rotterdam which are particularly affected by sediment carried in by the Rhine will certainly remain contaminated for the time being. It was anticipated that after the year 1999 some 3 million m^3 sludge would—initially at least—remain so contaminated that it would have to be stored in the Slufter.

Since the point sources along the Rhine and in the Port of Rotterdam area now (or in any case by the year 2000) usually treat their effluents using the most modern purification technology, we can no longer expect such spectacular reductions in contaminant loads from the point sources in the future as has been the case in the past. Where many substances are concerned, the future levels of contamination in the Rhine will largely be due to natural causes and by contamination from diffuse sources.

We cannot change the naturally occurring background concentration of the Rhine water. A reduction in diffuse sources can only be expected in the long term. The Rotterdam Municipal Authority cannot have such an effective impact on the diffuse sources as on the point sources.

From the lessons learned from the Rhine Research Project it may be recommended that due attention should be paid to the following preconditions in order to make a success of such a complicated and sensitive project:

- Have one or more powerful stakeholders who can put pressure on the (inter-)governmental management programmes;
- Have a sound legal framework in place and adopt sincere and well-coordinated public relations activities to support the legal actions;
- Ensure the right financial backing;
- Make an appeal to the public opinion on the basis of just and fair principles;

- Punish and penalize the non-sustainable use and encourage a more sustainable use of the river system;
- Create public support and political awareness to serve as a stimulant for the progress of the project;
- Develop a clear strategy, a consistent approach, a long-term plan, and a lot of patience and determination.

Following these recommendations, the model adopted by the Port of Rotterdam may hopefully lead to similar successes in other river catchments or other comparable complex environmental problems.

REFERENCE

Rotterdam Municipal Port Management, *Ten Years of the Rhine Research Project*, 1994.

CHAPTER 14

The Rhine Experience: The Development of Integrated River Basin Management in North-West Europe

KOOS WIERIKS

INTRODUCTION

For many, centuries the river Rhine has played an important role in the history of Europe and in the social, political, and economic developments of the continent. This role became even more important in the second half of the twentieth century. The multiple uses of the river, the conflicting interests, and especially the environmental and flooding problems in and alongside the river emphasize the importance of an integrated approach for solving the various problems associated with the Rhine.

The International Commission for the Protection of the Rhine (ICPR) was established in 1950. It was the first intergovernmental body set up for the management of transboundary waters. In more than half a century of its existence, the Rhine Commission has gained much experience in dealing with a broad range of river problems. The international management of the Rhine has continuously developed, since the 1950s and 1960s, a joint monitoring strategy, and has now emerged as a comprehensive integrated management system for the Rhine. The Commission now looks after all aspects of water quality, emission reduction measures, ecological restoration, and flood prevention measures. It was guided by a process of 'learning by doing' and was heavily influenced in its planning by some major catastrophes. The approach of the

Rhine Commission, and in particular the formula of the Rhine Action Programme, has proven to be very successful. Important lessons can be drawn from the experiences of the ICPR, lessons not only for the Rhine Commission itself, but also for other similar organizations dealing with transboundary water problems.

This chapter describes the various stages in the development of the management approach of the Rhine Commission until recent times and provides a guideline for the future course of the ICPR. Special attention is paid to the necessary prerequisites for a successful water-management strategy and to those factors that might be considered to be essential for its success.

THE RIVER RHINE: USES AND CONFLICTS

The Rhine is one of Europe's most well-known and important rivers. For many centuries it has not only been an important shipping lane, but also an important and valuable source of food and water, a crystallization point for human settlements, and a source of inspiration for poets and writers. Much of the industrial development in north-western Europe mainly took place on the banks of this river.

Originating in Switzerland, the Rhine flows through France, Germany, and the Netherlands into the North Sea. The Rhine is 1,320 km. long and has a catchment area of 200,000 sq. km. Already in Roman times many important cities had developed on its banks. Currently, more than 50 million people live in the Rhine catchment area. The river is Europe's most densely navigated waterway, connecting Rotterdam (one of the world's largest seaports) with Duisburg (the world's largest inland port). Vast industrial complexes have been built along the river, for example in the Ruhr, Main, and Rijnmond areas. Most of Europe's important chemical production plants can be found along the Rhine. Rhine water is used for industrial and agricultural purposes, for energy generation, for the disposal of municipal waste water, for recreational activities, and for the production of drinking water for more than 20 million people. Furthermore, the Rhine is the natural habitat for a diversity of plant life and many kinds of birds, fish, and other species. So many different claims on the river have led to various conflicts and problems concerning water quality, river ecology, and flooding.

WATER QUALITY

Although water quality problems in the Rhine had already been recognized in the fifteenth century, they continued until the end of the 1960s when the deteriorating quality of the river became very apparent. By that time, the pollution of the Rhine with organic substances led to acute oxygen problems, resulting in the disappearance of almost all forms of aquatic life from the river.

Other threats to the river ecosystem were caused by waste water discharges from industries, agriculture, traffic, and households. Large amounts of heavy metals, pesticides, hydrocarbons, and organic chlorine compounds were discharged into the river, causing further ecological problems such as the disappearance of indigenous species, deterioration of the water quality, and pollution due to sediments. More and more efforts were needed to produce good quality drinking water, and by the end of the 1960s the Rhine had the dubious reputation of being the sewer of Europe.

Severe water quality problems may also be caused by accidental pollution. Enormous amounts of dangerous substances are transported via the Rhine everyday, and many industrial complexes are located right on the banks of the river. The Sandoz accident in 1986 clearly illustrates the disastrous impact that accidental pollution can have on the whole river.

ECOLOGICAL PROBLEMS

To improve the shipping conditions and to enable the use of alluvial areas for agriculture and other purposes, the course of the river was drastically changed in the nineteenth century. Further 'corrections' in the riverbed followed in the twentieth century. As a result, the length of the river between Basel and the border of Hessen was reduced by more than 80 km. Meanders and alluvial areas were cut off, causing great changes in the river ecosystem. Other problems were the increase of flow velocity, the erosion of the riverbed, and the drop in groundwater levels.

Numerous dams and weirs were constructed on the Rhine and its tributaries for power generation. These physical barriers made it almost impossible for the most important fish species in the Rhine, the Atlantic salmon, to reach its spawning grounds. Catches of Rhine salmon decreased steadily from more than 280,000 tonnes around 1870 to zero in 1950.

FLOODING PROBLEMS

Flooding problems are as old as the river itself. Heavy rainfall or sudden snowmelts have always caused high water levels in the Rhine basin. However, human activities in and along the river have strengthened the negative impact of higher water. The risk of flooding has increased with the growing intensity of use of the river basin, erosion, urbanization, and changes in the use of water by households. At the same time, with more and more people settling down and economic activities expanding in these endangered areas, the threat of flooding of these areas by high water has increased. The 1993 and 1995 floods in the middle and lower Rhine clearly showed the need for a drastic change in both river management with respect to floods and risk management policies in potentially endangered areas. Exactly the same problems were experienced in the Oder and Elbe regions in 1997.

INTERNATIONAL COOPERATION

The above-mentioned problems can only be solved effectively in an international context. The sediment quality in the downstream Rotterdam area is directly influenced by upstream discharges. Salmon cannot return to their upstream spawning areas when downstream barrages and weirs block their long way up, and the magnitude of flooding is influenced by upstream and downstream riverbed conditions. International cooperation is therefore the first prerequisite for effective river management.

The main reason for the establishment of the Rhine Commission in 1950 was the declining quality of water in the river. To tackle this problem the states bordering the Rhine—Switzerland, France, Germany, Luxembourg, and the Netherlands—decided to cooperate in a new international forum: the International Commission for the Protection of the Rhine (ICPR). The basis for this cooperation was strengthened in 1963 with the signing of the Berne Convention on the Protection of the Rhine. The European Commission joined this Convention in 1976.

The first years of this cooperative effort were devoted to achieving a common understanding of the problems associated with the Rhine, and creating a legal and institutional basis for tackling them cooperatively. Some joint monitoring programmes were developed, and the first common measures to protect the river against the effects of organic pollution were taken around 1970. During the first two decades of the Rhine Commission, it was very important to create a positive atmosphere between the participating states. Confidence, trust, and understanding between the partners are very important for a cooperative international undertaking to be successful.

Between 1970 and 1985, some successful programmes were developed to reduce the inflow of polluted municipal and industrial waste water into the river. Between 1970 and 1990, more than 80,000 million DM were invested in the construction of communal and industrial treatment plants. The quality of the Rhine water slowly improved as a result of the work of the Commission. Oxygen levels improved steadily, and some improvements could also be observed with regard to other pollutants. In this period, the main efforts were focused on 'end-of-the-pipe' techniques, that is, the treatment of waste water rather than the adoption of preventive measures.

As a next step on the way to the further protection of the river, the Rhine Commission agreed, in 1976, on an International Convention on the Prevention of Chemical Pollution of the Rhine. A very important part of this Convention was the introduction of a system of black and grey lists aimed at reducing the influx of dangerous substances. One of the tasks under the Convention was the development of internationally binding limit values for extremely hazardous substances. Such limit values were developed for mercury, cadmium, and a number of organic substances, and national efforts were made to achieve them. However, it soon became evident that the problems concerning limit values for dangerous substances affected not only the Rhine catchment. For reasons of observing the rules of fair international competition, the Commission of the European Community was approached. Unfortunately, the European Community took very long to look into this problem. The slow progress also had an adverse effect on the ICPR. For a long time the EC-Commission discussed limit values only for selected hazardous substances while, in Germany, the sectoral approach had proved successful long before. Under this German approach, minimum requirements had been determined for all relevant pollutant parameters of a certain industrial sector. In individual cases they may be even more stringent but they may never be reduced. The different approaches were discussed in the ICPR but no clear decision could be taken.

This period proved to be a very difficult one for the International Rhine Commission. On the one hand, considerable international progress was made in reducing the load of organic pollutants into the Rhine. For the first time in the post-war period, biodiversity increased substantially. The monitoring programme, now considerably improved under the ICPR, records this progress. On the other hand, despite an internationally agreed upon Convention, pollution by dangerous chemical substances could not be tackled adequately. The prescriptive and detailed approach taken on the basis of the Chemical Pollution Convention blocked the realization of its own (positive) objectives.

This ambiguous situation considerably complicated and slowed down discussions within the ICPR. The situation lasted until 1986 and, as so often appears to be necessary in environmental decision making, a serious accident was needed to take the next step forward.

THE SANDOZ ACCIDENT: A TRIGGER FOR THE RHINE ACTION PROGRAMME

An accident in a Swiss warehouse, at the end of 1986, painfully showed how vulnerable the Rhine ecosystem still was and how the threat of such accidents had not been overcome. Due to this accident at the Sandoz plant near Basel, varieties of toxic substances flowed into the river and caused the death of almost all aquatic life downstream as far as Loreley near Coblence. The Sandoz incident triggered a wave of publicity in all the states bordering the river. Political attention was drawn to the problem and in a very short time not less than three ministerial conferences addressed the issue of Rhine pollution. Finally, all this activity resulted in in the Rhine Action Programme (RAP) of 1987.

The RAP set clearly-defined goals, to be reached by the year 2000:

- the ecosystem of the Rhine should be improved to such an extent that higher species, such as salmon and sea trout, could again become indigenous.
- the production of drinking water from the Rhine must be guaranteed in the future.

- the pollution of river sediments must be reduced to such an extent that sludge may at any time be used for land filling or be dumped into the sea.

Under the Rhine Action Programme, the ministers agreed on achieving some very challenging and ambitious targets like the 50 per cent reduction of inputs of dangerous substances between 1985 and 1995 and the return of the salmon by the year 2000.

In 1988, enormous algae blooms in the North Sea showed the close interrelationship between discharges into the river and the corresponding effects on the marine environment. Therefore, a fourth target was added to the RAP: the improvement of the ecological state of the North Sea.

The RAP was split into three phases. To begin with, the ICPR established a list of priority substances, analysed their sources, and determined the amounts discharged. Furthermore, the ICPR pressed for a working method based on an industrial-branch approach. In order to reduce pollution, both of water and suspended matter, the '*Sand der Technik*' (the best available technology) had to be determined for industrial production as well as urban waste water treatment. Additionally, the Commission developed well-defined measures limiting the risk of pollution of the Rhine in case of accidents like the Sandoz fire.

The second phase, up to 1995, aimed at the actual implementation of the agreed upon measures. The third phase (1995-2000) was to enable the further implementation, 'fine-tuning' and eventual development of additional measures to achieve the agreed upon goals.

An essential part of the RAP addresses the reduction of the continuous day-to-day input of dangerous substances into the river. The implementation of this part of the RAP has been very successful. All along the river measures have been taken to prevent pollution and in 1994, the ICPR could report that most of the reduction goals had been reached. In the field of industrial sources the 50 per cent target was almost completely met, and for many substances reductions up to 90 per cent were realized.

Difficulties in implementation are still being reported in the field of the diffuse sources of pollution, especially with respect to agricultural emissions of pesticides and nutrients. These emissions received further attention in the third phase of the Action Programme 1995-2000. Nevertheless, it can be stated that an enormous improvement in the water quality of the Rhine has taken place in a very short time. From being the sewer of Europe in 1970, the Rhine is now one of the cleanest transboundary rivers in Europe.

SALMON 2000: THE ECOLOGICAL DIMENSION OF THE RHINE ACTION PROGRAMME

Perhaps the most challenging goal before the RAP is the one to improve the Rhine ecosystem to such an extent that migratory species can return to their spawning grounds and become indigenous once again: the return of the salmon to the Rhine by the year 2000.

The ICPR decided on a range of measures to be implemented in a very short time. The improvement of the water quality of the river, as a result of the afore-mentioned emission reduction programme, was of course the first requisite step. Furthermore, extensive programmes were developed and implemented to enable the salmon to return to their spawning grounds. Parallel action was required in order to create a new stock of Rhine salmon. Salmon eggs were bought in Scotland and the south-west of France and were hatched in special fish hatcheries. Thousands of alevins were released into the river, and a programme was developed to follow the behaviour of the salmon.

The successful results of the Salmon 2000 programme can clearly be observed. Since 1990 salmon and sea trout have returned from the sea to the Rhine and its tributaries, and since 1992 natural reproduction has been recorded. In 1995 nine salmon were caught under the Iffezheim barrier, proving that salmon returning from the Atlantic can once again actually swim up the river more than 700 km. Another 40 were caught in 1996. The almost utopian political goal of 1987 appears to have been reached in a very short time. The success of the Salmon 2000 programme once more underlines the effectiveness of international cooperation between the states bordering the Rhine. Ambitious political goals on the international level have been translated into concrete measures and activities at the local level. All along the river, people are working to realize the same goals and are actually producing positive results.

THE 1993 AND 1995 FLOODINGS

In 1994, the Rhine ministers decided to broaden the scope of their cooperation even further by including the quantitative aspects of the river management system in their policies. This decision paved the way for the further integration of qualitative, quantitative, and ecological aspects of river management.

This concept of further integration received additional impetus and even more political commitment after the extreme floodings in the middle and lower Rhine areas in 1993 and 1995. On the basis of the very positive results of the RAP, the ministers involved charged the ICPR with the development of an international action programme for the management of flooding problems. The proven spirit of co-operation between the states bordering the Rhine and the efficient, holistic, and programmatic approach of the ICPR led to the belief that a comparable positive result could be achieved in reducing the problems of international flooding.

As in the case of the Sandoz disaster, it took a serious crisis to convince the Rhine states to take measures against flooding. Two enormous floods with disastrous consequences for many people seemed necessary before the Rhine Commission saw the necessity of taking adequate measures to prevent or at least to mitigate the impact of such accidents in the future. It may appear cynical to note that other river commissions do not seem to have learnt the necessary lessons from such past experiences and continue to neglect, for lack of realization, funds, or for whatever reason to take comprehensive and holistic preventive measures. The policy of 'management by accident' was probably the only possible way for river management in the early stages, but these days one might expect that the lessons of the past have been learnt and that suitable measures will be taken to prevent disasters.

After being charged with its new flood management task, the ICPR did not hesitate to integrate this new problem field into its mandate. At the end of 1995, the first international strategy for the management of flooding was agreed upon. The most important conclusion of this strategy document is that floodings cannot be prevented and that international action should aim at flood plain management rather than at flood management.

Ten guiding principles were formulated as the basis for a

structural multi-track approach. The ICPR also started to work on an inventory of all historically known and presently available natural flooding areas. This inventory, to be finalized by the end of 1996, was to form an important basis for later decisions concerning the actual international Action Programme and corresponding measures. A second inventory addresses the current international alarm-and-warning system along the Rhine in combination with the international network of weather forecast stations. The formulation of this international Action Programme on protection against flooding is a direct plea for transboundary solidarity. A narrow approach on the basis of only local, regional, or national activities will not be enough.

Technical solutions may offer some relief, but in the end downstream problems can only be solved by structural upstream measures. It will be a major task for the ICPR to initiate this process and to convince all the authorities, organizations, and people involved of the necessity of adopting a common approach. Another challenge for the ICPR is the close interrelationship between the Action Programme on protection against flooding and the current and future activities of the Commission with respect to the ecological restoration of the Rhine. The objectives of flood plain management and ecological restoration of the alluvial areas can be combined very well. Practical implementation on a project level, however, might appear to be more complicated.

Like the problem of pollution, the problem of the flooding of international rivers is not a problem exclusive to the Rhine. Similar problems are also faced, for example, in the Meuse and the Scheldt rivers. As a consequence, similar strategies and solutions are being developed for these rivers. At the same time, the European Community has taken the initiative to develop a programme for protection against floodings under the framework of the European spatial planning programmes. This very valuable European Community initiative offers the opportunity to take measures different from the 'classic measures' related to water management. Spatial planning instruments will help to keep the endangered areas free of construction activities; specific areas are designed for retention or overflow purposes and also to improve the storage capacity of the whole river basin.

It is obvious that so many initiatives, in so many different fora, demand well-balanced coordination between different groups and organizations. The above-mentioned examples of the complexity underlying international cooperation in solving water problems illustrate that there is a need for a high level coordination between different river-basins management organizations.

The experiences of the ICPR clearly point to the need for the adoption of a management approach on a river-basin scale. Such an approach is actually being applied to almost all important European rivers and seas. At the same time, the European Community is in the process of preparing an overall European water policy. Because of the enormous differences between different international water bodies, it would be wrong to try to develop one general prescriptive policy for all European waters. On the other hand, there is a need to create a structure of cooperation between the various management organizations handling different rivers and seas, aiming at an exchange of information and experiences and also ensuring a certain level harmonization of activities and goals between different areas. With respect to this harmonization of operations, it will of course be necessary to take into account the specific geographical, natural, and socio-economic conditions of the different basins.

THE PRESENT ORGANIZATION OF THE RHINE COMMISSION

The organization of the Rhine Commission has been changed several times to accommodate shifts in policy. Thanks to the integration of high-water problems in the Commission's work and the wish to optimize and streamline the work of the Commission, the structure of the ICPR was reorganized in 1995.

Ministerial conferences, held every two to three years, formulate the political goals of the Commission and are at the same time the platform for the assessment and evaluation of the activities carried out. The Commission itself, consisting of the highest officials from the different member states, meets annually and decides on working programmes, finances, and all formal procedures. A coordination group, meeting four times a year, is responsible for the actual planning and coordination of the work of the ICPR.

Three permanent working groups cover the areas of water quality, ecology, and emissions. Two non-permanent project groups, with a mandate limited in time, prepare the work on the new

Rhine Convention and on the Action Programme on Protection Against Flooding. A number of expert groups deal with specific problem areas related to the work of the working groups and the project groups. Each group consists of a number of national governmental experts. The work of the Commission is supported by a small international secretariat with a permanent base in Coblence, Germany.

A relatively new development in the framework of the ICPR is its positive attitude towards cooperation with non-governmental organizations (NGOs). The ICPR has decided to promote the exchange of information with international NGOs in order to create a better common understanding of problems and their necessary solutions. As its first initiative, the ICPR organized an international conference, 'Living with the Rhine'. This conference addressed all actual Rhine issues with a view to developing future Rhine policies. All relevant international NGOs took part in this conference and expressed their views on the future of the Rhine and the future tasks of the ICPR. This very successful conference was followed by an official hearing of NGOs organized by the Commission. The latest development is the introduction of special meetings between NGOs and the ICPR working groups. These meetings address specific issues such as the formulation of quality objectives, the development of monitoring programmes, etc. Exchange of information with NGOs is expected to be institutionalized as a regular part of the ICPR task.

RIVER-BASIN MANAGEMENT NOW AND IN THE FUTURE

The recent floods emphasize once more the importance of adopting and implementing an integrated approach to these problems at the level of the entire river basin. On the basis of international solidarity, measures must be taken in the whole catchment area of the Rhine. The ICPR has gained much experience in the development of such a river-basin management approach. Between 1950 and 1970, the main activities of the Commission were the development of a legal basis for cooperation and of conducting joint monitoring programmes. From 1970 onwards, many measures were developed to reduce inputs of pollutants into the river. The basic instrument was the construction of industrial and municipal waste water treatment plants. This 'end-of-the-pipe' approach improved the water quality of the river, but real progress was not made until the Rhine Action Programme of 1987 was agreed upon and launched.

The Rhine Action Programme tackled the water quality problems with the help of an integrated approach. Precaution and prevention became the basic principles, and for the first time actual emission reductions were realized. With the introduction of ecological goals for the river, the Rhine Action Programme broadened the scope of river management from water-quality management to water management. The integration of quantitative aspects into the work of the Commission after the 1993 and 1995 flooding problems further expanded the scope of integrated water management.

Due to these developments, the ICPR also has to change its methods in order to be able to realize its goals. Prevention of emissions of dangerous substances calls for more than just a licensing system. Process-integrated measures have to be developed for different branches of industry, product policies have to be changed in a European context, and even changes in consumption patterns may have to be considered. With respect to the flooding problems, it is obvious that close cooperation with spatial planning organizations is a prerequisite.

The ICPR is in the process of preparing a new Rhine Convention as a legal basis for the functioning of the integrated approach. This new convention will address and integrate all the elements that are necessary for the future sustainable use and development of the Rhine. An important element of the new convention will be the exchange of information and cooperation with NGOs. The successful implementation of future Rhine policies will only be possible on the basis of a general consensus about objectives and means. River management is not only a governmental responsibility but is also the responsibility of every citizen, farmer, community, and industry in the Rhine catchment area. The new Rhine Convention and the ICPR will have to provide the infrastructure for enabling discussions and agreements dealing with the implementation of Rhine policies.

On the basis of the experience gained during nearly five decades of cooperation, it may be suggested that future Rhine management, based on the concept of integrated water management on a river-basin scale, should leave some room for flexibility in the implementation of its policies. Detailed prescriptive policies take much more time to be developed, and will hinder rather than stimulate the implementation in different countries or areas. The same approach should be adopted in relation to the formulation of concrete measures to be taken by industries or other polluters. Decisions regarding technical measures, investments in new production processes, etc., can most effectively be taken by these very same target groups and not at a central inter-governmental level.

Openness is another important element which must be kept in mind when developing future policies. Political and public support are prerequisites for the successful formulation and implementation of such policies. This aim can only be realized when innovative, transparent, and goal-oriented policies are formulated and forwarded to a broad public for open discussion. The recent developments in computer and communication technology offer new possibilities for the effective implementation of the ICPR information strategy.

Although a lot has been done for the Rhine, much still remains to be done. In the years to come, the sustainable use and development of the Rhine will call for additional programmes and measures. The ICPR has the challenging task of demonstrating its capability to fulfil these responsibilities successfully. International cooperation, solidarity, flexibility, and practical implementation were and will continue to be the basis for the success of the Rhine Commission. The present process of the ongoing integration of all the relevant aspects of water management for the entire Rhine catchment area and future strategies aimed at the sustainable use and development of the river will further improve the Rhine, and that will be a new global example for water management on a river-basin scale.

The successful cooperation achieved under the framework of the ICPR has been an inspiring example for many other river-basin management organizations. It is of course not possible to project wholesale the Rhine approach on to any other river in the world. Therefore, it is very important to take proper note of the various steps of the process taken by the Rhine Commission and to learn from both the positive and negative experiences of the ICPR. In the same way, the ICPR can also learn from the various experiences in the management of other international rivers.

It is interesting to look at the current European Union initiative to draw up a new Water Resources Framework Directive. This directive will play an important role in future water and river management and might also create a mechanism for the exchange of information and experiences between different river-basin management authorities. A question remains as to how this directive will impact on existing river-basin management organizations like the ICPR.

CONCLUSION

International water problems can only be effectively solved and prevented on a river-basin scale. There is, however, a clear need for an exchange of information and sharing of experiences between the authorities dealing with different basins. The European Community can play an important role in creating a basis for such an exchange.

River-basin management requires an overall integrated approach to issues of water quality, water quantity, and all ecological aspects of the problem. An approach addressing only one aspect can only partially solve some problems and may even give rise to new problems in the long run. The sustainable use and development of the river should be the basis for formulating and implementing future policies for international rivers. That means the promotion of a balanced use of the river, respecting all interests and interest groups, now and in the future.

Precaution and prevention are the most important basic principles for river-basin management. Taking measures against damages after they occur is always more expensive than facing and preventing problems beforehand. Decision makers should learn their lessons from the Sandoz disaster and the recent floodings. International strategies and policies should leave room for flexible implementation at the national level. Working in the same direction does not automatically mean taking the same measures. Detailed, prescriptive policies take much more time to be drawn up and be agreed upon, and are often less effective in their implementation.

Important decisions concerning the solution of environmental problems cannot be taken without considering the opinions of all the groups involved. NGOs contribute positively to the quality of the decisions taken and therefore form a very important factor in the successful implementation of these decisions. Public and political support are also prerequisites for the successful formulation, and implementation of environmental policies. Open planning processes, access to data, and transparent and goal-oriented policies are essential for success.

Cooperation in river-basin management must be based on mutual confidence between all the parties involved. The negative and positive lessons learnt from the experiences of the ICPR may serve as an example for other international water-management authorities all over the world.

CHAPTER 15

Some Gender Dimensions in Managing Water Scarcity

LOES SCHENK-SANDBERGEN

INTRODUCTION AND RATIONALE

This chapter addresses some gender dimensions which pertain to the question: How to manage water in a context of scarcity? What are the experiences of, and prospects for, women in India? The analyses and observations offered here are mainly drawn from my own experiences as an anthropologist/gender expert in various water resource/irrigation projects in India. To understand the relativity of findings in the Indian context, some findings and observations regarding the role of women in the management of water scarcity in Laos, Vietnam, and Bangladesh are also presented.[1] The four countries represent a variety of religions, cultures, and structures of social organization: India with Hinduism, Laos with Buddhism, Vietnam with Confucianism, and Bangladesh as an Islamic state. These structures generate different gender relations and gender ideologies that have crucial consequences for the ownership of, access to, and control over water resources for women.

From a gender perspective, the subject of water-scarcity management in India, and also elsewhere in the world, should start with an analysis of kinship relations, post-marriage residence patterns, and (customary) inheritance patterns. These three crucial issues determine broadly the position and status of women. Therefore, these three crucial gender dimensions are not only specific to the problem of the management of water scarcity, but are also the basic dimensions for any other analysis of the rights of women to natural resources in general.

Nowadays, the starting point or basic assumption in almost all 'gender and development' debates, is a gender analysis focused on women's access to and control over water, and the decision-making role of women in matters relating to water and other natural resources such as land, trees, livestock, etc. But these are basically derivatives from the three basic arrangements of the prevailing social and cultural values.

'Why' is the participation of women in managing water scarcity important? This question is relevant because it addresses the subject of the importance of women's rights to natural resources. The belief that women should share these rights on an equal footing with men is now gaining widespread international acceptance (Rights of Women, 1997, p. 1). This chapter deals with the following dimensions of the relation between women and water:[2]

- Patriarchy and the absence of water rights for women.
- Gender conflicts over use of, access to, and control over water.
- Women's alternatives to cope with water scarcity in the cities.
- Participation of women in water users' associations.
- The feminine principle in water management.
- The role of women in construction, maintenance, and financing.

The main problems in each dimension will be indicated, followed by cases and examples of innovative and alternative approaches and strategies to improve the role of women in managing water scarcity as implemented in the projects in which I have participated.

THE PROBLEM

Managing water scarcity should be looked at from class and gender perspectives. Ownership of land and water provide the basis for immense power under existing Indian conditions. Therefore, we should be cautious not to focus only on water scarcity as a classless phenomenon, and instead we should ask: 'How to solve it?' First, we have to ask the question: 'Water scarcity, for whom?' Everybody knows that the middle-class housewife in an urban middle-class neighbourhood has a water tank on her roof, while her sister in the slum has to queue up for hours in the night to get a pot from the public tap. Real scarcity of water is the grim reality for members of the households of landless and marginal farmers, and not for the big landowner with his deep tube well which enables him to sell water to those who have little or none. The more severe and frequent the drought, the better for him and his family. Therefore, the main concern in the following pages will be with women (and men) of poor households. For the reader unfamiliar with the subject, a brief sketch of some of the problems faced by women of poor households regarding water scarcity may be useful.

According to the sexual division of labour, women are the water carriers and water managers, certainly at the household level. They have developed all kinds of informal strategies to secure access to water. They have a long experience of being deprived of water as a result of class and gender inequalities under which the wealthy upper castes and male officials appropriate this precious natural resource. Probably no other group is more affected by the destruction, depletion, pollution, and total neglect of traditional tanks, wells, and other water resources than women of poor households (Shiva, 1988; Venkateswaran, 1995; Agarwal and Narain, 1997). Fetching of water is very time consuming and tiresome. Water carrying is an extremely strenuous activity and consumes an enormous amount of women's energy and time (Page, 1996).

Lack of clean and sufficient drinking water affects the health of children most, but among adults, women are more exposed to the hazards of polluted water than men. What is even more serious is that there are indications that the reproductive health of women is endangered by water pollution.

The lowering of water tables means increasing work for women who then have to seek alternative sources. In particular for the women of the former untouchable castes this often implies that their labour is indiscriminately exploited, as they have to depend on the goodwill and willingness of the upper castes and their employers to use their wells to get water. The following field notes from my study in the Bidar district in Karnataka illustrates this situation (Schenk-Sandbergen and Mohan, 1996b, p. 17).

The constraints imposed by prevailing gender ideologies, and the unequal gender relations generated by the overwhelming patriarchal, patrilinear/patrilocal social organization of Indian society are the main causes that deny in particular poor women access to and control over water. In a more modern context, it prevents women from participating in the planning, implementation, operation, and maintenance of water-supply schemes and irrigation agencies and projects. Women face many constraints which hamper their public representation, participation, and empowerment. Patriarchy and male dominance are the basic underlying causes of the unequal gender decision-making power in water-resource management in India.

Additional constraints for tribal and scheduled caste women are the result of the hierarchical power structure shaped by class, caste, and poverty. Water-scarcity management problems for women in India go far beyond the stereotyped opinion that it is a question of different needs as a consequence of the sexual division of labour.

Women need water to produce food, to irrigate vegetable gardens and rice fields, to give drinking water to livestock, to raise fish, to raise saplings in nurseries, etc. But farm women and landless female agricultural labourers can use and draw water from wells on their husbands'/employers' land, only to work hard, to do the back-breaking work of irrigating the crops. But they cannot own, manage, or take decisions relating to the water resource.

This leads to negative consequences. The few studies on the impact of government and private initiated irrigation systems on women show the following effects:

- women are marginalized to subsistence production, and men produce commercial crops for the market (Boserup, 1970);
- more inequality in gender relations is the result as women become more dependent on the income of the men (Epstein, 1962; Stanbury, 1992);
- loss of productive resources such as access to land, or land rights and agricultural skills, more work load as drinking water sources are spoiled, and pasture land for grazing of cattle are diminished (Groverman and van Walsum, 1994; Van Wijk de Hange and Saunders, 1996);
- less formal public decision-making power, more indebtedness, and more pressure on women to earn cash (Schenk-Sandbergen, 1996a).

In the Tungabhadra canal irrigation project in Karnataka, it was found that the gap between poor and rich households had widened due to unequal water distribution. Consequently, farm women at the tail end of the irrigation sub-canals, although they themselves were small and marginal landowners, became increasingly dependent on access to land (leasing in), wage labour, and the credit of the head reach households (Schenk-Sandbergen and Mohan, 1996b).

PATRIARCHY AND THE ABSENCE OF GROUNDWATER RIGHTS FOR WOMEN

Experts rightly point to the intolerable pressures exerted as a consequence of the privileged access to groundwater enjoyed by the rich and wealthy landowners at the cost of the poor and marginal ones. However, it is mostly not mentioned that groundwater is also not accessible to women, except perhaps to some female-headed households. Women face gender constraints which *de facto* exclude them from groundwater rights. The practice is that groundwater rights are legally controlled by the holder of the land overlying the groundwater. Men inherit land, and in case where land is acquired through purchase, 'the head of the household', that is the male, automatically becomes the owner of the land. Formal land titles are adjudicated in his name.

Although women have legal rights to inherit and to own land, the reality is different (Sharma, 1980; Agarwal, 1994). Rahman and Van Schendel (1997) describe a similar situation in Bangladesh, where women are also disallowed from inheriting land. The existing post-marriage residence pattern in rural India and Bangladesh is that brides after marriage move to the village (house) of their bridegrooms' parents. That means a decisive cut-off from the house and land rights of their own parents.

For poor women, the ownership of a tube well or even a small hand pump to lift water is a distant dream. Therefore, some development projects have initiated efforts to give priority to women in the allocation of hand pumps. This is illustrated by the work of North Bengal Terai Project, a project with which I have been involved as a researcher and gender specialist for many years.

The North Bengal Terai Project

It is perhaps surprising that in the North Bengal Terai Project,[3] the male government staff of the Agricultural and Irrigation Department gladly accepted my recommendation to give preference to female-headed households (widows, single women, divorced, and abandoned women) in the selection of hand pump beneficiaries (Schenk-Sandbergen, 1991). The reason for the willingness of these men, in the relatively mild patriarchal context of west and north Bengal, to accept these recommendations is that the crucial pre-condition—that the 'land title' should be in the name of the female head of the household—is fulfilled. A woman without a husband is considered a kind of 'honourable male'. There was a firmly rooted attitude among the male implementing officers working on the project that only the registered person who is the owner of the land should be considered as a beneficiary, and in 95 per cent of the cases these were males. Women from female-headed households are mostly owners of land and therefore the criterion works out in a positive way in their favour. The situation became more difficult at a later stage of the project when it was suggested that hand pumps should also be given to a fixed quota of married women with the argument that they are the main users, and that it is therefore fair to put the pump in their name. This proposal, though similar to the notion of 'land to the tiller' which is widely accepted in the socialist political context of the Terai, was not embraced by the male staff.

However, what was not foreseen was that the project regulation—that women would get preference in the allotment of a hand pump if the land is in their name—has resulted in an unplanned, but very positive side effect. Marginal farmers transferred the ownership of their land to the name of their wives since they understood the pre-condition that women would get priority in the allotment of hand pumps when they are landowners. The transfer of the *patta* (land right document) costs only 10 rupees (Schenk-Sandbergen, 1994d). What does this reaction of the males tell us? The gender transfer of land might be perceived as just a paper formality, but it can also be seen as the first step towards the weakening hold of male dominance in control over groundwater resources. And, ultimately, it can work out very positively in case of the illness or death of the husband, or the divorce and abandonment of the wife.

Regarding the history of women, land, and water rights, it is useful to remember that for many women these rights were self-evident in the past. The important study of Agarwal (1994) shows that in the past matrilinear and bilinear societies in India, in which women inherited land and residence patterns were matrilocal, have been undermined by patriarchy and by the interventions of the colonial and post-colonial state. Agarwal illustrates the effects of these interventions with three examples of matri/bilinear societies. First, the ethnic community of the Garos in north-eastern India, second, the Nayyars in Kerala, in the south-western part of India, and third the Singhalese in Sri Lanka.

In her conclusion Agarwal writes (p. 192) that one of the striking features of the three case studies is 'the vulnerability of women's customary rights in land even in matrilineal and bilinear communities'. With the erosion of customary land rights, major groups of women in India also lost their groundwater rights. Gender relations and gender ideologies changed to the disadvantage of women; men of these communities have become only recently more powerful (Leacock, 1981; Etienne and Leacock, 1980).

In a few places in the world the inheritance of land (and house) rights for women is still the existing practice for large groups of women, as in Laos. But there are already alarming indications that the precious customary inheritance rights of women might be undermined by the consequences of the liberalization of the socialistic economy, unless timely action is taken (Schenk-Sandbergen, 1997). I found that in Buddhist Laos, characterized by a strong matrilinear and matrilocal cultural orientation, the participation and decision-making role of women in traditional community water-resource schemes is the best. The more patriarchal a society is, the less control women have over water resources and the less significant the role of women in management. The unique situation in Laos where women play a prominent role in the management and development of the traditional irrigation schemes is discussed later in this chapter.

It looks utopian, but the lesson learned from the Terai Project is that it is important to challenge the dominance of self-evident male ownership and make use of right of land and water to improve access to, and control over, groundwater resources for women. A crucial condition is to support and promote land rights for women of poor households.

GENDER CONFLICTS OVER USE OF, ACCESS TO, AND CONTROL OVER WATER

I have observed in several projects that competing gender demands over water always end up in the subordination of the water rights and interests of women. Men occupy strong power positions as decision makers in the local political institutions. Patriarchy hits women at various levels. The following three cases illustrate this aspect of gender-based conflicts over water.

Competing Gender Demands: Drinking and Irrigation Water

During my research in Gujarat some years ago, I was told that in a certain village an NGO had constructed bathing and washing places for women with water taps.[4] I visited the village and observed that these facilities could not be used. Women expressed their grief and sadness about the situation to me. It was decided by the males that as the water was scarce, it should be diverted to the irrigation channels, as for them irrigation was more important than meeting the women's interests and needs. The situation continued for months, also during those times when water was less scarce.

Closing Irrigation Canals: Gender Effects

The second example of competing gender demands relates to the general practice of 'closing the irrigation canals' widely followed by the officials of the irrigation department or the command area development boards.

An example is the Tungabhadra scheme in Karnataka. From the beginning of May, after the *rabi* harvest till the middle of July, the irrigation canals are closed as 'maintenance' work has to be carried out. Our gender study showed that 70 per cent of the women are dependent on the canal water for drinking and all other domestic uses (Schenk-Sandbergen and Mohan, 1996b). In the 40 years ever since the functioning of the canal irrigation system began, no concerted effort had been made to provide adequate drinking water sources for the duration of the closing of the canals. The decision to close the canals is taken by the fully male-dominated command area development authority.

The consequences have been disastrous. In 1996, a cholera epidemic broke out in Sindhanur, with 190 reported cases of death. In 1997 the situation was again very critical, and women's workload had increased to unacceptable proportions; they had to fetch water from the river (vakranis)[5] or from very far away places. The males in the area, more mobile, and possessing cycles and small transport lorries, saw commercial opportunities in the water scarcity, and began to deplete the scarce water resources in order to sell water. Women at the tail-end had to buy water for Rs. 1 to 2 per pot.

Male Choice of Technology

Another example of conflicting male and female interests in water use, access, and control relates to the choice of technology. In the formulation of the proposal of the North Bengal Terai Project, Phase II, the number of planned hand pumps was less, and the number of deep tube wells and river lift irrigation devices was more. A brief gender analysis conducted by interviewing various women in the villages revealed that the women did not want the bigger technology, but instead wanted hand pumps near to the house compound for multi-purpose use: irrigating their vegetable garden and for the easy fetching of drinking and domestic water. However, the male farmers wanted deep tube wells to grow commercial crops, and the staff of the irrigation department also had its own reasons to push for large-scale water devices. This conflict was finally and wisely resolved by increasing the number of hand pumps in the project.

A solution to this conflict between male and female interests in water use is to make concerted and coordinated efforts to combine irrigation and domestic water supply schemes; these are typically conceived as separate and independent projects and as belonging to different domains. Irrigation is considered as the male domain, and drinking water as the female domain. This artificial division between water for productivity and water for domestic use, that is, two spheres of water usage, is not realistic from the women's point of view. For rural women of poor households both spheres are very much related and therefore they should not be divided into different sectors (Centre for Women's Development Studies, 1990).

WOMEN'S ALTERNATIVES TO COPE WITH WATER SCARCITY IN THE CITIES

How feasible are the alternative ways of water harvesting designed to reduce water scarcity? Is roof-top water harvesting a viable option? In the following pages, I want to analyse the significance of women's alternatives in the area of water harvesting as I found them during my research in the slums of Indian cities.[6] But first, some remarks on roof-top harvesting of water. I had always wondered what makes Indian culture different from South-East Asian culture in this respect? This is not only a question of obviously better housing (bricks and tiles on the roof) in South-East Asia in comparison with India, where it is difficult to catch rainwater from a mud hut with thatched roofs. The fact that parts of India only get two months of monsoon does not entirely explain the differences in rain-pot customs.

In Vietnam, in rural villages in the Red River Delta, I observed that almost all peasants' houses have brick water containers in the compound, in which the women collect rainwater via bamboo or PVC pipes from the roof-tops. Even rice seedbeds are raised in the compound and on roof-tops as the water containers are directly within the reach of the women. In Laos, I visited villages where women construct huge water jars in which they collect rainwater from roof-tops as storage for periods of drought. The Lao Women's Union promotes this rainwater collecting method backed by UNICEF funds (Schenk-Sandbergen and Choulamany-Khamphoui, 1995b). In Thailand beautiful water pots in all shapes and colours can be seen around each and every house, clinic, shop, factory, temple, including slum areas where people do not always have 'hard' roof-tops. South-East Asian women catch water where it falls. Efforts to launch waterpot programmes by women's NGOs, or by women in NGOs, in India have, as far as I know, not been successful.

Women's Solutions to Cope with Scarcity in the Slums

Banerjee and Rohilla (Chapter 3 in this volume) have explored how to harvest more water to mitigate Delhi's water crisis. The existing strategies of women in slums to adapt and cope with water scarcity are (Schenk-Sandbergen, 1994b):

- Increase of water resources by private initiative and common management of women;
- Prevention of water wastage and reuse of water;
- Informal, tacit restriction of water users and quantity of water to be taken to ensure equal distribution and depletion;

Collective action by pressurizing authorities and development of countervailing power.

Increase of water resources by initiative and common management: In Bangalore I found that women initiated digging wells and collecting money to install water taps. Private wells are used by the community and they share in the costs for their maintenance and cleaning. I also found that women made a hole in the water supply line of the municipality to get water illegally but understandably. They dig water out of the sand with cups and filter it through cotton cloth in pots. In other places women steal water at night.

Prevent water wastage and reuse of water: From the roof of a house overlooking a slum in Bangalore, the public water taps were observed; there is one tap for 25 families. The water just flowed for four hours when the valve was opened. Taps are stolen, so when the water comes it just flows. We saw an old woman, who put a stick into a pipe to avoid wastage of water. When asked who has told her to do this, she replies, 'nobody'. She knows that otherwise there will be no water in a short time.

Women also have special skills and knowledge regarding multiple uses of water. They wash vegetables in clean water and then use the same water to soak dirty cloth. They sprinkle the same water when it is dirty on the road around the house to settle the dust. In fact, all kind of practices exist to reuse grey, and greyer water.

Restriction on water users and the quantity of water to be taken to ensure equal distribution: In the Somaswara slum in Bangalore, women follow the distribution system according to which the ten households which paid for the installation of the taps fetch water first, and afterwards other neighbours are allowed to take their share. At many places there are tacit rules which limit the water quantity to be fetched per trip; in general women are not allowed to fetch more than five water pots (the capacity of each plastic pot is 15 litres). There is an informal arrangement in the slum that when there is sufficient water in the well everybody is allowed to take a share, but as soon as the water scarcity starts in summer, the families looking after the well are only entitled to take water. We found the same regulations were applied in the case of the number of buckets that could be fetched from hand pumps. It is also the custom not to allow to water to be taken from public hand pumps of other neighbourhoods. There is no 'open' access to public water facilities.

Collective action by pressurizing authorities and building countervailing power: Action of people is very often associated with the mobilization of NGOs but in slums we came across many spontaneous campaigns of local people, and initiatives of groups of local women's organ-izations such as mahila mandals, which organized the women of the area to march to the houses of the councillors to request better latrines, pavements, the repair of hand pumps, etc.

Female Hand Pump Caretakers and Mechanics

Another effective way of increasing water quantity in cities is to train women to repair and maintain available but defunct water devices themselves in particular in slums. In my study of the slums of Bangalore I found that the duration of time during which hand pumps did not function varied from three weeks to three years. In most cases it took four to six months before the hand pumps were repaired. Hand pump repair is a political matter. The slum leader has to inform the councillor and the latter has to inform the maintenance department of the municipal corporation. The process of communication involves political support, bribes, and power.

In Bangalore I have initiated and implemented a programme to impart training in hand pump care taking and mechanic training for women. A long scouting tour along with many government officers was finally rewarded by the discovery that in the public health engineering department training and monitoring cell of the zilla panchayat a full-fledged training centre existed, with all the UNICEF manuals in the local language, audio-visual aids, and instructions to train hand pump caretakers and mechanics in rural areas. A helpful and enlightening talk with the trainers revealed that they were very interested in the hand pump caretaker and mechanic training also for a group of urban slum trainees.

Before the training started, data was collected on defunct water devices in some slums. The finding was alarming; many hand pumps, taps, and wells were out of order. Moreover, the figures showed that the distribution of the 'out-of-order' devices is very unevenly distributed. In some slums even the only available water facility was out of order. Additionally, there is hardly a correlation between the number of households in a slum and the number of water resources available depending on the 'encroachment' status of the slum. In a colony 2,000 households have to share the water of only one hand pump, which was, by the way, out of order during the survey. In another slum with a population of 4,200 households there are 22 hand pumps (7 out of order) and 35 water taps. The bad smell and the high iron content restrict the drinking water use even further. In the majority of slums there is no water during two days of the week. Water has often to be fetched in the night (at 1 or 2 o'clock). The result of these irregularities in the water supply is that many women have to wait in long queues in the middle of the night to fetch water.[7]

The experiments with the training of female hand pump caretakers and mechanics all over the country, but also in Bangalore, Kanpur, and Mirzapur, try to mitigate the problem by making women independent of male dominated maintenance departments. With the support of several NGOs, the hand pump training was given in Bangalore. A tool kit for repair was deposited in the offices of the NGOs. The programme was successful to a certain extent. The main problem was that hand pump caretakers were considered volunteers and were not rewarded with money. Poor women cannot afford to spend too much time on voluntary caretaker work as they have to earn income to feed their children.

However, one of the most important outcomes of the programme is that women become aware that when they organize, and are trained, they have the skills and capacity to improve their own living environment. The importance of community participation in the maintenance and repair of public amenities is also emphasized by the programme. Moreover, it is a first step towards strengthening the women involved to take up leadership roles in organizing women to demand their basic habitat and human rights at several levels of government institutions.

A lesson learned from the female hand pump caretaker and from the mechanical training of trainers is that in this approach also the self-evident male dominance was challenged: by training women as mechanics which is normally exclusively male work, and to reduce the decision-making power of the maintenance and repair institutions by facilitating women to do that themselves. The role of NGOs is crucial to make the impact and to effect the training course sustainable. They should encourage and stimulate the trained mechanics and caretakers to carry out their learned skills. Besides, they may plan and organize training courses themselves for training of hand pump caretakers and mechanics to involve more women and women's groups in the maintenance and repair of hand pumps. It is very encouraging that some NGOs in Bangalore have taken the initiative to give the training an enduring impact.

PARTICIPATION OF WOMEN IN WATER USERS' ORGANIZATIONS

Water development projects emphasize the urgent need to make reservations for women in water users' assiociations (WUAs), and women staff in government water management agencies (irrigation departments, CADAs, etc.).

In a previous section it was revealed that patriarchy and male dominance exclude women from decision making regarding the use of water. Internationally, the under-representation of women in WUAs gets much attention in the development of methodologies and approaches concerning the gender components in water projects. The efficiency argument is mostly the rationale behind this focus of attention, perhaps with a spark of awareness of gender equality as a measure of an equitable society.

Gender experts follow different approaches to get women involved in water committees: integrated empowerment, economic empowerment, and conscientization through a process of gender-awareness raising through training and participatory gender studies. If all these steps do not help, the only strategy left is to lobby for the reservation of a quota of seats for women in water-management organizations to ensure women's participation. Unfortunately, these approaches from the capitalist-, modernized-, patriarchal world have also become necessary nowadays in the matrilinear, socialist, economically 'opening-up' countries. Let us have a look at Laos.

Laos: An Example of Training Male Staff in Gender Awareness

During investigations in Laos it was found that many communities have a long tradition of voluntary labour to construct, operate, and maintain small local schemes. The people have inherited skills in local irrigation techniques and water management. This counts for the men, as much as for the women. However, we found that women hardly participate in the water users' organizations in the villages recently established and supported by the government.

According to a resolution of the Ministry of Agriculture, water users' organizations have to be formed to regulate the operation and maintenance of government-supported schemes. Procedures for water users' organizations before, during, and after the scheme construction are drawn up, and the role and duties of the organization formulated. The main duties and roles of the members of formal water users' organizations are to look after regular repair and maintenance work, canal excavation, weed cleaning in the scheme area, to give material and financial support, to collect water fees (either in cash or in kind) in accordance with the agreement of the association, and to protect the scheme against any type of destruction.

Our visits to irrigation schemes in six provinces showed that there are almost no women participating in the management and decision making of the 'new' irrigation scheme organizations with the exception of one scheme.[8] It was found that the 'modernization process' in formalizing and legalizing irrigation organizations has reduced the previously more equal gender relations in the management of irrigation schemes to the disadvantage of women. In the new concept of community participation, awareness has to be raised that women are important partners in agriculture and irrigation, operation and maintenance, whereas this had been a basic understanding in the context of the traditional system of irrigation.

In most of the Lao Lum communities, by far the largest group in the population of Laos, the post-marriage residence pattern is matrilocal. This means that the bridegroom lives in the house of his parents-in-law. It is mostly the youngest daughter who inherits her mother's/parents' land and house. She is the manager of the farm and participates in almost all agricultural operations. She is the main water user and manager at the local level. But the role of women also demands that grown-up daughters contribute to the heavy labour involved in the construction and maintenance of the schemes.

The irrigation investment, or water fees, are mainly paid by the productive labour of women and grown-up girls. Because of the daily use of water for domestic use (bathing, washing clothes, and irrigating small plot around the house or along the irrigation canal) it is clear that women and girls are highly interested in the planning, design, and maintenance of the scheme. Women are good at financial management and accounting because of social and economic experience at household finance and local economy: they are the sellers, buyers, traders, middle women, and entrepreneurs.

With the institutional formalization of government-supported schemes, there is a very real danger that women will lose control and access over 'their' land and water sources, leading to a process of 'disempowerment' of women.

The main reason for the exclusion of women is the intrusion of the urban-based and class-biased gender ideology that accords men a higher status as 'heads of households' and relegate women to an inferior position as caretakers of children and the household; this same kind of thinking regards irrigation as a male domain and not suitable work for women. These values seep through the whole hierarchy of levels, from the centre (Vientiane) to the village, as the process of mobilization and organization of water users' organizations has to be implemented through the male officers of the irrigation line departments. At the district level, the male heads of the villages are approached to organize meetings to discuss the formation, role, and duties of formal water users' committees. The head of the village will call the men of the households, as this matter relates to the public sphere of authority not specified by society as being the women's domain. In most cases, the head of the village will invite the official representatives of the village Lao Women's Union to be present at the meetings.

However, we came across a very large scheme where the representatives of the Lao Women's Union were not invited to attend the meetings and did not know that the male water users' organization had decided that all households should have to take a considerable credit to invest in the irrigation scheme. It was very embarrassing and painful for the village women when this point was brought up during a talk we had on the financing of the irrigation scheme. The consequence for the women was that under the possibilities for earning cash in that area the women have to earn most of the loans their husbands have taken.

In this male-dominated set-up it is no wonder that class and gender assumptions are reflected in the approach and conceptualization of the water users' gender composition. Additionally, in talks the argument was used that women are less educated and therefore less qualified to manage water facilities, and that women are too busy to attend meetings. The last argument may be true, but that does not justify their exclusion but should rather be a challenge to search for appropriate solutions to overcome this problem.

At present we can state that there is a strict demarcation line in perception and actual behaviour, whereby the water users' organizations are a male domain and the Lao Women's Union exclusively is for women. The formal creation of new institutions to operate and maintain government-supported irrigation schemes has created a new field of male authority and power as it originated in a social vacuum. Women are perceived as the followers and helpers of men, whereas in the daily activities of village life it is often the other way round.

Encouraging the appointment of women in water users' organizations at both the management level and also as members is an important part of irrigation development in which the general position of women can be strengthened. Therefore, raising gender awareness among the male staff of the water government agencies and the male farmers in the villages was one of the focus points of the gender component in the project in which I was involved. My Laotian counterpart and I wrote training manuals and reference books for the on-site training of male government staff to help them understand the importance of women's participation in the planning, design, implementation, and rehabilitation of water schemes (Schenk-Sandbergen and Choulamany-Khamphoui, 1995b). Most important was that we designed a set of very interesting visuals at the village level dealing with Lao village conditions regarding the division of labour on gender lines. The purpose is to enhance the understanding of the males of women's roles and status to facilitate the formation of and the participation of women in water users' associations. The objectives of the training/workshops are to raise gender awareness and to discuss ways to strengthen women's role in irrigation development.

Water Users' Associations and Reservation for Women

In India it is a general phenomenon that women are mostly excluded from membership in water users' associations, with the exception of some female-headed household representative. To understand this situation we have to understand the meaning of the concept 'participation in water management'. Usually the concept refers mainly to participation in official bodies, which are primarily meant for official decision making on operational activities such as the daily irrigation routine of opening and closing sluices and outlets, planning of maintenance, etc.[9] However, this forms only one aspect in which participation of water users can contribute to improved irrigation management.

The making of bunds, field channels on one's own land, facilitating the water to spread over the plot, the drainage, the choice of crops, and the investment in the inputs should also be defined as 'participation in water management' at the household level. Landless labourers and farm women play an important role at this level of water management, or have to bear the costs of these activities. However there is a lack of recognition of women as water users (Zwarteveen, 1997, p. 1346). This is the more surprising as there are certainly links between the two levels of water management. If the scheme level decides on night irrigation, it influences the household level, as women and children are mobilized to receive the water in the field at night. Other examples are the distribution days and rotation schedules. This affects the daily routine of all family members.

There are hardly any rules requiring that users should be informed about the availability of water and its supply dates in different parts of the command area. In Vietnam the brigade chiefs in the communes takes this responsibility, and inform the women in particular as they are the main labourers in scooping out the water from the canals into the fields.

In our study of the Tungabhadra scheme, we found that applying the narrow definition of 'participation in formal water management bodies' we had to admit that the offices of government agencies and formal jobs in irrigation are totally male-dominated. Farm women, as well as farm men, play hardly any role in water management as the activities are mostly in the hands of the male professionals of the Irrigation Department and the Command Area Development Authority. Even activities such as the construction of small field channels by the CADA and the maintenance of canals are subcontracted to contractors who hire outside labourers to do the work. The same practice is followed for the maintenance of the canals. This state of affairs reveals the urgent necessity for more community participation in water management.

In addition to several approaches to empower women in the command area, it seems that the extreme patriarchal context of the project warranted reservation of seats for women in decision-making water bodies. Reservation of one-third of the seats for women in WAUs has been proposed. This quota fits well into the general Karnataka government policies regarding the enhancement of the position of women. Karnataka (and also West Bengal) has made serious efforts to implement the seventy-third Amendment Act, 1992 of the Constitution. The Act requires that in the local government bodies of the panchayats, not less than one-third of the seats should be reserved for women. The intention of this policy is to involve women at the local level in the planning, decision making, and implementation of development activities. Karnataka has the reputation of a model state regarding the election of a high number of women members in panchayats and its effect on a successful involvement and participation of women in development activities.

After our illustrations of the conscientization and reservation approach to increase the participation of women in water management, let us now briefly look at the integrated approach of economic empowerment of women as implemented by the women's NGO, SEWA.

Women's NGOs and Pani Panchayats: an Integrated Approach

Often in community participation the interests of the women get lost. Therefore, women's NGOs have taken initiatives to organize women to manage water sources. One of them is the Self Employed Women's Association (SEWA) in Banaskantha. In Banaskantha district, in particular in the arid-zone *talukas* of Radhanpur and Santhalpur, the scarcity of fresh water and the depletion and degradation of the soil are such that people struggle for survival (Schenk-Sandbergen, 1995a).

The increasing salinity of the surface water and groundwater intruding from the Thar desert towards the north, and the Rann (Gujarati for 'desert') of Kutch on its northern border, have depleted and spoiled, over a short period of time, the last sources of fresh water available in the area. The indiscriminate over-drawing of irrigation water by rich farmers using deep tube wells in Mehsana district seems to be the main reason for the changes in the upper and under aquifers and for the lowering of the water table at the frightening rate of 3 m per year. Grain is a very profitable cash crop, and with irrigation the rich farmers of Mehsana can get three annual harvests but in doing so they have changed the whole underground water balance and composition.

Moreover, the Banaskantha area has suffered from years of continuous droughts since 1985 (1985-6, 1986-7, 1987-8, and recently 1991-2). This has exacerbated the desertification of the area and reduced the last resource on which the local people depended for their survival and the survival of their cattle. Despite relief measures such as cattle camps, cattle perished because of lack of fodder and water. In 1990 another instance of extreme weather hit the area: the monsoon rains were so heavy that a flood relief programme had to be launched by the government. Evidence of land degradation caused by the stagnant monsoon water was widely visible.

In the Banaskantha area, SEWA has worked for many years to improve the living conditions and the position of women. The main element of SEWA's economic empowerment strategy is to link women, water, and work in this arid environment. In this way, efforts are made to improve the socio-economic and ecological environment in order to reduce seasonal migration. This interesting process is described in their recent study on managing water for the people by the people (SEWA, 1997). The main strategy is to link women, water, and work in order to encourage women to play a crucial role in the pani panchayats (water committees) in order to ensure that the drinking water facilities which have been provided will be sustainable in the long run. Women, except possibly the Rajput women, are the water carriers and water managers in the two *talukas*. They are therefore the most concerned about and involved in water management. It is imperative that they participate on the same footing with men in the management of drinking water in the pani panchayats. The raising of the women's self-confidence and their empowerment is supported by training them in management skills and leadership; this training is a precondition for active participation in the pani panchayats which the project is actively encouraging.

Another link is that income generation should be related to the regeneration of the degraded environment by making optimal use of existing water resources. Eco-regenerative activities are experimental and require considerable effort which is hazardous and financially very risky in an arid region. This means that any failure resulting from climatic conditions can have a detrimental effect on the livelihood of the women involved. Aware of this enormous problem, SEWA nevertheless took up the challenge and selected a number of strategic activities such as raising nurseries, setting up fodder and fuel plantations, as well as social forestry, water conservation and water harvesting projects.

The SEWA case shows that the approach of linking women, water, and work is a successful one in the formation of pani panchayats. The lesson learned is that participation of women in water-managing institutions requires an integrated women and development approach. Economic empowerment should go hand-in-hand with social and cultural empowerment.[10]

THE FEMININE PRINCIPLE IN WATER MANAGEMENT

The core argument of this chapter leads me to reflect a little on women's role in the Chipko movement and the debate on the 'recovery of the *feminine principle* in water management'. The recovering of indigenous knowledge and the role of women and poor peasants and tribals as water managers is addressed here. Other studies show the importance of indigenous women's knowledge of water management in China and many parts of India (AIT Gender Studies, 1995).

According to the principles of the Chipko movement no high dam can provide the solution to the water problem. The recovery of the feminine principle involves the recognition that sustainable availability of water resources is based on participation in the water cycle, not on manipulation or mastery over it (Shiva, 1988, p. 215).

The Chipko movement started in 1973 at Mandal, Chamoli district where the organizers had been active for many years. In 1979, on World Environment Day, hundreds of women of belonging to Chipko movement protested with empty water pots (Shiva, 1988, p. 210). They were protesting against the deepening water scarcity as well as against the failure of official water supply schemes and against a model of science which saw metal pipes and concrete tanks as producers of water, and which regarded male engineers and technicians who fitted pipes and designed schemes as providers of that water. They said, 'If you want to solve our water problems please plan for water, not for pipes' (Shiva, 1988, p. 211). The problem will not be solved. On the contrary, the problems will increase. According to Vandana Shiva, it is imperative in water management to think and act ecologically: to 'think like a river' and to flow with the nature of water (Shiva, 1988, p. 183).

THE ROLE OF WOMEN IN CONSTRUCTION, MAINTENANCE, AND FINANCING

Women of poor households in Bangladesh play an important role in the maintenance of government-owned water works. For instance, since 1988 the concept of women's maintenance groups (WMGs) has been accepted in Bangladesh by the Bangladesh Water Board (BWDB). They have legal contract systems with the so-called women landless cooperative societies and women embankment maintenance groups (EMGs). These consist of small groups of female labourers recruited from among the poorest of the poor rural households. EMGs are contracted by the BWDB but differ from regular contractors. EMGs pursue a strategy of preventive maintenance of water management infrastructure.

The activities involved are the implementation of preventive maintenance, which provides an opportunity to intervene in the highly segregated labour market, by protecting this newly generated employment opportunity for women belonging to the poorest of the rural poor. Thus, while the main objective of EMGs is to ensure preventive maintenance, it also needs to be appreciated as a strategy designed to support the socio-economic development of poor women (Schenk-Sandbergen, 1994a). A two-day yearly training of women organized into groups on various technical, social, and gender issues is part of the set-up. Women are likely to be concerned with getting the maintenance work done properly. As Bangladesh, an Islamic country with many restrictions on the freedom of movement of women institutionalized in *purdah*, can organize these groups, it might be possible to do so in India too. The option has been proposed in the Tungabhadra project. It is too early to say anything regarding the results.

In Laos and Vietnam women are the constructors and the main operators and maintainers of water systems. As already mentioned, women are pump operators in the Red River Delta irrigation schemes. Most of the cleaning work on primary and secondary canals in Vietnam is carried out by women's labour. They even do this work voluntarily for the Vietnamese Women's Organization which receives the value of the work of the volunteers as a money contribution for their collective activities.

Liberalization and privatization policies hamper the handing over of the ownership and/or management of government-owned water devices to private groups of farmers. The underlying thought is that, if a service organization (WUAs) is financially dependent on fees from users, it is likely that the performance of that organization will improve. The assumption is controversial, but the questions here are: who will pay for the costs? And what social costs are involved? WUAs are generally dependent on the fees or contributions from their members, while state irrigation agencies are not.

The point here is that evidence reveals that a large share of the price to be paid for the privatization of water resources will fall on the shoulders of women and children. In Laos women and girls weave day and night to earn money for the running of the irrigation system. In Vietnam, with the so-called private company system of water-service stations, households have to pay incredibly high water fees to finance the enterprises. In the Red River Delta men migrate and the feminization of agriculture is evident. Unacceptable self-exploitation of women, dropping out of children from school, especially girls, as they have to help their mother in the rural/water managing work is already documented in many places around the world (Schenk-Sandbergen, 1996a).

These dimensions of the problem deserve our full attention when we talk of managing water scarcity.

SUGGESTIONS TO IMPROVE THE PARTICIPATION OF WOMEN IN THE MANAGEMENT OF WATER SCARCITY IN INDIA

In view of the above, the following suggestions are made to improve access to, control over, and participation of women in managing water scarcity in India:

1. Land rights for women are essential to gain access to and control over water resources. Project interventions which challenge the well-entrenched male dominance can be successful in improving the role of women in water resources development and management.
2. Water projects should make concerted efforts to combine irrigation with drinking and domestic water facilities. The separation of the two spheres into male and female domains should be diminished.
3. Practices such as 'water-pot' and 'catch the rain where it falls' followed in South-East Asia should also be promoted in India (even when it is for two months or less).
4. Lessons should be learned from the ways and means by which poor women cope with scarcity of water in urban slums including practices such as reuse of water, prevention of water wastage, restriction of quantity per household according to the season, etc.
5. The female hand pump caretaker and mechanic training programme should be supported and promoted.
6. It is essential to raise gender awareness among male employees of government line departments (CADA and DOI). This can be done by

   - organizing workshops on women (gender), water management, and development.
   - incorporation of gender issues in the training curriculum of colleges and universities related to water management, etc
   - support in the conducting of gender-sensitive socio-economic/monitoring action studies in planning/ formulation/evaluation phase of water projects.

7. More gender aware female personnel should be appointed in line with government regulations (one-third clause).
   Planning and design of water projects should be scrutinized and approved by a 'gender-effect reporting' carried out by a 'woman (gender) and development' specialist to be appointed at the state, district and block level. Close cooperation with NGOs and mahila mandals to mobilize and organize women for case studies, surveys and meetings should be pursued.

8. Reservation policies to enhance the participation of women in water management and community development efforts (water users' committees) should be stipulated in regulations on WUAs. Supportive gender training and other gender activities are imperative.
9. An integrated approach is necessary for the enhancement of women's representation in water-managing institutions. Socio/political empowerment has to go hand in hand with economic empowerment.
10. Indigenous knowledge and practices of women in the area of conservation and harvesting of water should be revived, strengthened, and supported.
11. The role of women (and men) of poor households in operation and maintenance activities should be institutionalized in cooperatives and societies, acknowledged by the government water agencies.
12. Women should be fully involved in the privatization of public water resources as the financial and labour aspects might have detrimental effects on them and their children (for example, an increase of child labour).

NOTES

1. The data in this chapter are based on my experiences working in the following water resources development cooperation, and academic research projects:
   - North Bengal Terai Development Project, West Bengal: small-scale irrigation
   - Banaskantha Women's Rural Development Project, Gujarat: SEWA project (desert area, drinking water)
   - Bangalore: Women and Water in the Slums
   - Kanpur/Mirzapur: Ganga Action Plan
   - Tungabhadra Irrigation Pilot Project, Karnataka: canal irrigation
   - Strengthening and Restructuring Irrigation Development Project, Women and Irrigation in Vientiane, Laos
   - TA 2, Women and Irrigation in the Red River Delta, Hanoi, Vietnam
   - Early Implementation Project and CCP-Tangail Project, Bangladesh
   - Land Rights for Women Project, Laos.
2. The choice of the selected gender dimensions in this contribution is an outcome of missing gender dimensions in the papers submitted for the IDPAD Water Seminar in October 1997.

3. The following overview shows the tangible results of the gender intervention strategty after five years of the implementation of Phase II:
   (a) 24,030 hand pumps have been sanctioned and are almost all installed (planned 13,500 in the preparatory documents: Master Plan Phase II).
   (b) Female-headed households belonging to marginal farmers have been given preference in getting a hand pump, and 10 per cent of the hand pumps have been given in the name of the woman of the household.
   (c) The sample survey shows that in 95 per cent of the cases, the site of the hand pump was selected in consultation with the women of the house.
   (d) Construction works of the project (pump houses, 4 ha. units, etc.) have provided wage labour to local landless women (and men).
   (e) Out of the planned 27 women extension workers (two per block), nine have been appointed in the Department of Agriculture.
   (f) Farm women of marginal and small landholders are trained to improve agricultural management and operations.
   (g) In some cases women are members of the water users' committees.
   (h) An in-depth follow-up study over four years has been conducted on the impact of the irrigation project on women.
4. The importance of bathing and washing places for women is grossly overlooked in many projects. The pioneering article by Anil Agarwal (1985) is one of the few studies which pay attention to the cultural taboos related to the subject.
5. Vakranis are small pits (1 to 2 feet deep) dug along the canal bed in which water is collected overnight, using cups or coconut shells. These pits are either owned by one individual family or by a few families depending on the quality and quantity of water that can be harvested in a pit.
6. For an excellent overview of the rise and fall of India's traditional water-harvesting systems, see Anil Agarwal and Sunita Narain (1997).
7. In 15 slums it was stated that the waiting time to collect water was as follows: half an hour three slums; one hour two slums; one and a half hours two slums; two hours six slums, two and a half hours two slums.
8. This is the Sustainable Irrigated Agriculture Project (SIRAP). The project has been established to provide support for irrigation scheme improvement (infrastructure, operation, and management) and agricultural development in seven provinces in north-east Thailand and three provinces in Lao PDR. It is funded by the Netherlands Government and implemented through the Mekong Secretariat. The emphasis of the SIRAP is on the involvement of government, farmers, and the

private sector in a development strategy based on a participatory approach. The strengthening of women's role in water users' organizations by providing gender-awareness training is the main focus of the two Women in Development experts working on the project.

9. The president of the Phu Tien Drainage and Irrigation Management Company, which is one of the 11 enterprises in the Bac Hung Hai scheme in the Red River Delta in Vietnam, is a woman. The Phu Tien DIMC covers 20,000 ha. (13,000 ha. cultivable) and 200,000 people. The land size per household is only 0.1 ha. There are 40 pumping stations (4 drainage, 36 for irrigation and drainage). The total staff is 186 people, of which 120 are pump station operators. To our surprise we were informed that 70 per cent of the operators of the pumping stations are women: married (80 per cent) and unmarried women (20 per cent). This is a unique gender representation in the irrigation sector, as all over the world this (technical) job is regarded as a typical male domain. The existence of female pump operators might be the legacy of the war period when women had to take over all occupations traditionally performed by men.
10. I am grateful to Dr. V.B. Eswaran who drew my attention to a similar approach adopted by a project of an NGO (Adithi) in Bihar. Poor women have successfully leased fishponds to raise fish.

REFERENCES

AIT Gender Studies, *Indigenous Women's Knowledge in Health Care and Water Management*, Workshop Report, 7-9 August 1995, Bangkok: Asian Institute of Technology, 1995.

Agarwal, Anil, 'Taboos Make Hygiene Difficult for Women', in *Gate*, no. 4, 1985, p. 29.

Agarwal, Anil and Sunita Narain (eds), 'Dying Wisdom: Rise, Fall and Potential of India's Traditional Water Harvesting Systems', in *State of India's Environment: A Citizens' Report, 4,* Delhi: Centre for Science and Environment, 1997.

Agarwal, Bina, *A Field of One's Own: Gender and Land Rights in South Asia*, Cambridge South Asian Studies 58, Cambridge: Cambridge University Press, 1994.

Baserup, E. *Womens Role in Economic Development*, London: Earthscan Publications, 1970.

Centre for Women's Development Studies, *National Colloquium on Role of Women in Water Resource Management*, Report, New Delhi: CWDS, 1990.

Epstein, Scarlett T., *Economic Development and Social Change in South India*, Manchester: Manchester University Press, 1962.

Etienne, M. and E. Leacock (eds), *Women and Colonization*, New York: Praeger, 1980.

Groverman, Verona and Edith van Walsum, 'India, Women may lose or gain: expected impact of irrigation projects', in Vera Gianotten, Verona Groverman, Edith van Walsum, Lida Zuiderberg (eds.), *Assessing the gender impact of development projects: Case studies from Bolivia, Burkina Faso and India*, Amsterdam: Royal Tropical Institute, 1994, pp. 71-94.

Leacock, Eleanor Burke, 'Women, development and anthropological facts and fiction', in *Myths of Male Dominance: Collected Articles on Women Cross-Culturally*, New York: Monthly Review Press, 1981, pp. 310-17.

Page, Ben, 'Taking the strain: the ergonomics of water carrying', in *Waterlines*, vol. 14, no. 3, January 1996.

Rahman, Md. Mahbubar and Willem van Schendel, 'Gender and the Inheritance of Land: Living Law in Bangladesh', in Jan Breman, Peter Kloos and Ashwani Saith (eds.), *The Village in Asia Revisited,* Delhi: Oxford University Press, 1997, pp. 237-76.

Rights of Women, *Rights of women to the natural resources land and water*, Working Paper 2, Women and Development, NEDA, Netherlands Development Assistance, Development Cooperation Information Department, Ministry of Foreign Affairs, The Hague, 1997.

Schenk-Sandbergen, Loes, 'Empowerment of women: its scope in a bilateral development project—a small-scale irrigation project in North Bengal', *Economic and Political Weekly*, vol. XXVI, no. 17, 27 April 1991.

———, *Strengthening of Women in CPP: Action Plan 1994-1995*, Bangladesh Flood Action Plan, Compartmentalization Pilot Project, FAP-20, Tangail, Bangladesh, Government of The Netherlands, Ministry of Foreign Affairs, Directorate General for International Cooperation, 1994a.

———, 'Women, Water and Sanitation in the Slums of Bangalore: A Case Study of Action Research', in Hans Schenk (ed.), *Living in India's Slums: The Case of Bangalore*, IDPAD Series, New Delhi: Manohar, 1994b.

———, 'Sustainability for Whom?: Lessons from the North Bengal Terai Small-Scale Irrigation Project', in Margaret M. Skutsch, J. Hans M. Opdam and Nico G. Schulte Nordholt (eds.), *Towards Sustainable Development: A Liber Amicorum for Enno W. Hommes,* Technology and Development Group, University of Twente, Enschede, 1994c.

———, *Main Findings of the Socio-eeconomic Impact Study: North Terai*

Development Project, Phase II, Euroconsult, Report No. 15 of the Research and Evaluation Unit, Arnhem, The Netherlands, 1994d.

———, *Women and Seasonal Survival: Migration in India, Series,* Indo-Dutch Alternatives in Development, New Delhi: Sage Publications, 1995a.

———, With Outhaki Choulamany-Khamphoui, *Women in Rice-Fields and Offices in Laos: Irrigation in Laos. Gender Specific Case-Studies in Four Villages,* Heiloo: Empowerment, 1995b.

———, *Rural Women in the Red River Delta: Gender, Water Management and Economic Transformation, A Study in Three Districts and Nine Communes,* Socialist Republic of Vietnam, Research Centre for Gender, Family and Environment in Development, Hanoi, Vietnam, Asian Development Bank, October 1996a.

———, With Shanta Mohan, *Rural Women in Tungabhadra: Gender, Water Management and Empowerment, A study in five villages and four camps in the command area of the D-36 Distributary,* TIPP-II, Indo-Dutch Development Cooperation Project, Ministry of Foreign Affairs, Directorate General for International Cooperation, Command Area Development Authority Tungabhadra Project, November 1996b.

———, *Report of the Evaluation of the Working Women's Forum,* Phase II, Madras, India, 1996c.

———, *Gender & Culture Issues Study: Gender, Culture and Land in Lao PDR: Towards a Gender Sensitive Land Titling: A Study in Districts, Villages and Households* (in cooperation with Hermien Rodenburg and Chansamone Phengkhay), AusAid, Vientiane, 1997.

SEWA (Self Employed Women's Association), *Managing Water for the People by the People.* A Participatory Assessment on SEWA's Experiences with Rural Water Resource Management in Gujarat, Ahmedabad, 1997.

Sharma, Ursala, *Women, Work and Property in Northwest India,* London: Tavistock Publishers, 1980.

Shiva, Vandana, *Staying Alive: Women, Ecology and Development,* New Delhi: Kali for Women, 1988.

Stanbury, Pamela, 'Women and Water: Effects of Irrigation Development in a North Indian Village', in Michael R. Dove and Carol Carpenter (eds.), *Sociology of Natural Resources in Pakistan and Adjoining Countries,* Lahore: Vanguard, 1992, pp. 372-99.

Van Koppen, Barbara and Simeen Mahmud, *Women and Water-pumps in Bangladesh: The impact of participation in irrigation groups on women's status,* Department of Irrigation and Soil and Water Conservation, Wageningen Agricultural University, Bangladesh Institute of Development Studies, Dhaka, 1995.

Van Wijk, Christine, Esther de Lange and David Saunders, 'Gender aspects in the management of water', *Natural Resources Forum*, vol. 20, no. 2, 1996, pp. 91-103.

Venkateswaran, Sandhya, *Environment, Development and the Gender Gap*, New Delhi: Sage, 1995.

Zwarteveen, Margreet Z., 'Water: From Basic Need to Commodity: A Discussion on Gender and Water Rights in the Context of Irrigation', *World Development*, vol. 25, no. 8, August 1997, pp. 1335-49.

CHAPTER 16

Gendered Poverty Alleviation Under Water Scarcity

BARBARA VAN KOPPEN

INTRODUCTION

The growing competition over water resources warrants a stronger focus on the access to water for resource-poor women and men by those governmental and non-governmental irrigation agencies and research institutes that have poverty alleviation as their ultimate aim. This paper elaborates some of the poverty alleviation issues that are indicated in other papers in this volume and complements these discussions with other experiences from Bangladesh, India, and elsewhere. Gendered water needs for domestic use and urban water supply have already been dealt with by several earlier papers and especially by Schenk-Sandbergen. This paper concentrates only on water for agricultural use. The main concern is how the access to irrigation water by resource-poor farmers who have limited, informal, or temporary land rights can be improved under current policies and interventions to manage water scarcity. This implies some further analysis of the relation between land rights and water rights.

Within the group of resource-poor farmers, women farmers and their access to water gets special attention. Women are a priority target group in poverty-alleviation efforts. They constitute the larger proportion of the world's poor. Improvement of women's status in both male-headed and female-headed households has also multiplier effects on the well-being of their dependents and on the future generations, which are lacking if only men's status was to improve (cf. Safiliou, 1991; Agarwal, 1994). The focus on irrigation water by definition excludes landless women and men who even lack some functional access to land for their own cultivation. The indirect effects of irrigation development under increasing water scarcity on the well-being of landless women and men, due to changes in on-farm and off-farm employment and food prices, are beyond the scope of this paper.

Access to water is operationalized as follows. This vague expression refers to a whole 'bundle of water rights'. This paper concentrates on the processes by which resource-poor women and men farmers are included or excluded as legitimate right holders under the prevailing pluralistic legal systems (Von Benda-Beckmann, 1991). The type of rights a person holds can vary largely, for example, from just use rights on the water entering the field to the authority to change rules or represent the water users' group to third parties (Von Benda-Beckmann et al., 1996). This variation is not further considered here. Rights to water conveyed by infrastructure below the point of water derivation from the flow or reservoir are primarily vested in those who invested in, own, and maintain the infrastructure (Coward, 1986). This historically and geographically widely accepted principle reflects the specific character of the natural resource of water, which is that water in the form as nature offers it is not fit for most human use as yet. In order to bring water in the right quantity and quality to the right site at the right moment the necessary infrastructure is needed. Depending on the main investor and the owner of the infrastructure, a distinction can be made between private and public irrigation. Inclusion and exclusion processes of the resource-poor as legitimate right holders differ accordingly.

In private irrigation, a farmer or a group of farmers invests in and owns the infrastructure and disposes of the water as they consider best. Private mechanized pump irrigation nowadays is the most important example. The usual failure of the state or NGOs to effectively regulate water extraction by private pump owners reflects these *de facto* water rights of those who invest in the infrastructure, in spite of statutory laws under which water is declared to be state property. Opportunities and constraints for resource-poor women and men to get access to water as private owners of infrastructure, and the roles of external support agencies are described in Section Two. Access to water by purchase from well owners is also briefly discussed.

Public irrigation is defined here as irrigation in which external governmental or non-governmental agencies bear most of the investment costs in infrastructure, although a certain contribution of the users towards the construction, operation, and maintenance is usual. In this situation, the ownership claims of the state are very effective. The state is the main decision maker in defining the division between its own rights and the rights of the users. Users' rights are mainly defined at two moments. During the design and construction or rehabilitation phase, also before the investments are made, the rights on the fruits of the investments are negotiated. If the state fails to do so explicitly, local people will do that themselves. During the subsequent use phase, changes in water rights are minor and gradual, unless major projects of irrigation management transfer are undertaken. One aspect of the (re-) definition of the rights and obligations of users is the (re-)definition of the legitimate 'user', or the membership criteria for water users' organizations. Opportunities and constraints to endow resource-poor women and men with water rights in public irrigation, especially under irrigation management transfer, are discussed in Section Three (minor irrigation) and Section Four (large-scale canal irrigation).

PRIVATE IRRIGATION

In private irrigation, in which owners of infrastructure are entitled to water, a critical issue for poverty alleviation is the access to own infrastructure and energy sources by the resource-poor. The current technology for private ownership tends to be biased towards the scale, cropping pattern, and financial means of larger farmers (cf. Chambers et al., 1989; Janakarajan, 2004).

In water-scarce regions, the access to increasingly expensive equipment is even more skewed. This intrinsic bias of private irrigation against the resource-poor can only be mitigated by improving the availability and distribution channels of technology and energy sources that fit the specific needs and potentials of the resource-poor. Where individual ownership is still not feasible, collective ownership, even with its complications (Janakarajan, 2004), may be a solution. The development of appropriate technology is promoted by external support agencies (cf. Datye et al., 1997).

Privatization of the market for irrigation equipment can also stimulate the availability of smaller and cheaper equipment and pump ownership by the smaller farmers. This has convincingly been shown for the spreading of shallow tube wells in Bangladesh since the 1980s (Kahnert and Levine, 1993; Mandal and Parker, 1995; Van Koppen and Mahmud, 1996). Mobile equipment, such as these low-discharge pumps to be used with bamboo pipes, brings mechanized equipment even within the reach of tenants and sharecroppers. In case of eviction they can take the equipment, and so most of their investments, with them. Thus, the usual land-boundedness of irrigation infrastructure ceases to be a disadvantage for the land-poor. As financial investments often remain substantial, middle- and long-term loan facilities are crucial. The loan conditions should assure the resource-poor against the risks involved.

In recent decades, the sale of water by private owners has provided access to water to millions of small holders in India, Bangladesh, and Pakistan. Although the own water interests of the pump owners still prevail (cf. Strosser and Meinzen-Dick, 1994), there is also a dependency of the water seller on the buyer. The increasing competition in the offer of water by multiple pump owners, and flat electricity rates lead to lower water prices and more adequate service (Shah, 1993). This is in the interest of all water buyers, but especially of small and marginal farmers who cannot afford their own equipment.

This commercial demand for water under private irrigation makes direct access to irrigation water attractive even for people who have very limited access to land to use water primarily for own production. In Bangladesh, where groundwater is readily available, some successful experiences of such NGO-supported male (Wood et al., 1990) and female water-selling groups (Van Koppen and Mahmud, 1996) are reported. The latter study revealed that women only take active part in pump management themselves if the supporting NGO targets not only their credit programme, but especially their irrigation training and organizational support.

Other NGOs target 'households', and either the man or the woman can become the NGO member and represent the household. If the women are members of the NGO they just sign for the irrigation loan, but their male relatives fully control pump management and benefits. This role of mediator only slightly increases women's status in most of the cases studied, but it does not vest water rights in women. Only women who are heads of household benefit in both women-managed and predominantly male-managed irrigation groups which take loans in women's names (Van Koppen and Mahmud, 1996). In fact, the common rule of NGOs in Bangladesh nowadays—that only one member per household can become a member of the NGO—stimulates this role of women as mediators. Women respond positively, and the proportion of women members of NGOs like the Rangpur Dinajpur Rural Service (RDRS) and the Bangladesh Rural Advancement Committee (BRAC) is increasing rapidly.

The crucial difference for women between so-called household-based membership and individual membership is discussed below. It is noted that in both the male and female NGO-supported water-selling enterprises landownership played a minor role. The site selection was also based upon topography and hydrology and, if needed, the small sites for the pumps could easily be leased in. In northern Bangladesh (Van Koppen and Mahmud, 1996) and in Bihar (Wood, 1995), the disconnection between land rights and water rights is further illustrated by the common practice of several pump owners to use their mobile pumps on one well, either for free or for a small compensation, and to shift from well to well, depending upon the sites to be irrigated.

TURNOVER IN PUBLIC MINOR IRRIGATION

Transfer of former state ownership of high-discharge water extraction mechanisms (WEMs) to users' groups is the current policy in both India (cf. Raju, 1997) and Bangladesh (cf. Kahnert and Levine, 1993). An important issue for poverty alleviation is the question to whom these pumps are to be transferred and whether the earlier skewness in control over this state-subsidized equipment can be redressed to some extent. In Bangladesh especially, BRAC has pioneered in specific targeting of such turnover to resource-poor women and men who are BRACs members (Van Koppen and Mahmud, 1996). Around 1990 state-subsidized deep tube wells (DTWs) were offered to BRACs members. They could become shareholders in a water-selling enterprise by signing up for a five-year capital loan and repaying this in weekly installments with a total interest of 50 per cent. BRAC itself kept 20 per cent of the shares and assured adequate irrigation management. Both male and female BRAC members responded favourably to this option to invest in a durable asset under conditions within their reach.

In 1991 more than 47,000 BRAC members, out of whom 52 per cent were women, participated in 642 DTW enterprises. However, it soon appeared that BRAC's feasibility studies had been grossly over-optimistic. The command areas, in particular, declined rather than increased under the encroaching shallow tube wells. For the shareholders the DTWs became 'their fate'. They found themselves trapped in the loan conditions that did not allow them to withdraw without losing the money already put in. Thus, the only way to get back at least some money for the DTWs was to finish all repayments and then try to sell the DTW on the free market, at price that a client would be willing to offer. By the end of 1995 BRAC admitted the failure of its programme.

These experiences show, in the first place, the capacity and willingness of the resource-poor to invest in productive assets if such an opportunity is offered to them. Second, they also underline the importance of realistic feasibility studies, the risks of imposed technology choice, and the need for longer-term credit systems that provide for risks.

TURNOVER IN PUBLIC LARGE-SCALE CANAL IRRIGATION

In management transfer of state-financed large-scale canal irrigation systems, the redefinition of the division of rights and obligations between the state and the users, or their representatives, is a much-debated topic (cf. Raju, 1997, Vermillion, 1996). Again, much less attention has been paid to the issue to whom rights and obligations will be transferred. Do membership criteria tend to include or exclude male and female resource-poor farmers? In most debates the characteristics of the potential users are hardly specified. Some 'universal farmer' is portrayed. This farmer uses water primarily for own production, which supposes some minimal access to land in the command area. It has often been assumed that this universal farmer is a man, although this bias is gradually weakening, at least in rherotic. Further, many turnover debates do not refer to the individual farmer but rather to groups of farmers beyond a certain outlet sharing one water source.

This thinking in collectivities with only their representatives interacting with the external agencies is unavoidable for the large numbers of farmers concerned with canal irrigation, but it entails serious risks. Any person who comes forward is easily defined as a 'representative' of the other household members or of the whole group (Sengupta, forthcoming). This practice of the external agencies served to strengthen the power and position of the local male elite and the male household heads who have the most contacts with state and administrative officials (cf. Von Benda-Beckmann, 1991).

When title criteria are specified in general law or in a particular management transfer project, two criteria are likely to be considered: household representation and land rights. These may entail far-reaching implications for the inclusion or exclusion of the resource-poor in general and of women in particular. These aspects are discussed below, as well as the issue of effective representation of the resource-poor in water users' groups. Household-based versus producer-based allocation of women's independent water rights are hardly on the academic, policy, and intervention agenda as yet. In an analogous debate on women's resource rights, namely land rights, women's needs for unconditional and independent rights have been much better articulated in the last decade. Agarwal (1994) documented the state of the art and the need for further action and research for South Asian women. African women voiced their need for land rights during the Beijing Women's Conference in 1995. In Latin America some governments recently even opted for preferential land title allocation to women heads of households irrespective of their conjugal status.

For irrigation development accompanied by expropriation and reallocation of the command area, the allocation of rights on irrigable land to women is also receiving quite some attention nowadays. Widespread practices of gender-skewed allocation have been severely criticized for settlement schemes like the Mahaweli Ganga Scheme (Kumar, 1987), the Mwea scheme in Kenya (Hanger and Morris, 1973), and the Office du Niger in Mali (Klaver and Van Koppen, forthcoming). Exclusive allocation to men has also been challenged in small-scale gravity-irrigation schemes, for example, in The Gambia (Dey, 1990; Carney, 1988) and Burkina Faso (Van Koppen, 1990). Experiments with more equal land allocation and women's inclusion as members of water users' organizations have also started to be implemented in Kenya (Povel, 1990; Hulsebosch and Ombara, 1995), The Gambia (Carney, 1994), and Senegal (Deuss, 1994), as well as in land reclamation schemes in Sumatra, Indonesia (Van Hussen, personal communication). In India successful efforts of women's groups to obtain access to irrigated land are reported (Datye et al., 1997). The importance of women's organized action to obtain land is also emphasized by Agarwal (1994).

For water rights under irrigation development in which land tenure does not change immediately, such as large-scale canal irrigation, women's needs for independent rights are still hardly acknowledged. In most academic circles and in intervention practice and policy, worldwide water rights are still seen as being vested in households. Implicitly this is often equalized to men's rights, with women at best having *de facto* derivative rights. In this specific domain, the unitary household concept is still adhered to, in spite of decades of strong criticism against the underlying assumption of this unitary household model, which is that the interests of the different household members are similar and that the male head of the household represents these interests (cf. Safiliou, 1986; Agarwal, 1995).

Nevertheless, evidence challenging this assumption of a unitary household has also started to accumulate in the field of water rights, especially in local legal systems. Illo's study (1988) on the Aslong scheme in the Philippines reveals that male farmers themselves pointed out to the community organizers of the National Irrigation Administration that women played an important role in irrigated agriculture and irrigation management and that more than only one member of the household should be represented in the water users' organization. In Bolivia women vest their independent water rights by participating in construction work (Prins, 1996; Noordholland de Jong, personal communication).

In Ecuador water rights are inherited by both sons and daughters (Krol, 1994). In south-west Burkina Faso, where women cultivate rice in the wetlands, it is even forbidden for male chiefs to enter these wetlands, because this would cause inundations (Van Koppen, forthcoming). On the other hand, in the traditional irrigation schemes in the highlands of Tanzania women are reported to be excluded from vesting water rights according to local law (Kitunga, 1989).

By now, evidence on gender differentiation in vesting water rights in public irrigation has especially been documented for allocation based upon participation in construction or maintenance activities. Several agencies in Africa, Latin America, and Asia are reported to have excluded women from vesting independent water rights by participating in construction in at least three ways.

The first way is to prohibit women from carrying out construction work. Either women are not informed about the possibility to vest resource rights in this way, as reported in Burkina Faso (Projet Sensibilisation, 1993), or women are not allowed to come, which occurred in Tanzania (Van der Grift, 1991). They may even be unceremoniously sent back when they have already come to the construction site, as Bolivian women experienced (Prins, 1996). Such exclusion is usually justified on the basis of middle-class norms that construction work is too heavy or is not decent for women, even in regions where such norms do not exist locally.

A second form of exclusion is that women do participate in construction work but that this is not counted. It may be counted in the name of the male head of the family, as in Nepal (Bruins and Heijmans, 1993), or tasks such as bringing food and drink to the construction site, which are often carried out by women, are not seen as work, even if they are considered to be so in local perception (Lynch, 1991; Pradhan, 1989).

Third, women may carry out work and this is registered, but then the rule is not applied on women in the same way as it is in the case of men. Women's investments do not lead to rights as the investments of men do. The author observed this in Burkina Faso.

Although these experiences are from other countries and concern the construction phases rather than the turnover phases, they highlight the general importance of membership criteria and representation open to both women and men producers and the importance of equality in the link between contributions in infrastructure maintenance or fee payment and rights.

Land Rights

Besides household relations that are considered in defining title criteria, land rights are also often considered in the definition of the 'user' and the specification of membership criteria for water users' organizations. As mentioned before some (potential) functional access to land is implicitly required if water is mainly used for irrigation of own crops. However, statutory law often interprets land rights as full ownership rights. Other regulations, on the other hand, mention tenants and sharecroppers as legitimate members of water users' organizations as occurs in several regions in Pakistan (Byrnes, 1992). Byrnes observes that many farmers are not even aware of the existence of such texts. Here, and elsewhere, local regulations prevail in which cultivators with longer-term use rights rather than absentee landlords are in reality the decision makers. Nevertheless, with growing scarcity a legal status as water right holder will increasingly influence the outcome of water disputes.

In the general perspective of poverty alleviation, membership criteria that suppose certain assets are likely to exclude the asset-poor. If only long-term ownership rights are considered to be valid, all land users with informal and temporary claims on land, or people with formal claims that are not officially registered yet, are excluded and further brought at the mercy of the landowners and administrators. If water rights are vested in the actual cultivator, on the other hand, this might improve his or her bargaining position in the tenancy or sharecropping arrangements. Such effective negotiation power is reported, for example, in Mexico.

Moreover, these water right holders gained in cash or kind from a temporary exchange of water rights with others till they needed the water themselves (Nederlof and Van Wayjen, 1996). Disconnecting land rights and water rights allows for seasonal changes in land use in head ends (*phad*) of reservoir-fed irrigation, as mentioned by Datye et al. (1997). Allocation of water to landless families who do not cultivate yet is also realized in the Sukhomajri case in Chandigarh (Malhotra, in Chambers, 1994).

There may be contexts in which landownership as a condition for membership criteria is more favourable for the resource-poor, like regions with inverse tenancy, or in situations in which tenants may be evicted if they are further empowered, or in the context described by Schenk-Sandbergen (1997). The condition of this project of women's landownership for pump installation strengthened the people's registration of land in women's names.

It is noted that the relation between land and water is also expressed by the water payment systems. If water is implicitly paid for by land tax, the role of the landowner is obvious. If water payment is based upon factual use of water for certain crops in a certain area, it does not matter who the owner of that area is. So the disconnection between land rights and water rights seems to contribute to poverty alleviation but this dynamic needs further reflection.

Another aspect of the link between land rights and water rights is the quantity of water to which one is entitled. If water allocation is proportional to land size, large farmers will automatically benefit more from public investments. Therefore, in the 1980s a water rights reform was proposed, which would allocate water to users in quantities that were sufficient for a certain area and crop only (Chambers et al., 1989). However, this proposal does not seem to have had practical consequences since. The experience in Bolivia teaches us that the establishment of ceilings in the quantities of water to which one is entitled is often difficult, as larger farmers can still obtain water rights by using the names of relatives (Prins, 1996). With increasing water scarcity and given the agencies' choices in stopping irrigation in certain areas, class and gender characteristics of those who are excluded need explicit attention.

Early Inclusion in Forums

One common characteristic of the resource-poor is that they are hardly organized as yet. For women this is even more valid although NGOs achieve considerable changes nowadays. Therefore, the suggestion made by Raju (1997: 25) that the law or government regulations should encourage and support catalyst agents for limited periods to bring farmers together initially and help them organize themselves and get official recognition is crucial. Datye et al. (1997) also emphasize the need to work with viable small groups and their federations. The definition and implementation of title criteria need not only explicit but also timely attention in these forums. The effective presence of the priority target groups in the decision-making process from the very start of any turnover project is a condition *sine qua non* to guarantee that the resource-poor not only get their legal rights but that they are also able to claim the rights they already have (Van Koppen, 1997).

CONCLUSION

With increasing water scarcity in both private and public irrigation, the political space for the resource-poor to claim their rights to water and to the technology to convey the water is likely to decrease. This has already happened in the case of their rights to land in many parts of South Asia, where land reform has met resistance that it would not have met if it had been implemented earlier. So it is all the more important to reconcile water management under scarcity and poverty alleviation as soon as possible.

REFERENCES

Agarwal, Bina, *A field of one's own: Gender and land rights in South Asia*, South Asian Studies 58, Cambridge: Cambridge University Press, 1994.

Bruins, Bert and Annelies Heijmans, 'Gender biases in irrigation projects', *Gender considerations in the rehabilitation of the Bauraha Irrigation System in the district of Dang, Nepal*, Kathmandu: SNV Nepal, 1993.

Byrnes, Kerry J., 'Water users' associations in World Bank-assisted irrigation projects in Pakistan', *World Bank Technical Paper*, no. 173. Washington: The International Bank for Reconstruction and Development, World Bank, 1992.

Carmen, Diana and Magdalena Léon, 'Women, land rights and the Latin American counter-reforms', paper prepared for presentation at the XX International Congress of the Latin American Studies Association (LASA), Guadalajara, Mexico, 17-19 April 1997.

Carney, Judith, 'Struggles over land and crops in an irrigated rice scheme: The Gambia', in Jean Davison (ed.), *Agriculture, women and land: The African experience,* Boulder, Colorado: Westview Press, 1988, pp. 59-78.

———, 'Gender and the sustainability of irrigated farming in The Gambia', in Engstrom, Ingrid, Patricia Jeffery, Kenneth King, Camilla Toulmin (eds.), *Gender and environment in Africa; Perspectives on the politics of environmental sustainability*, Centre of African Studies, University of Edinburgh, 1994.

Chambers, Robert, 'Irrigation against rural poverty', in R.K. Gurjar, *Socio-economic dimensions and irrigation*, Jaipur: Printwell, 1994.

Chambers, Robert, N.C. Saxena and Tushaar Shah, *To the hands of the poor*, London: Intermediate Technology Publications, 1989.

Coward, Walter E., Jr., 'State and locality in Asian irrigation development: The property factor', in K.C. Nobe and R.K. Sampath (eds.), *Irrigation management in developing countries: Current issues and approaches, Proceedings of an Invited Seminar Series sponsored by the International School for Agricultural and Resource Development* (ISARD), *Studies in Water and Policy Management*, No. 8. Boulder and London: Westview Press, 1986.

Datye, K.R., Suhas Paranjape, Seema Kulkarni and K.J. Joy, 'Krishna Valley Development: An Approach to Equitable and Efficient Use of Water', Chapter 8 in this volume.

Deuss, Marleen, 'Do women's gardens hold water?: Gender relations and the introduction of irrigation systems at the Ile a Morphil in Senegal', M.Sc. thesis, Department of Irrigation and Soil and Water Conservation, Wageningen Agricultural University and Third World Centre University of Nijmegen, Occasional Paper 42, Nijmegen: Third World Centre, Catholic University of Nijmegen, 1994.

Dey, Jennie, 'Gender issues in irrigation project design in Sub-Saharan Africa', Contribution to International Workshop Design for Sustainable, Farmer-Managed Irrigation Schemes in Sub-Saharan Africa, Department of Irrigation and Soil and Water Conservation, Wageningen Agricultural University, 1990.

Hanger, Jane and Jon Morris, 'Women and the household economy', in Robert Chambers and Jon Morris (eds.), *Mwea: An irrigated rice settlement in Kenya*, Munchen: Weltforum Verlag, 1973.

Hulsebosch, Joitske and Doris Ombara, 'Towards gender balance in irrigation management: Experiences in Kenya South-west Kano Project', in *Irrigation and Drainage systems* 9, Netherlands: Kluwer Academic Publishers, 1995, pp. 1-14.

Illo, Jeanne Frances 1, Susan E. Leones, Grace C. Ignacio, Karen H. Jacob and Victoria R. Pineda, 'The Philippine Communal Irrigation Program', in Jeanne Frances 1. Illo (ed.), *Gender issues in rural development: A workshop report,* Institute of Philippine Culture, Quezon City, Ateneo de Manila University, 1988.

Janakarajan, S., 'Conflict Over the Use of Groundwater: Some Evidence from Tamil Nadu, Chapter 5 in this volume.

Kahnert, Friedrich and Gilbert Levine (eds.), *Groundwater irrigation and the rural poor: Options for development in the Gangetic Basin*, A World Bank Symposium, Washington D.C.: The World Bank, 1993.

Kitunga, Demere, *The role of women in traditional irrigation in Same and Mwanga districts. Research Report*, Dar-es-Salaam: Traditional Irrigation Improvement Programme, 1989.

Klaver, Dieuwke and Barbara van Koppen, *Changing alliances between the state, male tenants and female farmers: The struggle for land in the settlement scheme,* Office du Niger, Mali, forthcoming.

Krol, Marjon, 'Irrigatie is mannenwerk: Genderverhoudingen in een kleinschalig irrigatieproject in de Ecuadoriaanse Andes', Doctoraalscriptie voor de Vakgroepen Vrouwenstudies in de Landbouw en Irrigatie aan de Landbouw universiteit, Wageningen, 1994.

Kumar, Shanti P., 'The Mahaweli Scheme and Rural Women in Sri Lanka', in N. Heyzer (ed.), *Women Farmers and Rural Change in Asia: Towards equal access and participation,* Kuala Lumpur, Malaysia: Asian and Pacific Development Centre (APDC), 1987.

Lynch Deutsch, Barbara, *Natural Resources,* vol. 4, 1991

Mandal, M.A.S. and D.E. Parker, *Evolution and implications of decreased public involvement in minor irrigation management in Bangladesh*, Colombo, Sri Lanka: International Irrigation Management Institute: Short report series on locally managed irrigation, no. 11, 1995.

Nederlof, Marc and Eric van Wayjen, 'Religion and local water rights versus landowners and state irrigation in Izucar de Matamoros (West Bank Mexico, in Geert Diemer and Frans Huibers (eds.), *Crops, people and irrigation: Water allocation practices of farmers and engineers*, London: Intermediate Technology Publications, 1996.

Povel, S.A.M.T., Participatory development of a women's irrigation scheme case: The Nyandusi Women's Horticultural Scheme, Nyanza Province, Kenya, Contribution to the International Workshop 'Design for Sustainable Farmer Managed Irrigation Schemes in Sub-Saharan Africa', Department of Irrigation and Soil and Water Conservation, Wageningen Agricultural University, 1990.

Pradhan, Naresh, 'Gender participation in irrigation system activities in the hills in Nepal', in Proceedings of the second annual workshop on Women in Farming Systems, September 1989, Rampur: Institute of Agriculture and Animal Science, and Kathmandu, Nepal: USAID, 1989.

Prins, Djura, 'La dinamica de los derechos de agua en el contexto de la intervencion 'el Proyecto Multiple Laka Laka' en Bolivia. Un estudio sensitivo hacia el papel de la mujer en la intervencion', M.Sc. thesis, Department of Irrigation and Soil and Water Conservation, Wageningen Agricultural University, The Netherlands, 1996.

Projet Sensibilisation et Formation des paysans autour des barrages, Attribution des parcelles aux femmes dans les périmetres en aval des barrages: Possibilités et limites, Ministere de l'Agriculture et des Ressources Animales, Ouagadougou, Burkina Faso, 1993.

Raju, K.V., 'Irrigation rules and laws for water distribution and management transfer', paper prepared for the IDPAD Seminar 'Managing Water Scarcity: Experiences and Prospects', Amersfoort, 13-17 October 1997.

Safiliou, Constantina, 'Agricultural strategies and programmes: The status of women and fertility', Background Paper for the International Seminar on Women in Agriculture and Rural Development in Asia, Huangxian, China (FAO/ESH/ A86/3), 1986.

———, 'Gender and rural poverty in Asia: implications for agricultural project design and implementation', in *Asia-Pacific Journal of Rural Development*, vol. 1, no. 1, July 1991.

Schenk-Sandbergen, Loes, 'Some Gender Dimensions in Managing Water Scarcity', Chapter 15 in this volume.

Sengupta, Nirmal, 'Negotiating with an under-informed bureaucracy: The case of water rights on system tanks of Bihar', in Bryan, Bruns, and Ruth Meinzen-Dick (eds.), 'Negotiating water rights', unpublished manuscript.

Shah, Tushaar, *Groundwater markets and irrigation development: Political economy and practical policy*, Bombay: Oxford University Press, 1993.

Strosser, Pierre and Ruth, Meinzen-Dick, 'Groundwater markets in Pakistan: An analysis of selected issues', in Marcus Moench (ed.), *Selling Water: Conceptual and policy debates over groundwater markets in India*, Vikram Sarabhai Centre for Development Interaction (VIKSAT), Pacific Institute for Studies in Environment, Development and Security, Natural Heritage Institute, 1994.

Van der Grift, Eveline W., 'Gender relations in traditional irrigation in Malolo, Tanzania', M.Sc. thesis, Department of Irrigation and Soil and Water Conservation, Wageningen Agricultural University, in collaboration with SNV Tanzania, 1991.

Van Koppen, Barbara, 'Women and the design of farmer-managed irrigation schemes: Experiences provided by two projects in Burkina Faso', in *Contributions to the International Workshop Design for Sustainable Farmer-Managed Irrigation Schemes in Sub-Saharan Africa*, February 1990, Wageningen, The Netherlands: Wageningen Agricultural University.

———, 'Allocation and poverty alleviation: A gendered analysis of inclusion and exclusion of the resource-poor in vesting water rights', paper prepared for the Workshop on Women and Water, International Irrigation Management Institute, 15-19 September 1997, Colombo, Sri Lanka.

———, 'Negotiating water rights in public irrigation: The case of rice valleys in Burkina Faso', in Bryan Bruns and Ruth Meinzen-Dick (eds.), 'Negotiating Water Rights', unpublished manuscript.

Van Koppen, Barbara and Simeen Mahmud, *Women and Water-pumps: The impact of participation in irrigation groups on women's status*, London: Intermediate Technology Publications, 1996.

Vermillion, Douglas L. (ed.), *The privatization and self-management of irrigation*, Final Report, Colombo, Sri Lanka: International Irrigation Institute, 1996.

Von Benda-Beckmann, Keebet, 'Development, law and gender skewing: An examination of the impact of development on the socio-legal position of women in Indonesia, with special reference to the Minangkabau', in LaPrairie and Els Baerends (guest editors), *The socio-legal position of women in changing society, Journal of Legal Pluralism and Unofficial Law*, nos. 30 & 31, 1990-1, Groningen,

The Netherlands: Foundation for the Journal of Legal Pluralism, 1991.

Von Benda-Beckmann, Keebet, Mirjam de Bruijn, Han van Dijk, Gerti Hesseling, Barbara van Koppen and Lyda Res, 'Rechten van vrouwen op de natuurlijke hulpbronnen land and water', Literature review for The Special Program Women and Development of the Ministry of Foreign Affairs, Department of International Co-operation, Wageningen, 1996.

Wood, Geoffrey D., 'Private provision after public neglect: Opting out with pumpsets in North Bihar', paper presented at the International Conference on the Political Economy of Water in South Asia: Rural and Urban Action and Interactions, Madras Institute of Development Studies, January 1995. Occasional Paper 01195, Bath: Centre for Development Studies, University of Bath, 1995.

Wood, Geoffrey D., Richard Palmer-Jones, Q.F. Ahmed, M.A.S. Mandal and S.C. Dutta, *The water sellers: A cooperative venture by the rural poor*, Connecticut: Kumarian Press, 1990.

Participants in the Seminar

CONVENERS

A. VAIDYANATHAN, currently Emeritus Professor in the Madras Institute of Development Studies, has worked in the Indian Planning Commission, non-governmental research institutions, and also international organizations. Water resource planning and management are major areas of his research. His publications include the books *Water Resource Management: Institutions and Irrigation Development in India* and *Tanks of South India*.

H.M. OUDSHOOM, a civil engineer, has held senior positions as manager, adviser, and policy maker in the Netherlands Ministry of Transport, Public Works and Water Management. He is a member of the UN Committee on Natural Resources, which advises Economic and Social Council (ECOSOC) on water related issues. Currently he is Emeritus Professor of the Delft University of Technology.

KEYNOTE SPEAKERS

ANIL AGARWAL, founder of the Centre for Science and Environment, was a prominent and influential figure in the movement for environmental protection both in India and abroad. He played a major role in spreading awareness of environmental issues among the public, and campaigning for their redressal through the media, the government, and the courts. His edited collections *Dying Wisdom: State of India's Environment Report* and *Making Water Everybody's Business* are widely acclaimed. He passed away in 2002.

V.B. ESWARAN, a retired civil servant, has been Secretary to several departments in the state government of Gujarat and in the Government of India. For more than 10 years now he has been engaged in promoting or advising non-governmental organizations, working in different aspect of natural resources management. He has chaired several special government committees to advise on energy watershed and environment.

ANIL C. SHAH was Secretary for Rural Development in the Gujarat government for nearly 30 years. After retirement, he served as Chief Executive of the Aga Khan Rural Support Programme before moving to his current position as Chairman of the Development Support Centre, Ahmedabad, India. He has played an active role in promoting a participatory approach to the development of natural resources. He has also written extensively on various aspects of watershed, irrigation, and forest management. A collection of his papers was published in 2001.

CONTRIBUTORS

PAUL P. APPASAMY, currently Director of the Madras School of Economics, Chennai, is also an engineer turned environmental economist, interested in urban water supply and environmental management. His major studies include pollution in the Noyyal River Basin and urban water problems of Chennai. His research papers on these subjects are being put together in book form.

VISHWA BALLABH, an agricultural economist, is currently Professor in the Institute of Rural Management in Anand. The management of natural resources, including water and forests, are his main area of research. He has co-edited two books: one on *Cooperative Management of Natural Resources* (with Katar Singh) and another on *Farm Forestry in South Asia* (with N.C. Saxena).

KATAR SINGH, an agricultural economist, has done extensive research on environmental issues, especially management of common pool resources. He retired recently after serving several years as Director of the Institute of Rural Management, Anand. His publications include *Managing Common Pool Resources: Principles and Case Studies*; *Rural Development: Principles, Policies and Management*, and co-edited volumes on *Cooperative Management of Natural Resources*; and *Designing and Managing Rural Development Organisations in India*.

ASHIS BANERJEE is Secretary and Executive Director of India Trust for Art and Cultural Heritage and head of its Natural Heritage Division. He is an active member of International Union for Conservation of Nature (IUCN) and is on the governing bodies of several non-governmental organizations (NGOs) and educational institutions in India.

SURESH K. ROHILLA is an urban and regional planner. At the time of the conference, he was Director of the National Heritage Division in INTACH. Currently he is researching on the role of groundwater in urban development in Delhi, on leave from the Ministry of Urban Development, Government of India. He has co-authored a book on Delhi's water and solid waste management (2000).

KANCHAN CHOPRA is currently Professor in the Institute of Economic Growth, Delhi, and member of the Executive Committee of the Indian Society of Environmental Economics and the International Society of Environmental Economics. Her areas of research include sustainable development and institutional aspects of water and forest management. Besides several journal articles, she has published four books, three of them as co-author, *Operationalising Sustainable Development: Ecological Modeling for Developing Countries* (1999); and *Migration, Common Property and Environmental Degradation* (2000) being the most recent of them.

SEEMA BATHLA, an economist, is currently a Research Associate in the Institute of Economic Growth. Her research interests cover common property management, sustainable development, and resource accounting. Besides several papers, she co-authored a book *Joint Forest Management: India Country Study—Policy that Works for Forests and People* (2000).

S. JANAKARAJAN, an economist, has done a number of field studies of agrarian institutions, management of surface and groundwater and water pollution in south India. He has been on the faculty of the Madras Institute of Development Studies since 1987 and is currently a Professor. Besides a number of papers in journals and edited volumes, he was co-author of *Study of Surface and Groundwater Irrigation in Parambikulam Aliyar, Palar and Vaigai Basins.*

He is a co-author of a forthcoming book *Rural India Facing the 21st century: Essays on Long Term Village Change and Recent Development Policy.*

TRIYUGI PRASAD has done extensive studies on international river basins and conjunctive use of surface and groundwater for irrigation. He is currently Professor and Head of Department of Civil Engineering and also Director of Centre for Water Resource Status in Patna University. His publications include *Cooperation for International River Basin Development: The Case of the Kosi Basin*; *Sustainable Water Resources Management for Agricultural Development in the Humid Alluvial Region of Eastern India*; and *Planning and Design of Conjunctive Irrigation Systems in Alluvial Plains: Principles and Practices.*

K.V. RAJU, currently Professor in the Institute for Social and Economic Change (ISEC), Bangalore, has worked on water management issues in a wide variety of contexts and institutions as researcher and as consultant to projects. Besides several papers, he has co-authored a monograph 'Rejuvenating Tanks: A Socio-Ecological Approach' (2003). He is currently with the Ecological Economics of the ISEC, Bangalore.

K.R. DATYE, a civil engineer, worked for many years in government construction and management of large surface water projects. For the last 15 years he has been involved in rural development, developing and testing innovative technologies, water, infrastructure and energy sectors, and promoting participatory approaches to equitable management of water. He is co-author of two, major monographs: 'Farmer Managed Irrigation Systems: Indian Experience', and 'Sustainable Agriculture in Semi-Arid Regions: Opportunities for Small and Marginal Farms'. His most recent publication is a book titled *Banking on Biomass: A New Strategy for Sustainable Prosperity Based on Renewable Energy and Dispersed Industrialisation.*

SUHAS PARAJAPE is working in the Society for People's Participation in Eco-System Management (SOPPECOM), an NGO in Pune.

SEEMA KULKARNI is working in the Society for People's Participation in Eco-System Management (SOPPECOM), an NGO in Pune.

K.J. Joy is working in the Society for People's Participation in Eco-System Management (SOPPECOM), an NGO in Pune.

Tushaar Shah, an economist, was Director of the Institute of Rural Management at Anand for several years. Currently he leads International Water Management Institute's (IWMI) global research programme on sustainable groundwater management. His extensive writings on the subject include two books: *Groundwater Markets and Irrigation Development: Political Economy and Practical Policy* (1993) and (as co-author) *To the Hands of the Poor: Water and Trees* (1987).

Himanshu Thakkar, an engineer, has been actively involved with the work of the World Commission on Dams (WCD) right from its inception. He wrote a paper 'India Irrigation Options Study' for the WCD besides making many submissions to it. He earlier worked as senior policy analyst in the Centre for Science and Environment, New Delhi. He was actively involved in the Narmada Bachao Andolan (Save Narmada Campaign). He is currently with South Asia Network on Dams, Rivers and People and edits a periodical '*Dams, Rivers and People*'.

Arthur van Norden headed the IWT Foundation from 1989 until 1995. Since the IWT Foundation had become inactive, he joined Friends of the Earth International where he worked as a senior consultant from 1995 until 1999. Thereafter he decided to say goodbye to the international environmental arena to devote all his time to writing children's books of which a vast number have been published in the Netherlands.

Wim van Leussen, a civil engineer, is Professor of River Basin Management in the Faculty of Engineering of the Twente University and Deputy Coordinator for the implementation of the European Water Framework Directive for the rivers Rhine and Meuse in the Netherlands. He has researched on dredging and estuarine processes and co-ordinated several European research projects. His publications include *Erosion Sedimentation Cycles in the Ems Estuary*; *Estuarine Macroflocs and their Role in Fine-grained Sediment Transport*; and *Multi-level Approach to Flood Control in the Dutch Part of the River Meuse*. He has co-edited a book *Physical Processes in Estuaries*.

TIEDO VELLINGA is a civil engineer from the Delft University of Technology. At present he is general manager, Maritime Development in the Rotterdam Port Management. He is also active in international committees on maritime environment and sedimentation problems.

KOOS WIERIKS, a Dutch spatial planning engineer, is a specialist on international cooperation and river basin management. He was Secretary General of the International Commission for the Protection of the Rhine and is currently working on capacity building projects for river basin management and planning in Indonesia. Some of his publications are *Integrated Water Management for the Rhine River Basin*; *Trans-boundary Cooperation in Shared River Basins*; *The Rhine Experience*; and *Rejuvenating the Rhine, the Long Road to Integrated River Management.*

LOES SCHENK-SANDBERGEN is an anthropologist, currently Associate Professor of the University of Amsterdam, specializing in gender (women) and development issues. She has researched and published extensively on gender-related issues, linking them with poverty and empowerment. Some of her recent publications are: (as co-author) *Women in Rice Fields and Offices: Irrigation in Laos, Gender Specific Case-studies in four villages* (1995), *Gender, Culture and Land Rights in Rural Laos* (1998), and (as co-author) *From Heroines to Beneficiaries: From Beneficiaries to Heroines?* (2003) A forthcoming publication is *The Impact of a Small-scale Irrigation Project in West-Bengal, Terai, India.*

BARBARA VAN KOPPEN specializes in studies of poverty, gender, and water law and institutions. She has taught at Wagnngen University, been technical advisor to the government of Burkina Faso, and is currently Senior Researcher at the International Water Management Institute (IWMI), Africa Regional Program. Her publications include *A Gender Performance Indicator for Irrigation. Concepts, Tools, and Applications* and *Poverty Dimensions of Irrigation: Management Transfer in Large-scale Canal Irrigation in Andhra Pradesh and Gujarat, India*, both published by IWMI in its research report series.

DISCUSSANTS

RAMASWAMY R. IYER, formerly Secretary, Ministry of Water Resources in the Government of India, is currently Honorary Research Professor at the Centre for Policy Research, New Delhi. He has been a member of several government committees, the latest being the National Commission on Integrated Water Resources Planning (1996-9). He participated in a collaborative inter-country study of the integrated development of the Himalayan rivers and co-edited two volumes based on this research: *Harnesssing the Eastern Himalyan Rivers* (2000) and *Converting Water into Wealth* (1994). His book *Water: Perspectives, Issues, Concerns* has recently been published (2003).

JACOBUS H. DE GOEDE is currently Director of a firm of consultants on human resources in the Netherlands. He has taught in the Asian Institute of Technology, Bangkok, and worked with UN Commission for Habitat Studies (UNCHS). He has also been involved in several water supply projects in India, including an environmental project under the Ganga Action Plan.

JAN HOEVENAARS, a graduate of Wageningen Agricultural University, has worked as a water management expert in Wageningen, and in Delft. He is now an independent consultant for the same subject. Mr Hoevenaars is a regular advisor to the Netherlands Embassy in Delhi. He is co-author of the (forthcoming) World Bank publication: *Reclaiming Drainage: Towards an Integrated Approach*.

JAN VAN DER LAAN is currently a freelance consultant on land and water related issues. He has taught in Wageningen and Mozambique universities and was for sometime specialist advisor in the Netherlands Embassies in India and Bangladesh and has been consultant to several countries in Asia and Africa.

N. MADHAVA MENON, formerly Director of the National Law School in Bangalore, is currently Vice-Chancellor of the National University of Juridical Sciences, Kolkata. He has served on several national level expert committees on legal and judicial reform and has been a member of the Law Commission of India. Water law is one of his special interests.

P. MOLLINGA, a graduate of Wageningen University is an irrigation engineer with special interest in 'Political Economy of Irrigation'. He does comparative research on water management in large scale irrigation systems in South Asia, as an example of politically contested resource use and the social shaping of irrigation technology. He has recently edited a book *Water for Food and Development on the Water Front: Water Distribution, Technology and Agrarian Change in South Indian Canal Systems* (2000).

LINDEN F. VINCENT taught in East Anglia and worked with the Overseas Development Institute before becoming Professor of Irrigation and Water Engineering at Wageningen Agricultural University. Her special interest is research on water scarcity contexts and their socio-technical dimensions. She has done research and consultancy work related to water development in several Indian states.

JAN TEUN VISSCHER, an engineer, is currently Acting Director of the IRC International Water and Sanitation Centre based in the Hague, the Netherlands. He has co-authored several IRC publications on better water management, pre-treatment alternatives for drinking water supply systems, renewable energy for rural water supply systems, and water harvesting.

PETER BURY is a Programme Officer at the IRC International Water and Sanitation Centre based in the Hague, the Netherlands. A geographer and planner, he specializes in decentralized, integrated land use and rural development planning and natural resources management. He has advised several African governments on these issues.

LINDEN VINCENT worked at the University of East Anglia, and with the Overseas Development Institute, London, before becoming Professor of Irrigation and Water Engineering at Wageningen University, the Netherlands, in 1994. Her research interests include agro-ecology (particularly in hill and drought-prone areas), and the socio-technical dimensions of irrigation. She has done research and consultancy work related to water development in several Indian states.

Index